CLASSICAL FIELD THEORY

CLASSICAL FIELD THEORY

ELECTROMAGNETISM AND GRAVITATION

Francis E. Low

A Wiley-Interscience Publication

JOHN WILEY & SONS, INC.

New York ■ Chichester ■ Weinheim ■ Brisbane ■ Singapore ■ Toronto

Library of Congress Cataloging in Publication Data:

Low, Francis E. (Francis Eugene), 1921–
    Classical field theory : electromagnetism and gravitation /
    Francis E. Low.
        p.   cm.
    Includes index.
    ISBN 0-471-59551-9 (cloth : alk. paper)
    1. Electromagnetic fields.   2. Gravitational field.   I. Title.
    QC665.E4L69   1997
    530.1′41—dc20   96-25448
                                                              CIP

Printed in the United States of America

10 9 8 7 6 5 4 3 2

Contents

Preface

It is hard to fit a graduate course on electromagnetic theory into one semester. On the other hand, it is hard to stretch it to two semesters. This text is based on a two-semester MIT ccurse designed to solve the problem by a compromise: Allow approximately one and a half semesters for electromagnetic theory, including scattering theory, special relativity and Lagrangian field theory, and add approximately one-half semester on gravitation.

It is assumed throughout that the reader has a physics background that includes an intermediate-level knowledge of electromagnetic phenomena and their theoretical description. This permits the text to be very theory-centered, starting in Chapters 1 and 2 with the simplest experimental facts (Coulomb's law, the law of Biot and Savart, Faraday's law) and proceeding to the corresponding differential equations; theoretical constructs, such as energy, momentum, and stress; and some applications, such as fields in matter, fields in the presence of conductors, and forces on matter.

In Chapter 3, Maxwell's equations are obtained by introducing the displacement current, thus making the modified form of Ampère's law consistent for fields in the presence of time-dependent charge and current densities. The remainder of Chapters 3–5 applies Maxwell's equations to wave propagation, radiation, and scattering.

In Chapter 6, special relativity is introduced. It is also assumed here that the reader comes with prior knowledge of the historic and experimental background of the subject. The major thrust of the chapter is to translate the physics of relativistic invariance into the language of four-dimensional tensors. This prepares the way for Chapter 7, in which we study Lagrangian methods of formulating Lorentz-covariant theories of interacting particles and fields.

The treatment of gravitation is intended as an introduction to the subject. It is not a substitute for a full-length study of general relativity, such as might be based on Weinberg's book.[1] Paralleling the treatment

---

[1]Steven Weinberg, *Gravitation and Cosmology*, New York: John Wiley & Sons. 1972.

of electromagnetism in earlier chapters, we start from Newton's law of gravitation. Together with the requirements of Lorentz covariance and the very precise proportionality of inertial and gravitational mass, this law requires that the gravitational potential consist of a second-rank (or higher) tensor.

In complete analogy with the earlier treatment of the vector (electromagnetic) field, following Schwinger,[2] we develop a theory of the free tensor field. Just as Maxwell's equations required that the vector field be coupled to a conserved vector source (the electric current density), the tensor field equations require that their tensor source be conserved. The only available candidate for such a tensor source is the stress-energy tensor, which in the weak field approximation we take as the stress-energy tensor of all particles and fields other than the gravitational field. This leads to a linear theory of gravitation that incorporates all the standard tests of general relativity (red shift, light deflection, Lense–Thirring effect, gravitational radiation) except for the precession of planetary orbits, whose calculation requires nonlinear corrections to the gravitational potential.

In order to remedy the weak field approximation, we note that the linear equations are not only approximate, but inconsistent. The reason is that the stress-energy tensor of the sources alone is not conserved, since the sources exchange energy and momentum with the gravitational field. The remedy is to recognize that the linear equations are, in fact, consistent in a coordinate system that eliminates the gravitational field, that is, one that brings the tensor $g_{\mu\nu}$ locally to Minkowskian form. The consistent equations in an arbitrary coordinate system can then be written down immediately—they are Einstein's equations. The basic requirement is that the gravitational potential transform like a tensor under general coordinate transformations.

Our approach to gravitation is not historical. However, it parallels the way electromagnetism developed: experiment → equations without the displacement current; consistency plus the displacement current → Maxwell's equations. It seems quite probable that without Einstein the theory of gravitation would have developed in the same way, that is, in the way we have just described. Einstein remarkably preempted what might have been a half-century of development. Nevertheless, I believe it is useful, in an introduction for beginning students, to emphasize the field theoretic aspects of gravitation and the strong analogies between gravitation and the other fields that are studied in physics.

The material in the book can be covered in a two-term course without crowding; achieving that goal has been a boundary condition from the start. Satisfying that condition required that choices be made. As a conse-

[2]J. Schwinger, *Particles, Sources and Fields*, Addison-Wesley, 1970.

quence, there is no discussion of many interesting and useful subjects. Among them are standard techniques in solving electrostatic and magneto-static problems; propagation in the presence of boundaries, for example, cavities and wave guides; physics of plasmas and magnetohydrodynamics; particle motion in given fields and accelerators. In making these choices, we assumed that the graduate student reader would already have been exposed to some of these subjects in an earlier course. In addition, the subjects appear in the end-of-chapter problems sections.

My esteemed colleague Kenneth Johnson once remarked to me that a textbook, as opposed to a treatise, should include everything a student *must* know, not everything the author *does* know. I have made an effort to hew to that principle; I believe I have deviated from it only in Chapter 5, on scattering. I have included a discussion of scattering because it has long been a special interest of mine; also, the chapter contains some material that I believe is not easily available elsewhere. It may be omitted without causing problems in the succeeding chapters.

The two appendices (the first on vectors and tensors, the second on spherical harmonics) are included because, although these subjects are probably well known to most readers, their use recurs constantly throughout the book. In addition to the material in the appendices, some knowledge of Fourier transforms and complex variable theory is assumed.

The problems at the end of each chapter serve three purposes. First, they give a student an opportunity to test his or her understanding of the material in the text. Second, as I mentioned earlier, they can serve as an introduction to or review of material not included in the text. Third, they can be used to develop, with the students' help, examples, extensions, and generalizations of the material in the text. Included among these are a few problems that are at the mini-research-problem level. In presenting these, I have generally tried to outline a path for achieving the final result. These problems are marked with an asterisk. I have not deliberately included problems that require excessive cleverness to solve. For a teacher searching for a wider set of problems, I recommend the excellent text of Jackson,[3] which has an extensive set.

One last comment. I have not hesitated to introduce quantum inter-pretations, where appropriate, and even the Schröedinger equation on one occasion, in Chapter 3. I would expect a graduate student to have run across it (the Schröedinger equation) somewhere in graduate school by the time he or she reaches Chapter 3.

Finally, I must acknowledge many colleagues for their help. Special thanks go to Professors Stanley Deser, Jeffrey Goldstone, Roman Jackiw, and Kenneth Johnson. I am grateful to the late Roger Gilson and to Evan Reidell, Peter Unrau, and Rachel Cohen for their help with the

---

[3]J. D. Jackson, *Classical Electrodynamics*, New York: John Wiley and Sons. 1962.

manuscript, and to Steven Weinberg and David Jackson for their excellent texts, from which I have freely borrowed.

FRANCIS E. LOW

Cambridge, Massachussetts

CHAPTER 1

Electrostatics

1.1. COULOMB'S LAW

In the first half of the eighteenth century, the basic facts of electrostatics were sorted out: the existence of two signs of transferable electric charge; the additive conservation of that charge; the existence of insulators and conductors. The process is described in a lively way by Whittaker.[1] In the next half-century, the quantitative law of repulsion of like charges was determined by Priestley and extended to charges of both signs by Coulomb. By 1812, with the publication of the famous memoir of Poisson,[2] the science of electrostatics was understood almost in its present form: potentials, conductors, etc. Of course, the specific knowledge of the nature of the carriers of electric charge awaited the experimental discoveries of the late nineteenth and early twentieth century.

The resultant formulation of electrostatics starts from Coulomb's law for the force between two small particles, each carrying a positive or a negative charge. We call the charges q_1 and q_2, and their vector positions $\mathbf{r}_1$ and $\mathbf{r}_2$, respectively:

$$\mathbf{F}(2 \text{ on } 1) = q_1 q_2 \frac{\mathbf{r}_1 - \mathbf{r}_2}{|\mathbf{r}_1 - \mathbf{r}_2|^3} \tag{1.1.1}$$

and

$$\mathbf{F}(1 \text{ on } 2) = -\mathbf{F}(2 \text{ on } 1). \tag{1.1.2}$$

Like charges repel, unlike attract. Most important, the forces are linearly

---

[1] *A History of the Theories of Aether and Electricity*, Tomash Publishers (American Institute of Physics, New York), 1987 (1st edition, 1910).

[2] *Mém. de l'Institut*, 1811.

additive. That is, there are no three-body electrostatic forces.[3] Thus, with three charges present, the total force on 1 is found to be

$$\mathbf{F}(\text{on } 1) = q_1 q_2 \frac{\mathbf{r}_1 - \mathbf{r}_2}{|\mathbf{r}_1 - \mathbf{r}_2|^3} + q_1 q_3 \frac{\mathbf{r}_1 - \mathbf{r}_3}{|\mathbf{r}_1 - \mathbf{r}_3|^3}. \tag{1.1.3}$$

If $\mathbf{r}_2$ and $\mathbf{r}_3$ are close together, the form of (1.1.3) goes over to (1.1.1) with $q_{2+3} = q_2 + q_3$. Thus, charge is additive. It is also conserved. That is, positive charge is never found to appear on some surface without compensating positive charge disappearing or negative charge appearing somewhere else.

Equation (1.1.1) serves to define the electrostatic unit of charge. This is a charge that repels an equal charge 1 cm away with a force of 1 dyne.

It is useful to define an electric field at a point $\mathbf{r}$ as the force that would act on a small test charge δq at $\mathbf{r}$ divided by δq, where the magnitude of δq is small enough so that its effect on the environment can be ignored. Thus, the field, a property of the space point $\mathbf{r}$, is given by

$$\mathbf{E}(\mathbf{r}) = \frac{\mathbf{F}(\text{on } \delta q \text{ at } \mathbf{r})}{\delta q} \tag{1.1.4}$$

and, by (1.1.3) generalized to many charges,

$$\mathbf{E}(\mathbf{r}) = \sum_i q_i \frac{\mathbf{r} - \mathbf{r}_i}{|\mathbf{r} - \mathbf{r}_i|^3}. \tag{1.1.5}$$

We can generalize (1.1.5) to an arbitrary charge distribution by defining a charge density at a point $\mathbf{r}$ as

$$\rho(\mathbf{r}) = \frac{\delta q}{\delta \mathbf{r}} \tag{1.1.6}$$

where δq is the charge in the very small three-dimensional volume element $\delta \mathbf{r}$. The sum in (1.1.5) turns into a volume integral:

$$\mathbf{E}(\mathbf{r}) = \int d\mathbf{r}' \frac{\mathbf{r} - \mathbf{r}'}{|\mathbf{r} - \mathbf{r}'|^3} \rho(\mathbf{r}') \tag{1.1.7}$$

where $d\mathbf{r}'$ represents the three-dimensional volume element. Note that in

---

[3]This statement does not hold at the microscopic or atomic level. For example, the interactions between atoms (van der Waals forces) include three-body forces. These are, however, derived from the underlying two-body Coulomb interaction.

spite of the singularity at $\mathbf{r}' = \mathbf{r}$, the integral (1.1.7) is finite for a finite charge distribution, even when the point $\mathbf{r}$ is in the region containing charge. This is because the volume element $d\mathbf{r}'$ in the neighborhood of a point $\mathbf{r}$ goes like $|\mathbf{r}' - \mathbf{r}|^2$ for small $|\mathbf{r} - \mathbf{r}'|$, thereby canceling the singularity.

We can return to the form (1.1.5) by imagining the charge distribution as consisting of very small clumps of charge q_i at positions $\mathbf{r}_i$; the quantity

$$q_i = \int_{i\text{th clump}} d\mathbf{r}' \, \rho(\mathbf{r}') \tag{1.1.8}$$

is the charge q_i in the ith clump. Its volume must be small enough so that $\mathbf{r}'$ in (1.1.7) does not vary significantly over the clump.

The mathematical point charge limit keeps the integral $\int_{\text{clump}} \rho \, d\mathbf{r}' = q$ constant as the size of the clump goes to zero. It is useful to give a density that behaves this way a name. It is called the delta function, with the properties

$$\delta(\mathbf{r} - \mathbf{r}') = 0, \qquad \mathbf{r} \neq \mathbf{r}' \tag{1.1.9}$$

and

$$\int d\mathbf{r}' \, \delta(\mathbf{r} - \mathbf{r}') = 1 \tag{1.1.10}$$

provided the $\mathbf{r}'$ integration includes the point $\mathbf{r}$. Of course, $\delta(\mathbf{r})$ is not a real function; however, as we shall see repeatedly, its use leads to helpful shortcuts, provided one takes care not to multiply $\delta(\mathbf{r})$ by functions that are singular at $\mathbf{r} = 0$.

Evidently, the fields of surface and line charge distributions can be written in the form (1.1.7), with the charge density including surface and line charge (i.e., one- and two-dimensional) delta functions. When the dimensionality of the delta function is in doubt, we add a superscript, thus $\delta^3(\mathbf{r}_3)$ for a point charge, $\delta^2(\mathbf{r}_2)$ for a line charge, and $\delta^1(\mathbf{r}_1)$ for a surface charge; here, $\mathbf{r}_3$, $\mathbf{r}_2$, and $\mathbf{r}_1$ represent three-, two-, and one-dimensional vectors, respectively. Note that δ^3, δ^2, and δ^1 can be expressed as products of one-dimensional delta functions. Thus, for example, $\delta^3(\mathbf{r}_3) = \delta^1(x)\delta^1(y)\delta^1(z)$, $\delta^2(\mathbf{r}_2) = \delta^1(x)\delta^1(y)$, and $\delta^1(\mathbf{r}_1) = \delta^1(x)$.

Given the charge distribution $\rho(\mathbf{r})$, (1.1.7) tells us how to calculate the electric field at any point by a volume integral—if necessary, numerically. We might therefore be tempted to terminate our study of electrostatics here and go on to magnetism. There are, however, a large number of electrostatic situations where we do not know $\rho(\mathbf{r})$, but are nevertheless able to understand and predict the field configuration. In order to do that, however, it is necessary to study the differential equations satisfied by the electric field.

We start by observing from (1.1.7) that the electric field can be derived from a potential $\phi(\mathbf{r})$. That is,

$$\mathbf{E}(\mathbf{r}) = -\nabla\phi(\mathbf{r}), \tag{1.1.11}$$

where

$$\phi(\mathbf{r}) = \int \frac{d\mathbf{r}'\, \rho(\mathbf{r}')}{|\mathbf{r} - \mathbf{r}'|}. \tag{1.1.12}$$

Equation (1.1.7) follows from (1.1.11) and (1.1.12) since

$$\nabla \frac{1}{r} = \left(\hat{\mathbf{e}}_x \frac{\partial}{\partial x} + \hat{\mathbf{e}}_y \frac{\partial}{\partial y} + \hat{\mathbf{e}}_z \frac{\partial}{\partial z} \right) \frac{1}{r} \tag{1.1.13}$$

(where $\hat{\mathbf{e}}_x$, $\hat{\mathbf{e}}_y$, and $\hat{\mathbf{e}}_z$ are unit vectors in the three coordinate directions) and

$$-\frac{\partial}{\partial x} \frac{1}{r} = -\frac{\partial}{\partial x} \frac{1}{\sqrt{x^2 + y^2 + z^2}} = \frac{x}{(x^2 + y^2 + z^2)^{3/2}} \tag{1.1.14}$$

so that, with similar equations for y and z,

$$-\nabla \frac{1}{r} = \frac{\hat{\mathbf{e}}_x x}{r^3} + \frac{\hat{\mathbf{e}}_y y}{r^3} + \frac{\hat{\mathbf{e}}_z z}{r^3} = \frac{\mathbf{r}}{r^3} \tag{1.1.15}$$

and

$$-\nabla \frac{1}{|\mathbf{r} - \mathbf{r}'|} = \frac{\mathbf{r} - \mathbf{r}'}{|\mathbf{r} - \mathbf{r}'|^3}. \tag{1.1.16}$$

From (1.1.11) we learn that

$$\nabla \times \mathbf{E} = 0 \tag{1.1.17}$$

since

$$\nabla \times \nabla \phi = 0 \tag{1.1.18}$$

identically for any ϕ. Of course, we could have derived (1.1.17) directly by taking the curl of (1.1.7).

On the other hand, given (1.1.17), we can derive the existence of a potential. We define

$$\phi(\mathbf{r}) = - \int_{\mathbf{r}_0}^{\mathbf{r}} \mathbf{E}(\mathbf{r}') \cdot d\mathbf{l}', \qquad (1.1.19)$$

where $\int d\mathbf{l}'$ represents a line integral along an arbitrary path from the point $\mathbf{r}_0$ (where ϕ is defined to be zero) to the point $\mathbf{r}$. We show that ϕ in (1.1.19) is independent of the path by calculating the difference of ϕ defined by two paths, P_1 and P_2:

$$\phi_1(\mathbf{r}) - \phi_2(\mathbf{r}) = - \int_{P_1} \mathbf{E} \cdot d\mathbf{l} + \int_{P_2} \mathbf{E} \cdot d\mathbf{l}$$

$$= - \oint_C \mathbf{E} \cdot d\mathbf{l} \qquad (1.1.20)$$

where $\oint_C \mathbf{E} \cdot d\mathbf{l}$ represents the line integral around a closed path C, given by going from $\mathbf{r}_0$ to $\mathbf{r}$ along P_1 and back from $\mathbf{r}$ to $\mathbf{r}_0$ along P_2.

By Stokes theorem,

$$\oint_C \mathbf{E} \cdot d\mathbf{l} = \int_S \nabla \times \mathbf{E} \cdot d\mathbf{S}, \qquad (1.1.21)$$

where $d\mathbf{S}$ is any oriented surface S bounded by C. Thus, since $\nabla \times \mathbf{E} = 0$, $\phi_1 = \phi_2$ and the integral defining ϕ is independent of the path from $\mathbf{r}_0$ to $\mathbf{r}$.

We note that changing $\mathbf{r}_0$ corresponds to adding a constant to ϕ:

$$\phi_{\mathbf{r}_0} = - \int_{\mathbf{r}_0}^{\mathbf{r}} \mathbf{E} \cdot d\mathbf{l}, \qquad (1.1.22)$$

and

$$\phi_{\mathbf{r}_1} = - \int_{\mathbf{r}_1}^{\mathbf{r}} \mathbf{E} \cdot d\mathbf{l},$$

so that

$$\phi_{\mathbf{r}_1} = \phi_{\mathbf{r}_0} - \int_{\mathbf{r}_1}^{\mathbf{r}_0} \mathbf{E} \cdot d\mathbf{l} \qquad (1.1.23)$$

with $- \int_{\mathbf{r}_1}^{\mathbf{r}_0} \mathbf{E} \cdot d\mathbf{l}$ the additive constant (it is independent of $\mathbf{r}$).

Finally, it is clear that $-\nabla\phi$ defined by (1.1.19) is the electric field. We show this for the x component: Let

$$d\mathbf{r} = \hat{\mathbf{e}}_x\,dx$$

and choose the path to $\mathbf{r} + d\mathbf{r}$ as $\mathbf{r}_0$ to $\mathbf{r}$ followed by $d\mathbf{r}$. Then

$$-\frac{\partial\phi}{\partial x} = -\frac{\phi(\mathbf{r} + \hat{\mathbf{e}}_x\,dx) - \phi(\mathbf{r})}{dx}$$

$$= \frac{\int_{\mathbf{r}_0}^{\mathbf{r}}\mathbf{E}\cdot d\mathbf{l} + E_x\,dx - \int_{\mathbf{r}_0}^{\mathbf{r}}\mathbf{E}\cdot d\mathbf{l}}{dx}$$

$$= E_x \qquad\qquad (1.1.24)$$

in the limit $dx \to 0$.

In general, a vector function of position (which goes to zero sufficiently rapidly as $\mathbf{r} \to \infty$) is completely determined by its curl and its divergence. In our case, a charge density confined to a finite region of space will—according to (1.1.7)—gives rise to an electric field that goes to zero like $1/r^2$; this is fast enough for the theorem to hold. (See Problem A.21.) We therefore turn to the calculation of $\nabla\cdot\mathbf{E}$.

For this purpose, we consider the field of a single point charge at the origin,

$$\mathbf{E} = q\,\frac{\mathbf{r}}{r^3}. \qquad\qquad (1.1.25)$$

$\nabla\cdot\mathbf{E}$ would appear to be given by

$$\nabla\cdot\mathbf{E} = \frac{\partial}{\partial x}\left(\frac{x}{r^3}\right) + \frac{\partial}{\partial y}\left(\frac{y}{r^3}\right) + \frac{\partial}{\partial z}\left(\frac{z}{r^3}\right)$$

$$= \left(\frac{1}{r^3} - \frac{3x^2}{r^5}\right) + \left(\frac{1}{r^3} - \frac{3y^2}{r^5}\right) + \left(\frac{1}{r^3} - \frac{3z^2}{r^5}\right) = 0. \qquad (1.1.26)$$

Equation (1.1.26) clearly holds for $r \neq 0$. The singular point $\mathbf{r} = 0$ presents a problem: Consider the electric flux through a closed surface S enclosing the charge at the origin, that is, the surface integral of the electric field over a surface S,

$$I = \int_S d\mathbf{S}\cdot\mathbf{E}, \qquad\qquad (1.1.27)$$

with the vector $d\mathbf{S}$ defined as the outward normal from the closed surface. The integral (1.1.27) is independent of the surface, provided the displacement from one surface to the other does not cross the origin. Thus,

$$I_1 - I_2 = \int_{S_1} d\mathbf{S}_1 \cdot \mathbf{E} - \int_{S_2} d\mathbf{S}_2 \cdot \mathbf{E}$$

where $d\mathbf{S}_1$ and $d\mathbf{S}_2$ are outward normals viewed from the origin. The two surface vectors $d\mathbf{S}_1$ and $-d\mathbf{S}_2$ are the outward normals of the surface bounding the volume contained between S_1 and S_2, provided S_1 is outside S_2. Thus,

$$I_1 - I_2 = \int d\mathbf{S} \cdot \mathbf{E} \tag{1.1.28}$$

and by Gauss' theorem

$$I_1 - I_2 = \int d\mathbf{r}\nabla \cdot \mathbf{E} = 0 \tag{1.1.29}$$

since the space between the surfaces does not include the singular point at the origin.

Consider first the integral (1.1.27) with the origin inside the surface. We choose the surface to be a sphere about the origin and find

$$I = \int_S \mathbf{E} \cdot d\mathbf{S} = q \int \frac{\mathbf{r}}{r^3} \cdot \hat{r}\, d\Omega\, r^2 \tag{1.1.30}$$

where $d\Omega$ is the solid angle subtended by $d\mathbf{S}$. Thus,

$$I = 4\pi q. \tag{1.1.31}$$

If S encloses several charges, we can calculate the contribution of each charge to I separately (since the fields are additive), yielding Gauss' law:

$$\int \mathbf{E} \cdot d\mathbf{S} = 4\pi \sum_i q_i \tag{1.1.32}$$

where the sum is over all the charges inside the surface S.

If the surface has no charges inside it, the integral $\int \mathbf{E} \cdot d\mathbf{S}$ is zero by

Gauss' theorem:

$$\int \mathbf{E} \cdot d\mathbf{S} = \int d\mathbf{r} \nabla \cdot \mathbf{E} = 0 \qquad (1.1.33)$$

since $\nabla \cdot \mathbf{E} = 0$ away from charges. Clearly, however, $\nabla \cdot \mathbf{E}$ cannot equal zero everywhere, since, if it did, the integral (1.1.32) would be zero instead of $4\pi \Sigma_i q_i$.

We can find the equation for $\nabla \cdot \mathbf{E}$ by considering finite charge density $\rho(\mathbf{r})$. Then (1.1.32) tells us that for any closed surface, the flux through the surface is equal to 4π times the total charge inside the surface:

$$\int_S \mathbf{E} \cdot d\mathbf{S} = 4\pi \int_V d\mathbf{r}\, \rho, \qquad (1.1.34)$$

where the integral $d\mathbf{r}$ is over the enclosed volume. Gauss' theorem applied to (1.1.34) gives

$$\int_V d\mathbf{r}\left(\nabla \cdot \mathbf{E} - 4\pi\rho\right) = 0 \qquad (1.1.35)$$

for any volume V. Thus, the integrand must be zero and we have the equation for the divergence of $\mathbf{E}$:

$$\nabla \cdot \mathbf{E} = 4\pi\rho. \qquad (1.1.36)$$

The special case of a point charge at the origin, for which $\rho = q\delta(\mathbf{r})$ and $\mathbf{E} = q(\mathbf{r}/r^3)$, shows that $\nabla \cdot (\mathbf{r}/r^3)$ acts as if

$$\nabla \cdot \frac{\mathbf{r}}{r^3} = 4\pi\,\delta(\mathbf{r}). \qquad (1.1.37)$$

Equation (1.1.36) yields an equation for the electrostatic potential ϕ:

$$\nabla \cdot \mathbf{E} = -\nabla \cdot \nabla\phi = 4\pi\rho \qquad \text{or} \qquad \nabla^2\phi = -4\pi\rho. \qquad (1.1.38)$$

This is known as Poisson's equation. In a portion of space where $\rho = 0$, (1.1.38) becomes

$$\nabla^2\phi = 0, \qquad (1.1.39)$$

which is called Laplace's equation. A function satisfying Laplace's equation is called harmonic.

As we remarked earlier, given the charge density ρ, the potential ϕ is determined (up to a constant) by the integral (1.1.12). We have given the subsequent development in (1.1.13–1.1.39) for three reasons.

First, the integral form (1.1.32) can be a useful calculational tool in situations where there is sufficient symmetry to make the flux integration trivial. These applications are illustrated in the problems at the end of this chapter.

Second, the differential equation (1.1.38) can be used when the actual charge distribution is not known and must be determined from boundary conditions, as in the case of charged conductors and dielectrics.

Third, the Coulomb law does not correctly describe the electric field in nonstatic situations, where we shall see that $\nabla \times \mathbf{E}$ is no longer zero. However, the divergence equation does continue to hold.

1.2. MULTIPOLES AND MULTIPOLE FIELDS

The electrostatic multipole expansion, which we take up in this section, provides an extremely useful and general way of characterizing a charge distribution and the potential to which it gives rise. Analogous expansions exist for magnetostatic and radiating systems [discussed in Chapter 2 (Section 2.3) and Chapter 5 (Section 5.10), respectively].

As shown in Appendix B, the electrostatic potential outside of an arbitrary finite charge distribution can be expressed as a power series in the inverse radius $1/r$:

$$\phi = \sum_{l=0}^{\infty} \frac{1}{r^{l+1}} F_l(\theta, \varphi).$$

The lth term in the series is called a multipole field (or potential) of order l; it can, in turn, be generated by a single multipole of order l, which we now define, following Maxwell.

A monopole is a point charge Q_0; it gives rise to a potential [choosing $\phi(\infty) = 0$]

$$\phi_0 = \frac{Q_0}{|\mathbf{r} - \mathbf{r}_0|}, \tag{1.2.1}$$

where $\mathbf{r}_0$ is the location of the charge.

A point dipole consists of a charge q at position $\mathbf{r}_0 + \mathbf{l}$ and a charge $-q$ at $\mathbf{r}_0$, where we take the limit $\mathbf{l} \to 0$, with $\mathbf{l}q = \mathbf{p}$ held fixed. $\mathbf{p}$ is called the electric dipole moment of the pair of charges. The potential of a point dipole is given by

$$\phi_1 = \text{Lim}\left\{\frac{q}{|\mathbf{r} - \mathbf{r}_0 - \mathbf{l}|} - \frac{q}{|\mathbf{r} - \mathbf{r}_0|}\right\}$$

$$= -\mathbf{l}q \cdot \nabla \frac{1}{|\mathbf{r} - \mathbf{r}_0|} = -\mathbf{p} \cdot \nabla \frac{1}{|\mathbf{r} - \mathbf{r}_0|}. \tag{1.2.2}$$

We separate $\mathbf{p}$ into a unit vector $\hat{l}$ and a magnitude Q_1 with $\mathbf{p} = Q_1\hat{l}$.

We define higher moments by iterating the procedure: A quadrupole is defined by displacing equal and opposite dipoles, etc. Thus, the 2^l'th pole gives rise to a potential

$$\phi_l = Q_l(-1)^l\hat{l}_1 \cdot \nabla\hat{l}_2 \cdot \nabla \cdots \hat{l}_l \cdot \nabla \frac{1}{|\mathbf{r} - \mathbf{r}_0|}. \tag{1.2.3}$$

The potential ϕ_l is specified by $2l + 1$ numbers: the polar angles θ_i and azimuths φ_i of the l unit vectors, and the magnitude Q_l.

On the other hand, an arbitrary charge distribution $\rho(\mathbf{r})$ generates an electrostatic potential

$$\phi(\mathbf{r}') = \int \frac{d\mathbf{r}\,\rho(\mathbf{r})}{|\mathbf{r} - \mathbf{r}'|} \tag{1.2.4}$$

which can, for $\mathbf{r}'$ outside the charge distribution, be expanded in two equivalent ways. The first is

$$\phi(\mathbf{r}') = \sum_{l=0}^{\infty} \frac{(-1)^l}{l!} \int \rho(\mathbf{r}) P^{(l)}_{i_1\ldots i_l}(\mathbf{r})\,d\mathbf{r} \frac{\partial}{\partial x'_i} \cdots \frac{\partial}{\partial x'_{i_l}} \frac{1}{r'} \tag{1.2.5}$$

where the harmonic polynomials $P^{(l)}_{i_1\ldots i_l}$ are defined in Appendix B:

$$P^{(l)}_{i_1\ldots i_l}(\mathbf{r}) = x_{i_1} x_{i_2} \ldots x_{i_l} - \text{(traces times Kronecker deltas)} \tag{1.2.6}$$

where the traces are subtracted to make the tensor $P^{(l)}_{i_1\ldots i_l}$ traceless. The expansion (1.2.5) is then

$$\phi(\mathbf{r}') = \sum_{l=0}^{\infty} \frac{(-1)^l}{l!} Q^{(l)}_{i_1\ldots i_l} \phi^{(l)}_{i_1\ldots i_l}(\mathbf{r}') \tag{1.2.7}$$

where the potential $\phi^{(l)}_{i_1\ldots i_l}$, defined in (B.2.3), is

$$\phi_{i_1 \ldots i_l}^{(l)}(\mathbf{r}') = \frac{\partial}{\partial x'_{i_1}} \cdots \frac{\partial}{\partial x'_{i_l}} \frac{1}{r'} \tag{1.2.8}$$

and the Cartesian lth rank tensor $Q_{i_1 \ldots i_l}^{(l)}$ is

$$Q_{i_1 \ldots i_l}^{(l)} = \int d\mathbf{r}\, \rho(\mathbf{r})\, P_{i_1 \ldots i_l}^{(l)}. \tag{1.2.9}$$

We call $Q_{i_1 \ldots i_l}^{(l)}$ the 2^lth pole moment of the charge distribution. Since $Q_{i_1 \ldots i_l}^{(l)}$ is an lth rank, traceless, symmetric tensor in three dimensions, the number of independent $Q_{i_1 \ldots i_l}^{(l)}$'s is $2l + 1$, as shown in (B.2).

The second equivalent expression for (1.2.4) is

$$\phi(\mathbf{r}') = \sum_{l=0}^{\infty} \int d\mathbf{r}\, \rho(\mathbf{r})\, P_l(\hat{r} \cdot \hat{r}') \frac{r^l}{r'^{l+1}} \tag{1.2.10}$$

$$= \sum_{l=0}^{\infty} \frac{4\pi}{2l+1} \sum_{m=-l}^{l} \frac{Y_{l,m}(\theta', \varphi')}{r'^{l+1}} Q_{l,m} \tag{1.2.11}$$

where the 2^lth pole moments $Q_{l,m}$ are given by

$$Q_{l,m} = \int d\mathbf{r}\, \rho(\mathbf{r})\, Y_{l,m}^*(\theta, \varphi)\, r^l. \tag{1.2.12}$$

Note that here also the number of independent $Q_{l,m}$'s for each l is $2l + 1$. An obvious question to ask is whether the general potential given by (1.2.11) can be reproduced by a series of Maxwell multipoles, one for each l. The answer is yes; the proof was given by Sylvester and can be found in that source of all wisdom, the 11th edition of the *Encyclopedia Britannica*; look for it under harmonic functions. We do not give the proof here. It is not trivial. Try it for $l = 2$. (See Problem 1.18.)

The number $2l + 1$ for the number of independent $Q_{l,m}$'s is slightly deceptive, since the $Q_{l,m}$'s depend on the coordinate system in addition to the intrinsic structure of the charge distribution. Since a coordinate system is specified by three parameters—for example, the three Euler angles with respect to a standard coordinate system—the number of intrinsic components is, in general, $2l + 1 - 3 = 2l - 2$. This fails to hold for $l = 1$ or 0. Since rotations about a vector leave the vector invariant, the number for $l = 1$ is $2l + 1 - 2 = 2l - 1 = 1$, as it must be: the magnitude of the vector. For $l = 0$, the number is 1, since the charge is invariant to all rotations. The full effect of the freedom of rotations shows up for the first time for $l = 2$. Here, it is convenient to define a coordinate system that diagonalizes the Cartesian tensor $Q_{ij}^{(2)}$. In this coordinate system, the tensor $Q_{ij}^{(2)}$ vanishes for $i \neq j$; it has, in general, three nonvanishing components $Q_{11}^{(2)}$, $Q_{22}^{(2)}$, and $Q_{33}^{(2)}$, with zero trace, that is,

$$Q_{11}^{(2)} + Q_{22}^{(2)} + Q_{33}^{(2)} = 0. \tag{1.2.13}$$

Any two of the three $Q_{ii}^{(2)}$'s (no sum over i) characterize the intrinsic quadrupole structure of the charge distribution.

1.3. ENERGY AND STRESS IN THE ELECTROSTATIC FIELD

The work done in bringing a small charge δq_i from far away to a point $\mathbf{r}_i$ is

$$\delta W_i = - \int_{\infty}^{\mathbf{r}_i} d\mathbf{l} \cdot \mathbf{E} \delta q_i = [\phi(\mathbf{r}_i) - \phi(\infty)] \delta q_i \tag{1.3.1}$$

where we conventionally take $\phi(\infty)$ to be zero for a system whose charges are all contained in a finite volume.

If we bring up several charges δq_i, each to a position $\mathbf{r}_i$, we have, to lowest order in δ_q,

$$\delta W = \sum_i \delta W_i = \sum \phi(\mathbf{r}_i) \delta q_i$$

and for a continuous distribution (with E for electric)

$$\delta W_E = \int_{\text{all space}} d\mathbf{r}\, \phi(\mathbf{r}) \delta\rho(\mathbf{r}). \tag{1.3.2}$$

This is the work done, to first order in $\delta\rho$, in changing $\rho(\mathbf{r})$ to $\rho(\mathbf{r}) + \delta\rho(\mathbf{r})$ and $\mathbf{E}$ to $\mathbf{E} + \delta\mathbf{E}$, where $\nabla \cdot \delta\mathbf{E} = 4\pi\delta\rho$. Thus,

$$\delta W_E = \frac{1}{4\pi} \int d\mathbf{r}\, \phi(\mathbf{r})\nabla \cdot \delta\mathbf{E}$$

and, integrating by parts (i.e., dropping a surface integral at ∞), we have

$$\delta W_E = \frac{1}{4\pi} \int d\mathbf{r} \, \mathbf{E} \cdot \delta \mathbf{E}$$

$$= \frac{1}{8\pi} \delta \int d\mathbf{r} \, \mathbf{E}^2 \tag{1.3.3}$$

all to first order in $\delta\rho$ and $\delta\mathbf{E}$.

Equation (1.3.3) can be integrated: The total work done is

$$\frac{1}{8\pi} \int d\mathbf{r} \, \mathbf{E}_f^2 - \frac{1}{8\pi} \int d\mathbf{r} \, \mathbf{E}_0^2 \tag{1.3.4}$$

where $\mathbf{E}_f$ is the field after the work has been done, $\mathbf{E}_0$ before.

If the initial charge configuration is a uniformly distributed finite charge over a very large volume $\int d\mathbf{r} \, \mathbf{E}_0^2/8\pi$ goes to zero.

If, however, we are bringing together small clumps of charge, then $\int d\mathbf{r} \, \mathbf{E}_0^2/8\pi$ will be different from zero for each clump and must be subtracted in the above formula.

Assuming the first case, we can write

$$W_E = \frac{1}{8\pi} \int d\mathbf{r} \, \mathbf{E}^2 = \frac{1}{2} \int d\mathbf{r} \, \phi(\mathbf{r}) \, \rho(\mathbf{r}) \tag{1.3.5}$$

or

$$W_E = \frac{1}{2} \int d\mathbf{r} \, d\mathbf{r}' \, \rho(\mathbf{r}) \frac{1}{|\mathbf{r} - \mathbf{r}'|} \rho(\mathbf{r}') \tag{1.3.6}$$

for the work done in assembling the charge density ρ. Going to the limit of point charges (i.e., charges with radii small compared to the distance between them) we find that

$$W_E = \frac{1}{2} \sum_{i \neq j} q_i q_j / |\mathbf{r}_i - \mathbf{r}_j| \tag{1.3.7}$$

is the work done in bringing all the charges q_i from $\mathbf{r} = \infty$ to $\mathbf{r}_i$. [The missing terms with $i = j$ are left out because they would have been included in the initial energy of the separated charges. Of course, the point charge approximation could not be made for such terms, since the integral (1.3.6) would be infinite.]

The electrostatic energy W in (1.3.7) has the property that, together with the kinetic energy of the charges q_i,

$$T = \sum_i m_i \frac{\mathbf{v}_i^2}{2}, \tag{1.3.8}$$

it is conserved. That is,

$$\frac{d}{dt}(T + W_E) = 0 \tag{1.3.9}$$

provided the forces on the charges are purely electrostatic and given by Coulomb's law.

We shall see later that a similar calculation can be made for a static (really, a slowly changing) magnetic field:

$$W_M = \frac{1}{8\pi} \int_{\text{all space}} d\mathbf{r}\, \mathbf{B}^2(\mathbf{r}). \tag{1.3.10}$$

Although (1.3.9) and (1.3.10) will have been derived for slowly changing fields, it turns out remarkably, as we shall see later, that the conservation law

$$\frac{d}{dt}(T + W_E + W_M) = 0 \tag{1.3.11}$$

still holds for rapidly changing fields. This appears to be a lucky accident, since it holds for electrodynamics, but does not hold for other field theories, in which an explicit interaction term appears in the conservation law. An example is discussed in Section 7.4.

We turn next to stress in the electrostatic field. We calculate the total electrical force on the charge inside a surface S. Introducing the summation convention we have

$$F_i = \int d\mathbf{r}\, \rho(\mathbf{r})\, E_i(\mathbf{r}) \tag{1.3.12}$$

$$= \frac{1}{4\pi} \int d\mathbf{r}\, E_i(\mathbf{r}) \nabla \cdot \mathbf{E}(\mathbf{r})$$

$$= \frac{1}{4\pi} \int d\mathbf{r}\, E_i \frac{\partial E_j}{\partial x_j} \tag{1.3.13}$$

$$= \frac{1}{4\pi} \int d\mathbf{r} \left[\frac{\partial}{\partial x_j} (E_j E_i) - E_j \frac{\partial E_i}{\partial x_j} \right]. \tag{1.3.14}$$

But,

$$(\nabla \times \mathbf{E})_i = \epsilon_{ijk} \frac{\partial}{\partial x_j} E_k \tag{1.3.15}$$

$$= \frac{1}{2} \epsilon_{ijk} \left(\frac{\partial}{\partial x_j} E_k - \frac{\partial}{\partial x_k} E_j \right). \tag{1.3.16}$$

Therefore, since $\nabla \times \mathbf{E} = 0$, $\partial E_i / \partial x_j = \partial E_j / \partial x_i$ and (1.3.14) becomes

$$F_i = \frac{1}{4\pi} \int d\mathbf{r} \left(\frac{\partial}{\partial x_j} (E_j E_i) - \frac{\partial}{\partial x_i} \frac{E_j^2}{2} \right)$$

$$= \frac{1}{4\pi} \int d\mathbf{r} \frac{\partial}{\partial x_j} \left(E_j E_i - \delta_{ji} \frac{E_k^2}{2} \right). \tag{1.3.17}$$

Gauss's theorem leads to

$$F_i = \int dS_j \, T_{ji} \tag{1.3.18}$$

where T_{ji}, the Maxwell stress tensor, is given by

$$T_{ji} = \frac{E_j E_i - \frac{1}{2} \delta_{ji} \mathbf{E}^2}{4\pi}. \tag{1.3.19}$$

Equation (1.3.18) tells us that the force on charges inside an arbitrary surface S may be thought of as coming from a stress through that surface, where $-T_{ji}$ is the ith component of the force transmitted in the j direction per unit area into the surface. The minus sign exists because dS_j in (1.3.18) is the outward normal.

A simple example: Two charges are shown in Figure (1.1). If both are positive, as in Figure (1.1a), the normal component of the field E_n at the surface equidistant from the two charges is zero, so that the first term in T_{ji} gives zero force through that surface; the second term is negative and, hence, corresponds to a force into the surface, and hence a repulsion. This is as if the lines of force repel each other.

For one positive and one negative charge, as in Figure (1.1b), the situation is different: The parallel component of the field at the surface $E_{\parallel} = 0$, $E_n \neq 0$. Hence, the first term in T_{ji} is twice the second term, and the sign of the force changes, corresponding to an attraction. The lines of force are under tension along their length.

Note that there is no contradiction between a right pointing force on

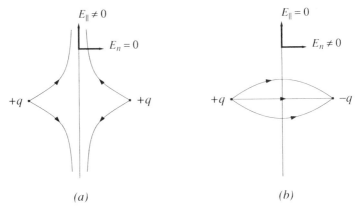

Figure 1.1.

the object on the right and simultaneously a left pointing force on an object on the left. This is, in fact, demanded by Newton's third law.

1.4. ELECTROSTATICS IN THE PRESENCE OF CONDUCTORS: SOLVING FOR ELECTROSTATIC CONFIGURATIONS

The electrostatic field in a conductor must be zero. Otherwise, current would flow, and we would not be doing electrostatics. Therefore, the potential difference between two points in or on the conductor must be zero, since

$$\phi_{12} = -\int_{1}^{2} \mathbf{E} \cdot d\mathbf{l}.$$

Therefore, the surface of the conductor is an equipotential, and the field at the conducting surface is normal to it. It then follows from Gauss' law that the outgoing normal field at the surface, E_n, will be given by

$$E_n = 4\pi\sigma \qquad (1.4.1)$$

where σ is the surface charge density. Note that there can be no volume charge density in the conductor, since $\nabla \cdot \mathbf{E} = 0$ there. Of course, σ cannot be chosen arbitrarily for a conducting surface. Only the total charge Q (if it is insulated) or the potential ϕ (if it is connected to a battery) can be so chosen. The surface charges will adjust themselves to make the conducting surfaces equipotentials. The basic calculational problem of electrostatics is to find out how the charges have adjusted themselves and to calculate the potential (and fields) they generate after doing so.

We show first that given a charge density ρ and a set of conducting surfaces S_i. with either Q_i or ϕ_i known, the electric field is uniquely determined.

Let ψ_1, ψ_2 be two presumed different solutions for the potential. Then

$$I = \int_V d\mathbf{r}(\nabla(\psi_1 - \psi_2))^2 = \int_V d\mathbf{r}\nabla(\psi_1 - \psi_2) \cdot \nabla(\psi_1 - \psi_2)$$

$$= \sum_i \int_i d\mathbf{S}_i(\psi_1 - \psi_2) \cdot \nabla(\psi_1 - \psi_2) - \int_V d\mathbf{r}(\psi_1 - \psi_2)\nabla^2(\psi_1 - \psi_2) \quad (1.4.2)$$

where V is the space contained between the conductors. Both terms are zero. Since $\psi_1 - \psi_2$ is constant over the conducting surface, the first term is proportional to $\Sigma_i \Delta Q_i \Delta \phi_i = 0$. The second term is zero because both ψ_1 and ψ_2 satisfy the Poisson equation $\nabla^2 \psi = -4\pi\rho$ with the same charge density ρ. Therefore, I is zero so that $(\nabla(\psi_1 - \psi_2))^2$ is zero, and ψ_1 and ψ_2 differ at most by a constant. Thus, the electric field is uniquely determined by the boundary conditions and Poisson's equation. Note that if any set of conductors is joined by batteries, with given potential differences between them and given total charge shared among them, the expression $\Sigma_i \Delta Q_i \Delta \phi_i$ is still zero.

The general electrostatic problem can therefore be formulated as follows: Given a set of conducting surfaces, the (appropriately specified) potentials and charges on the surfaces, and a given fixed charge distribution $\rho(\mathbf{r})$ in the space outside of the conducting surfaces, find the potential everywhere.

There is no general method for solving this problem. For certain geometries, however, there are available specific methods, with which we assume the reader is familiar. These include the method of images, the use of special coordinate systems appropriate to the geometry, and the use of analytic functions of a complex variable for two-dimensional problems. Examples of all these are given in the problems at the end of the chapter.

We wish to take up briefly two very general methods that are of use in many areas of physics. These are, first, the method of Green's functions and, second, the use of variational principles.

Green's functions make it possible to reduce to quadratures a class of problems with given potentials or charges on conducting surfaces, and arbitrary spatial charge distribution. The formulation is as follows: given potentials ϕ_c on conducting surfaces S_c and total charges Q_b on conducting surfaces S_b.[4] The Green's function $G(\mathbf{r}, \mathbf{r}_1)$ is the potential produced by a unit point charge at $\mathbf{r}_1$, with zero potential on the S_c's and zero charge on the S_b's. The potential is referred to zero at infinity. Thus, $\nabla^2 G(\mathbf{r}, \mathbf{r}_1) = -4\pi\delta(\mathbf{r} - \mathbf{r}_1)$, $G(\mathbf{r}, \mathbf{r}_1) = 0$ with $\mathbf{r}$ on each S_c and is constant on each S_b with $\int d\mathbf{S}_b \cdot \nabla G(\mathbf{r}, \mathbf{r}_1) = 0$. Let ψ be the actual potential for given ρ, ϕ_c, and Q_b. Consider

$$I(\mathbf{r}_1) = \sum_{i=b,c} \int \{G(\mathbf{r}_i, \mathbf{r}_1) \nabla\psi(\mathbf{r}_i) - \nabla G(\mathbf{r}_i, \mathbf{r}_1) \psi(\mathbf{r}_i)\} \cdot d\mathbf{S}_i \quad (1.4.3)$$

with $d\mathbf{S}_i$ the inward normal to each conducting surface. On S_c, the first integral vanishes, and the second is

$$-4\pi\phi_c \cdot Q_c(\mathbf{r}_1)$$

where $Q_c(\mathbf{r}_1)$ is the charge on S_c for the Green's function boundary condition. On S_b, the second integral vanishes, and the first is $4\pi Q_b \cdot \phi_b(\mathbf{r}_1)$, where $\phi_b(\mathbf{r}_1)$ is the potential on S_b for the Green's function boundary condition.

Now use Gauss' theorem:

$$I(\mathbf{r}_1) = -4\pi \int d\mathbf{r}\, \rho(\mathbf{r})\, G(\mathbf{r}, \mathbf{r}_1) + 4\pi \int d\mathbf{r}\, \delta(\mathbf{r} - \mathbf{r}_1)\, \psi(\mathbf{r}). \quad (1.4.4)$$

Combining (1.4.4) and (1.4.3), we have

$$\psi(\mathbf{r}_1) = \int d\mathbf{r}\, \rho(\mathbf{r})\, G(\mathbf{r}, \mathbf{r}_1) + \sum_b Q_b \phi_b(\mathbf{r}_1) - \sum_a Q_c(\mathbf{r}_1)\, \phi_c, \quad (1.4.5)$$

so that ψ is given by integrals over presumed known functions.

We show now that the Green's function is symmetric: $G(\mathbf{r}, \mathbf{r}_1) = G(\mathbf{r}_1, \mathbf{r})$. Consider

$$I(\mathbf{r}_1, \mathbf{r}_2) = \sum_i \int d\mathbf{S}_i \cdot [G(\mathbf{r}_i, \mathbf{r}_1) \nabla G(\mathbf{r}_i, \mathbf{r}_2) - G(\mathbf{r}_i, \mathbf{r}_2) \nabla G(\mathbf{r}_i, \mathbf{r}_1)].$$

$$(1.4.6)$$

---

[4]Note that given ϕ_i corresponds to Dirichlet boundary conditions, but given Q_i does *not* correspond to Neumann boundary conditions, since only the total charge on a surface is given. Nevertheless, given Q is the physically interesting case and by the uniqueness theorem determines the solution and the Green's function.

Clearly, $I = 0$, since on each surface G is constant, and either zero, or such that $\int \nabla G(\mathbf{r}, \mathbf{r}') \cdot d\mathbf{S} = 0$. So, using Gauss' theorem, we obtain

$$0 = \int d\mathbf{r}\left[G(\mathbf{r}, \mathbf{r}_1) \nabla^2 G(\mathbf{r}, \mathbf{r}_2) - \nabla^2 G(\mathbf{r}, \mathbf{r}_1) G(\mathbf{r}, \mathbf{r}_2) \right]$$

$$= 4\pi[G(\mathbf{r}_2, \mathbf{r}_1) - G(\mathbf{r}_1, \mathbf{r}_2)]. \qquad (1.4.7)$$

Although there is no general exact method for finding the field in the presence of a given configuration of conductors and charges, there is an exact variational principle that applies to a general electrostatic problem and can be used to generate approximate solutions.

Suppose we have a given set of conductors, with label c, on which the potential ϕ_c is given, another set of conductors, with label b, on which the charge Q_b is given, and a given spatial charge density $\rho(\mathbf{r})$. Then, as we have shown, the field $\mathbf{E}(\mathbf{r})$ is determined, as is the potential $\phi(\mathbf{r})$ to within a constant. The variational principle we consider here is for the quantity[5]

$$I = \frac{1}{2} \int d\mathbf{r}(\nabla \psi)^2 - 4\pi \int d\mathbf{r}\rho\psi - 4\pi \sum Q_b\psi(b) \qquad (1.4.8)$$

and states that I is an absolute minimum when the variational function $\psi(\mathbf{r})$ equals the correct potential everywhere. In the variation of ψ about the minimum, it must take on the assigned values ϕ_c on the c conductors. In (1.4.8), the function $\psi(b)$ signifies the constant value of the function $\psi(\mathbf{r})$ on the surface of conductor b (where the potential ϕ_b is *not* given).

To prove the principle, we let

$$\psi = \phi + \delta\psi \qquad (1.4.9)$$

where ϕ is the exact solution of the electrostatic problem. Then

$$I(\psi) = I(\phi) + \int d\mathbf{r} \nabla\phi \cdot \nabla\delta\psi - 4\pi \int d\mathbf{r}\rho\delta\psi$$

$$-4\pi \sum_b Q_b\delta\psi(b) + \frac{1}{2} \int d\mathbf{r}(\nabla\delta\psi)^2 \qquad (1.4.10)$$

---

[5] I is called a functional of ψ. A functional is a number whose value depends on a function. We shall encounter this concept often.

and, after a partial integration,

$$I(\psi) - I(\phi) = \sum_b \int \nabla\phi\,\delta\psi(b) \cdot d\mathbf{S}_b - \int d\mathbf{r}\,\delta\psi\nabla^2\phi$$

$$-4\pi \int d\mathbf{r}\rho\,\delta\psi - 4\pi \sum_b Q_b\delta\psi(b) + \frac{1}{2}\int d\mathbf{r}(\nabla\delta\psi)^2. \quad (1.4.11)$$

The surface element $d\mathbf{S}_b$ points into the surface of conductor b. Conductors c do not contribute, since $\delta\psi_c = 0$. Thus,

$$I(\psi) - I(\phi) = \frac{1}{2}\int d\mathbf{r}(\nabla\delta\psi)^2, \quad\quad\quad (1.4.12)$$

since $\nabla^2\phi = -4\pi\rho$ and

$$\int \nabla\phi \cdot d\mathbf{S}_b = 4\pi Q_b.$$

Therefore, $I(\phi)$ is an absolute minimum for $\delta\psi = 0$.

1.5. SYSTEMS OF CONDUCTORS

Suppose we have a set of conductors, each carrying a charge Q_i. The potential on each conductor will be a linear function of the Q_i's:

$$\phi_i = \sum_j p_{ij}Q_j. \quad\quad\quad (1.5.1)$$

This follows from the linearity of the equations for the fields. Thus, a charge Q_1 on conductor 1, with boundary condition $Q = 0$ on the other conductors, leads to a potential $\phi^{(1)}(\mathbf{r})$ that takes on the value $\phi_i^{(1)} = p_{i1}Q_1$ on the ith conductor. Similarly, a charge Q_2 on 2 with $Q = 0$ on the other conductors leads to a potential $\phi^{(2)}(\mathbf{r})$ that takes on the value $\phi_i^{(2)} = p_{i2}Q_2$ on the ith conductor. If both (1) and (2) carry charges Q_1 and Q_2, the potential is clearly the sum of these two, since the Laplace equation and boundary condition are satisfied. The generalization is (1.5.1). The p_{ij}'s are called coefficients of potential.

The energy of the configuration can be calculated in two ways:

1. We bring charge δQ_i up to the ith conductor. The differential work done is

$$\delta W = \sum_i \phi_i \delta Q_i = \sum_{i,j} p_{ij} Q_j \delta Q_i. \qquad (1.5.2)$$

2. We know, in general, that

$$W = \frac{1}{2} \sum_i \phi_i Q_i = \frac{1}{2} \sum p_{ij} Q_j Q_i \qquad (1.5.3)$$

giving

$$\delta W = \sum_{i,j} \frac{1}{2} (Q_i \delta Q_j + Q_j \delta Q_i) p_{ij} = \frac{1}{2} \sum Q_j \delta Q_i (p_{ij} + p_{ji}) \quad (1.5.4)$$

so that (since Q_j, δQ_i are arbitrary)

$$\frac{p_{ij} + p_{ji}}{2} = p_{ij} \qquad \text{or} \qquad p_{ij} = p_{ji}$$

and the matrix p is symmetric.

We have not considered a charge density $\rho(\mathbf{r})$ here. Clearly, one would take such a charge density into account by first solving the problem of all neutral conductors with the given charge density ρ. Then if $\phi^\rho(\mathbf{r})$ is that solution, the expression (1.5.1) becomes [with $\phi_i^\rho = \phi^\rho(\mathbf{r}$ on $i)$]

$$\phi_i - \phi_i^\rho = \sum_j p_{ij} Q_j \qquad (1.5.5)$$

with the p_{ij} the same as before. $\phi_i - \phi_i^\rho$ is the potential produced on the conductor by the charges Q_i alone.

Returning to (1.5.1), we may solve for the Q's as functions of the ϕ's. This is possible, since we know that

$$W = \frac{1}{2} \sum_{ij} Q_i p_{ij} Q_j > 0$$

unless all Q's are zero. Therefore, p_{ij} has an inverse, c_{ij}, such that $\sum_i p_{ij} c_{jk} = \delta_{ik}$ and

$$Q_i = \sum_j c_{ij} \phi_j. \qquad (1.5.6)$$

The c_{ij} are called coefficients of capacitance. Of course, c_{ij} is also symmetric.

To calculate the generalized force F_i on the ith conductor, we displace it at constant Q:

$$F_i = -\left(\frac{\partial W}{\partial \xi_i}\right)_Q.$$ (1.5.7)

The generalized force F_i is defined by the requirement that $-F_i \delta \xi_i$ be the work done in the displacement $\delta \xi_i$. Thus, F_i can be a force or a torque, depending on whether $\delta \xi_i$ is a translation or a rotation. From (1.5.7) and (1.5.3), we find

$$F_l = -\frac{1}{2}\sum_{i,j} Q_i Q_j \frac{\partial p_{ij}}{\partial \xi_l}.$$ (1.5.8)

Note: Differentiating at constant ϕ would give the wrong answer. In fact,

$$\tilde{F}_l = -\frac{1}{2}\sum_{ij} \phi_i \phi_j \frac{\partial c_{ij}}{\partial \xi_l} = -F_l!$$ (1.5.9)

This follows since, in matrix notation [ϕ, Q are vectors, p, c are matrices, and (f, g) is a (real) vector inner product],

$$\tilde{F} = -\frac{1}{2}\left(\phi, \frac{\partial c}{\partial \xi}\phi\right)$$

$$= -\frac{1}{2}\left(pQ, \left(\frac{\partial}{\partial \xi}\frac{1}{p}\right)pQ\right)$$

$$= -\frac{1}{2}\left(Q, p\left(\frac{\partial}{\partial \xi}\frac{1}{p}\right)pQ\right)$$

(since p is symmetric)

$$= \frac{1}{2}\left(Q, \frac{\partial p}{\partial \xi}Q\right).$$ (1.5.10)

The difference in sign comes about because in a displacement keeping the ϕ's constant charge will flow, and the batteries holding the ϕ's constant will be doing work. It follows that the derivative that gives $\tilde{F}_l$ in (1.5.9) includes, in $\tilde{F}_l \delta \xi_l$, both mechanical and electrical work. A correct account can be kept. The charge transported to the ith conductor is, from (1.5.6),

$$\delta Q_i = \sum_j \delta c_{ij} \phi_j \tag{1.5.11}$$

and the work done by the batteries is

$$\delta W_B = \sum_i \delta Q_i \phi_i = \sum_{i,j} \phi_j \delta c_{ij} \phi_i. \tag{1.5.12}$$

The total work done at constant ϕ is

$$\delta W_T = \frac{1}{2} \sum \phi_i \delta c_{ij} \phi_j = \delta W_B + \delta W_M.$$

where δW_M is the mechanical work:

$$\delta W_M = \delta W_T - \delta W_B = -\frac{1}{2} \sum \phi_i \delta c_{ij} \phi_j$$

$$= \frac{1}{2} \sum Q_i \delta p_{ij} Q_j$$

in agreement with (1.5.8).

The capacitance of a capacitor can be calculated from either set of coefficients. A capacitor consists of two conductors carrying equal and opposite charge. So, with $Q_1 = -Q_2 = Q > 0$, we have

$$\phi_1 = p_{11} Q_1 + p_{12} Q_2 = (p_{11} - p_{12})Q$$
$$\phi_2 = p_{21} Q_1 + p_{22} Q_2 = (p_{12} - p_{22})Q$$

and

$$\phi_1 - \phi_2 = (p_{11} + p_{22} - 2p_{12})Q$$

and

$$C \text{ (the capacitance)} = \frac{Q}{\Delta \phi} = \frac{1}{p_{11} + p_{22} - 2p_{12}}.$$

Note that $C > 0$, since $p_{11} + p_{22} > |2p_{12}|$; otherwise, the energy,

$$W = \frac{1}{2}(Q_1^2 p_{11} + Q_2^2 p_{22} + 2Q_1 Q_2 p_{12}),$$

could become negative.

1.6. ELECTROSTATIC FIELDS IN MATTER

We wish here to study macroscopic electrostatics in the presence of matter. We will make the assumption, following Lorentz, that the macroscopic equations we have been using

$$\nabla \cdot \mathbf{E} = 4\pi\rho \qquad (1.1.36)$$

and

$$\nabla \times \mathbf{E} = 0 \qquad (1.1.17)$$

hold microscopically, that is, at the atomic level. Thus, our basic equations are

$$\nabla \cdot \mathbf{e} = 4\pi\rho_m \qquad (1.6.1)$$

and

$$\nabla \times \mathbf{e} = 0 \qquad (1.6.2)$$

where we use lowercase letters to denote microscopic fields. The symbol ρ_m stands for microscopic charge density. Evidently, $\mathbf{e}$ and ρ_m will fluctuate over atomic scale distances. We eliminate these fluctuations by considering average fields and charge densities, where we average over a region containing many atoms. We then try to obtain equations for the averaged fields.

A subtle issue arises here: Can a description of the interaction of fields and matter that does not make use of quantum mechanics be correct? The answer is yes and no. No, obviously, because ordinary matter and its atomic constituents cannot be accounted for by the laws of classical mechanics. Yes, because in many cases, once the basic structure of the system has been determined by quantum mechanics, interactions with electric fields can be characterized by a few parameters, in addition to macroscopic currents and charges. Examples are the dipole moment per unit volume $\mathbf{P}$ and the dielectric constant ϵ, which we discuss in the following; the magnetic dipole moment per unit volume $\mathbf{M}$ and magnetic permeability μ, which we discuss in Section 2.4; and in addition all of the above as functions of frequency, which we discuss in Chapter 3 on time-dependent fields and currents.

A subtler issue has to do with the validity of classical equations for the electric field. Discussion of this question of course requires the use of quantum field theory. The emission of a single photon by a single atom can not in general be described clasically. However, the multiple photon emission by many atoms, each emitting one photon at a time, and their subsequent absorption, can in many cases be described classically, even though the radiation itself is not in a classical state. This is largely a consequence of the linearity of the field equations. The source of the radiation, the charge and current densities of the radiating system, must be correctly described, classically or quantum mechanically as appropriate. The quantum behavior of matter may be taken into account, either by cautious phenomenology (the nineteenth century method) or by correct theory (current condensed matter physics). We will stick mostly to the nineteen century way, with the exception of the case of a dilute gas, where simple quantum mechanical calculations of the dielectric properties can be carried out.

We proceed by averaging (1.6.1) and (1.6.2) over a region that contains a large number of atoms, but that is small compared to the scale of spatial variation of the fields. We average with a smooth function $f(\mathbf{x})$ such that

$$\int d\mathbf{x}\, f(\mathbf{x}) = 1$$

and such that the characteristic size Δ of f has the two properties

$$\Delta \ll \lambda \tag{1.6.3}$$

where λ is the scale of distance variation we hope to describe, and

$$n\Delta^3 \gg 1 \tag{1.6.4}$$

where n is the number of atoms per unit volume. Δ might be defined, for example, by

$$\Delta^2 = \int d\mathbf{x}\, f(\mathbf{x})\mathbf{x}^2 .$$

We, of course, choose $f(\mathbf{x})$ to be isotropic, that is, a function of $\mathbf{x}^2$. A simple model might be $f = 3/4\pi R^3$ for $r < R$ and $f = 0$ for $r > R$, with some smoothing at the boundary. This model evidently gives $\Delta^2 = (3/5)R^2$. The averages are calculated as

$$E_i(\mathbf{x}) = \int d\mathbf{x}' f(\mathbf{x} - \mathbf{x}')e_i(\mathbf{x}') , \tag{1.6.5}$$

etc. This way of averaging has the advantage that it commutes with differentiation. That is,

$$
\frac{\partial}{\partial x_j} E_i = \int d\mathbf{x}' \frac{\partial f(\mathbf{x} - \mathbf{x}')}{\partial x_j} e_i(\mathbf{x}')
$$

$$
= - \int d\mathbf{x}' \frac{\partial f(\mathbf{x} - \mathbf{x}')}{\partial x_j'} e_i(\mathbf{x}')
$$

$$
= \int d\mathbf{x}' f(\mathbf{x} - \mathbf{x}') \frac{\partial e_i(\mathbf{x}')}{\partial x_j'}
$$

$$
= \overline{\frac{\partial e_i}{\partial x_j}}. \tag{1.6.6}
$$

Thus we find, from (1.6.1) and (1.6.2), for $E_i = \bar{e}_i$:

$$
\nabla \times \mathbf{E} = 0 \tag{1.6.7}
$$

and

$$
\nabla \cdot \mathbf{E} = 4\pi\bar{\rho} \tag{1.6.8}
$$

where $\bar{\rho}$ is the average charge density.

In order to determine $\bar{\rho}$, we divide the charge density into two classes:

$$
\rho_f = \rho_{\text{free}} \quad \text{and} \quad \rho_b = \rho_{\text{bound}}.
$$

One may think of ρ_f as the charge density of charged atomic scale bodies, such as electrons or ions on the surface of a conductor. However, the division is not unique. For example, the induced "bound" surface charge on a dielectric sphere placed in an external field is, for a large dielectric constant, almost identical to the "true" (or "free") surface charge induced on a conducting sphere. (See Problem 1.27).

Our problem is to find a useful way of expressing the space averages of ρ_f and ρ_b. We call

$$
\bar{\rho}_f = \rho, \tag{1.6.9}
$$

the macroscopic charge density, which we presume to be independent of the applied field (except for its distribution on the surface of a conductor). There remains $\bar{\rho}_b$. Evidently, different material systems will behave quite differently, and a separate analysis is really required for each one. In order to fix our ideas, it is convenient to consider the simplest possible system: a set of spherically symmetric atoms whose separations from each other are large compared to their common radius, that is, a dilute gas.

We consider an applied field E_0 that is small compared to the internal fields of the atoms, that is,

$$E_0 \ll \frac{e}{a_B^2} \sim \frac{26 \text{ Volts}}{a_B} \sim 5 \times 10^9 \text{ Volts/cm} \qquad (1.6.10)$$

where e is the electron's charge and a_B the Bohr radius:

$$a_B = \frac{\hbar^2}{me^2} \sim \frac{1}{2} \times 10^{-8} \text{ cm}. \qquad (1.6.11)$$

Here, $\hbar$ is Planck's constant divided by 2π and m is the electron mass.

In view of (1.6.10), the effect of the applied field on the matter will be small, and we can confine ourselves to the linear approximation in an expansion in powers of the field. There are, of course, systems where the required inequality fails to apply, for example, in molecules with large permanent dipole moments as discussed in Section 2.4, or in highly excited atoms. Since the atoms in our model are far apart, the interaction between them will be largely governed by the multipole moments produced by the applied field.

Since the atoms are neutral, the largest effect will come from the induced electric dipole moment. This moment will be proportional to the local electric field $\mathbf{E}_l$ at the position of the atom. For our model of widely separated atoms, we will have approximately $\mathbf{E}_l = \mathbf{E}$, the average electric field, so that

$$p_i = \alpha E_{li} \cong \alpha E_i. \qquad (1.6.12)$$

The polarizability α has the dimensions of a volume; the atomic unit of volume is a_B^3, so that we expect α to be of order a_B^3. The field produced by the polarized atoms will then be of order

$$\delta E \sim \alpha E_0 \sum_i \frac{1}{|\mathbf{r} - \mathbf{r}_i|^3} \qquad (1.6.13)$$

so that

$$\frac{\delta E}{E_0} \sim \frac{\alpha N}{R^3} \qquad (1.6.14)$$

where N is the number of atoms in the sample creating the field δE, and R is a mean separation of the atoms. Thus, very crudely,

$$\frac{\delta E}{E_0} \sim \frac{\text{total atomic volume}}{\text{total occupied volume}}; \qquad (1.6.15)$$

for a gas, this ratio is $\sim d/d_s$, where d is the gas density and d_s the liquid or solid density of the same atom. For air at normal temperature and pressure, the ratio (1.6.15) is about 10^{-3}, so the vacuum field is not appreciably perturbed by the presence of the gas, and the effect of atoms on each other will be small. The local field $\mathbf{E}_l$ that polarizes the individual atom will be approximately equal to the average field $\mathbf{E}$. We will return to the question of atomic polarizability in Chapter 3. (See also Problem 1.36.)

A quadrupole moment can also be present; in an isotropic atom, however, the tensor quadrupole Q_{ij} can only be induced by a tensor field:

$$Q_{ij} = \alpha_Q \left(\frac{\partial E_i}{\partial x_j} + \frac{\partial E_j}{\partial x_i} \right). \qquad (1.6.16)$$

The quadrupole polarizability α_Q in (1.6.16) has the dimensionality L^5, so we expect α_Q for an atom to be $\sim a_B^5$. The field generated by the induced moment, in analogy with (1.6.13), will be

$$\delta E_Q \sim \alpha_Q \frac{\partial E_0}{\partial x} \sum_i \frac{1}{|\mathbf{r} - \mathbf{r}_i|^4} \qquad (1.6.17)$$

or

$$\delta E_Q \sim \frac{E_0}{R} \times \frac{N\alpha_Q}{R^4} \qquad (1.6.18)$$

so that

$$\frac{\delta E_Q}{E_0} \sim \frac{\text{total atomic volume}}{\text{total occupied volume}} \times \frac{a_B^2}{R^2}. \qquad (1.6.19)$$

This is smaller than the dipole effect by about 10^{-16} and is thus completely negligible. Higher moments clearly make even smaller contributions.

If we modify our model by bringing the atoms closer together, for example as a dense gas, or a solid or liquid, it will no longer be possible to ignore the interatomic interactions. It will still be true that the effective field δE generated by the atomic dipoles will be of the order of magnitude given by (1.6.15); however, the ratio in (1.6.15) can now be of the order of unity, so that the effect of the atomic polarization will be not only large but also not simply calculable. (For a quite successful way of estimating $\mathbf{E}_l$ for a denser system, see Problem 1.36.) We still expect the atomic polarization to be a linear function of the average field in the neighborhood of the atom, and we still expect the higher multipole moments to make negligible contributions. However, in addition to the density dependence, a significant change will be that the relation between **p** and **E** may, in general, be tensorial:

$$p_i = \alpha_{ij} E_j \qquad (1.6.20)$$

where the tensor α_{ij} would depend on the symmetry of the material. A locally isotropic material, or a crystal with cubic symmetry, would revert to the scalar relation with $\alpha_{ij} = \alpha \delta_{ij}$.

Before proceeding, we observe that material not locally isotropic can also possess electric moments even in the absence of an applied field. A crystal without reflection symmetry, for example, can have a permanent electric moment. Such a material is called ferroelectric or pyroelectric. We can obtain an order-of-magnitude estimate of the field produced outside a material whose atoms are permanently polarized with a dipole moment p_0. It will be

$$E_{p0} \sim \frac{N p_0}{R^3}$$

where N is the number of atoms in the sample and R a mean distance to the field point. Following the reasoning used to arrive at (1.6.15), we find

$$E_{p0} \sim \frac{p_0}{a_B^3}.$$

With $p_0 \sim e a_B \xi$, where ξ is a number of very rough order-of-magnitude unity, we have

$$E_{p0} \sim \xi \frac{e}{a_B^2} \sim 26\xi \frac{\text{Volts}}{\text{Å}},$$

which is a very large macroscopic field. This field is reduced by two effects. First, the parameter ξ turns out to be quite small since the energetics of the quantum states mitigates against a large dipole moment. Second, since the conductivity of the material is never exactly zero, the dipole moment of the sample tends to be canceled by a migration of electrons to the surface. Similar reasoning shows that permanent quadrupole moments can generate macroscopic fields of rough order, Volts/cm. Although these permanent fields are of considerable interest, they do not require further discussion here, since they play the role of fixed applied fields in our discussion of electrostatics.

The final result: Only the dipole field is important in most macroscopic electrostatics. For it, the average potential will be given by the average dipole moment per unit volume [which we call $\mathbf{P}(\mathbf{r})$] by the dipole formula

$$\phi = -\int d\mathbf{r}' \, \mathbf{P}(\mathbf{r}') \cdot \nabla \frac{1}{|\mathbf{r} - \mathbf{r}'|}. \tag{1.6.21}$$

To see this unambiguously, we calculate the averaged value of the potential ϕ_p arising from the atomic dipole moments:

$$\phi_p(\mathbf{R}) = \int d\mathbf{r} \, f(\mathbf{R} - \mathbf{r}) \sum_i \mathbf{p}_i \cdot \frac{\mathbf{r} - \mathbf{r}_i}{|\mathbf{r} - \mathbf{r}_i|^3} \tag{1.6.22}$$

where we have legitimately used the formula for the field of a point dipole, since in the averaging process, most of the range of $\mathbf{r}$ will be far from the dipole $\mathbf{r}_i$.

Now let $\mathbf{r} - \mathbf{r}_i = \mathbf{R} - \mathbf{R}'$ with $\mathbf{R}'$ the new integration variable. Then (1.6.22) becomes

$$\phi_p(\mathbf{R}) = \int d\mathbf{R}' \left(\sum_i f(\mathbf{R}' - \mathbf{r}_i) \mathbf{p}_i \right) \cdot \frac{\mathbf{R} - \mathbf{R}'}{|\mathbf{R} - \mathbf{R}'|^3}. \tag{1.6.23}$$

We see that $\sum_i f(\mathbf{R}' - \mathbf{r}_i) \mathbf{p}_i$ essentially counts all dipoles inside the averaging distance of f, so that $\sum_i f(\mathbf{R}' - \mathbf{r}_i) \mathbf{p}_i = \mathbf{P}$, the dipole moment per unit volume. Returning to (1.6.21), we see that it can be rewritten

$$\phi = \int d\mathbf{r}' \left[\nabla' \cdot \left(\frac{\mathbf{P}(\mathbf{r}')}{|\mathbf{r} - \mathbf{r}'|} \right) - \frac{\nabla' \cdot \mathbf{P}(\mathbf{r}')}{|\mathbf{r} - \mathbf{r}'|} \right]. \tag{1.6.24}$$

We now have a choice. We clearly will use Gauss' theorem, but we may either treat the dielectric boundary as a continuous (but rapid) change from finite $\mathbf{P}$ to $\mathbf{0}$, in which case we would have no surface term, and a potential

$$\phi = \int \frac{d\mathbf{r}'}{|\mathbf{r} - \mathbf{r}'|} \left(-\nabla' \cdot \mathbf{P}(\mathbf{r}') \right), \tag{1.6.25}$$

corresponding to an average bound charge density

$$\bar{\rho}_b = -\nabla \cdot \mathbf{P}; \tag{1.6.26}$$

or, we could treat the integral as confined to the dielectric with a sharp surface S, in which case we would have

$$\phi = \int \frac{\mathbf{P}(\mathbf{r}') \cdot d\mathbf{S}'}{|\mathbf{r} - \mathbf{r}'|} + \int_{V \subset S} \frac{d\mathbf{r}'\left(-\nabla' \cdot \mathbf{P}(\mathbf{r}')\right)}{|\mathbf{r} - \mathbf{r}'|} \qquad (1.6.27)$$

corresponding to a $\bar{\rho}_b = -\nabla \cdot \mathbf{P}$ and a surface charge $\bar{\sigma}_b = P_{\text{normal}}$ on the surface. Obviously, if there is a sharp boundary, (1.6.25) must produce (1.6.27) in the limit. This clearly comes about because $\nabla \cdot \mathbf{P}$ approximates a delta function on the surface.

We can now return to our original averaged equations. We call $\bar{\rho}_f = \rho$, $\bar{\rho}_b = -\nabla \cdot \mathbf{P}$, and find, from (1.6.26), (1.6.8), and (1.6.7),

$$\nabla \times \mathbf{E} = 0 \qquad (1.6.7)$$

and

$$\nabla \cdot \mathbf{E} = 4\pi\rho - 4\pi\nabla \cdot \mathbf{P}. \qquad (1.6.28)$$

We are moved to define an electric displacement

$$\mathbf{D} = \mathbf{E} + 4\pi\mathbf{P} \qquad (1.6.29)$$

which satisfies the equation

$$\nabla \cdot \mathbf{D} = 4\pi\rho. \qquad (1.6.30)$$

Equations (1.6.7) and (1.6.30) determine how we should treat a sharp boundary. From (1.6.7) we find, using Stokes' theorem on the rectangle shown in Figure 1.2a, and $\Delta x / \Delta l \rightarrow 0$, that $E_{\text{tangential}}$ must be continuous across the surface. For the $\mathbf{D}$ boundary condition, we use Gauss' theorem and the pillbox as shown in Figure 1.2b, with $\Delta x / \sqrt{\Delta S} \rightarrow 0$, to find $\Delta D_n = 4\pi\sigma$, where σ is the free surface charge density.

Equations (1.6.7) and (1.6.30) require a relation between $\mathbf{E}$ and $\mathbf{D}$ to

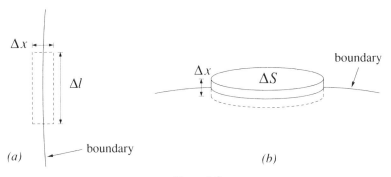

Figure 1.2.

determine the configuration. Since the system is linear, the earlier arguments of this section show that **P** must be proportional to **E**, that is,

$$P_i = \chi_{ij} E_j \qquad \text{(remember the summation convention)} \qquad (1.6.31)$$

where for a dilute gas $\chi_{ij} = n\alpha_{ij}$, with n the number of dipoles per unit volume. Therefore,

$$D_i = (\delta_{ij} + 4\pi\chi_{ij})E_j = \epsilon_{ij} E_j, \qquad (1.6.32)$$

where a symmetric α_{ij} will produce a symmetric ϵ_{ij}. ϵ_{ij} is called the dielectric tensor. An isotropic material as will be the case here would require $\epsilon_{ij} = \epsilon \cdot \delta_{ij}$ and $\mathbf{D} = \epsilon\mathbf{E}$. ϵ is called the dielectric constant.

1.7. ENERGY IN A DIELECTRIC MEDIUM

From the basic equations (1.6.7) and (1.6.30), we learn that there exists a potential ϕ such that $\mathbf{E} = -\nabla\phi$. Therefore, the work done in bringing an infinitesimal charge δq to a point $\mathbf{r}$ is

$$\delta W = \delta q\,\phi(\mathbf{r}), \qquad (1.7.1)$$

where we have set $\phi(\infty) = 0$, as usual. For a distributed charge,

$$\delta W = \int d\mathbf{r}\,\delta\rho(\mathbf{r})\phi(\mathbf{r})$$

$$= \frac{1}{4\pi}\int d\mathbf{r}\,\delta(\nabla \cdot \mathbf{D})\phi(\mathbf{r})$$

$$= -\frac{1}{4\pi}\int d\mathbf{r}\,\delta\mathbf{D} \cdot \nabla\phi, \qquad (1.7.2)$$

or

$$\delta W = \frac{1}{4\pi}\int d\mathbf{r}\,\delta\mathbf{D} \cdot \mathbf{E}. \qquad (1.7.3)$$

In the integration by parts, we have ignored surface terms on conducting surfaces. These are implicitly included in (1.7.2); if made explicit, the surface term in the Gauss' theorem integration by parts would have canceled the $\delta Q\phi$ surface contribution to δW, so that (1.7.3) continues to hold in the presence of conductors.

We consider here only linear media, for which

$$D_i = \epsilon_{ij} E_j .$$

Then

$$\delta W = \frac{1}{4\pi} \int d\mathbf{r} \epsilon_{ij}(\delta E_j) E_i \qquad (1.7.4)$$

$$= \frac{1}{4\pi} \int d\mathbf{r} \left(\epsilon_{ij}^S + \epsilon_{ij}^A \right) \delta E_j \cdot E_i \qquad (1.7.5)$$

where $\epsilon_{ij}^S = (\epsilon_{ij} + \epsilon_{ji})/2$ is the symmetric part of ϵ_{ij}; $\epsilon_{ij}^A = (\epsilon_{ij} - \epsilon_{ji})/2$ is the antisymmetric part.

The first term in (1.7.5) can be integrated since

$$\epsilon_{ij}^S \delta E_j E_i = \frac{\epsilon_{ij}^S \delta(E_j E_i)}{2} . \qquad (1.7.6)$$

However, the integral of $\delta W^A = \epsilon_{ij}^A \delta E_j E_i$ depends on the path of integration. In particular, a closed path in $\mathbf{E}$ space will not, in general, integrate to zero. Take a path in the E_z, E_y-plane. Then by Stokes' theorem

$$\delta W^A = \epsilon_{zy}^A \oint (E_z \, dE_y - E_y \, dE_z)/8\pi$$

$$= \epsilon_{zy}^A \frac{\text{(area enclosed by the path)}}{4\pi} . \qquad (1.7.7)$$

Thus, a static, antisymmetric component in the dielectric tensor indicates a medium that absorbs or produces energy.[6] An energy-conserving system must therefore have a symmetric dielectric tensor, $\epsilon_{ij} = \epsilon_{ji}$, to which case we restrict ourselves here. The work done in charging the system is then

$$W = \frac{1}{8\pi} \int d\mathbf{r} \, \mathbf{E} \cdot \mathbf{D} \qquad (1.7.8)$$

and may be identified with the electrostatic contribution to the free energy of the system.

Equation (1.7.8) has a remarkable property. It appears to express the electrostatic energy

$$W_m = \frac{1}{8\pi} \int d\mathbf{r} \, \mathbf{e}^2 \qquad (1.7.9)$$

[6]We will discuss absorption for time-dependent fields in Chapter 3.

(where the subscript m indicates that W_m is calculated from the microscopic field strength) as a quadratic functional of the averaged field strength and polarization, $\mathbf{E}$ and $\mathbf{P} = (\mathbf{D} - \mathbf{E})/4\pi$; that is, it appears to ignore fluctuations of the microscopic field.

Actually, in the absence of conductors, the quadratic dependence is only apparent. We can see this by writing the field $\mathbf{E}$ as

$$\mathbf{E} = \mathbf{E}_0 + \mathbf{E}_p \tag{1.7.10}$$

where $\mathbf{E}_0$ is the field produced by the macroscopic charge density ρ and $\mathbf{E}_p$ the field produced by the polarization charge density $\rho_p = -\nabla \cdot \mathbf{P}$. The electrostatic energy change produced by introducing the dielectric medium is

$$\Delta W = \frac{1}{8\pi} \int d\mathbf{r} (\mathbf{E} \cdot \mathbf{D} - \mathbf{E}_0^2) \tag{1.7.11}$$

$$= \frac{1}{8\pi} \int d\mathbf{r} \left[(\mathbf{E} + \mathbf{E}_0) \cdot (\mathbf{D} - \mathbf{E}_0) - 4\pi \mathbf{P} \cdot \mathbf{E}_0 \right], \tag{1.7.12}$$

where $\mathbf{P}$ is the polarization density. In the absence of conductors, the first term in (1.7.12) vanishes after an integration by parts, since $\nabla \cdot (\mathbf{D} - \mathbf{E}_0) = 0$ and $(\mathbf{E} + \mathbf{E}_0) = -\nabla(\phi + \phi_0)$. There remains

$$\Delta W = -\frac{1}{2} \int d\mathbf{r}\, \mathbf{P} \cdot \mathbf{E}_0 \tag{1.7.13}$$

which is a linear functional of the microscopic polarization and bilinear only in the applied field.

Equation (1.7.13) also holds in the presence of conductors, although the proof is more complicated.

It should be noted that following (1.7.8), we referred to W as the electrostatic contribution to the free energy (as opposed to internal energy) of the system. This is because in calculating the work done in electrifying our system, we kept the dielectric constant, and hence the matter density and temperature, fixed.

The change in free energy F is defined by

$$\delta F = \delta(U - TS) \tag{1.7.14}$$

where U is the internal energy, T the absolute temperature, and S the entropy. Therefore,

$$\delta F = \delta U - T\delta S - S\delta T$$

$$= \delta W - S\delta T \tag{1.7.15}$$

where δW is the work done on the system.

The work δW includes mechanical work, such as the familiar mechanical work done on a gas:

$$\delta W_m = -P\delta V \tag{1.7.16}$$

and the electrical work done on a dielectric medium:

$$\delta W_e = \frac{1}{4\pi} \int d\mathbf{r}(\mathbf{E} \cdot \delta \mathbf{D} - \mathbf{E}_0 \cdot \delta \mathbf{E}_0). \tag{1.7.17}$$

As before, $\mathbf{E}_0$ is the applied field, that is, the field that would be present in the absence of the dielectric sample. We see thus that at constant T and V, $\delta F = \delta W_e$.

The arguments leading to (1.7.13) also lead to

$$\delta W_e = -\int d\mathbf{r}\, \mathbf{P} \cdot \delta \mathbf{E}_0, \tag{1.7.18}$$

showing that the independent variables in the free energy should be volume (or density), temperature, and $\mathbf{E}_0$.

We then learn, from (1.7.15), that

$$S = -\frac{\partial F}{\partial T}\bigg|_{\rho, \mathbf{E}_0} \tag{1.7.19}$$

and, from (1.7.18), that

$$\mathbf{P}(\mathbf{r}) = -\frac{\delta F}{\delta \mathbf{E}_0(\mathbf{r})}, \tag{1.7.20}$$

where (1.7.20) and (1.7.18), in fact, define $\delta F/[\delta \mathbf{E}_0(\mathbf{r})]$ as the functional derivative of F with respect to $\mathbf{E}_0(\mathbf{r})$. Here ρ is the matter density of the material.

The internal energy is given by

$$U = F + TS$$

$$= F - T\left(\frac{\partial F}{\partial T}\right)_{\rho, \mathbf{E}_0}. \tag{1.7.21}$$

The electrical contribution to the internal energy is therefore, from (1.7.13) for ΔW and $\Delta W = \Delta U$ in (1.7.21),

$$U_{el} = -\frac{1}{2} \int d\mathbf{r}\, \mathbf{E}_0(\mathbf{r}) \cdot \left[\mathbf{P}(\mathbf{r}) - T\frac{\partial \mathbf{P}(\mathbf{r})}{\partial T} \right]. \tag{1.7.22}$$

Forces and stresses may be obtained by calculating the change in F resulting from appropriate displacements. For example, the total force or torque F_i on a dielectric would be calculated as

$$F_i = -\frac{\partial F}{\partial \xi_i} \tag{1.7.23}$$

where ξ_i is the appropriate conjugate variable: for force, a fixed coordinate in the dielectric, for example, the center of mass, for torque an infinitesimal angle of rotation about a coordinate axis. If there are conductors present, the arguments of Section 1.5 show that the total charge on each conductor must be held fixed in (1.7.23).

The calculation of internal stresses in a dielectric is harder, but can be carried out by considering internal displacements. A straightforward treatment of stresses in a fluid dielectric is given in Panofsky and Phillips.[7]

CHAPTER 1 PROBLEMS

Application of Gauss' and Coulomb's law to simple systems

1.1. Consider a spherically symmetric charge distribution, $\rho = \rho(r)$. Assuming the resultant electric field $\mathbf{E}$ to be radial, $\mathbf{E} = \hat{r}E_r(r)$, show that $E_r(r)$ is given by

$$E_r(r) = \frac{1}{r^2} Q(r)$$

where $Q(r)$ is the charge inside r.

1.2. Consider a cylindrically symmetric charge distribution, that is, $d = d(\rho)$, where z, ρ, and φ are cylindrical coordinates, and d is now the charge density. Assuming the field to be radial, $\mathbf{E} = \hat{e}_\rho E_\rho(\rho)$, show that $E_\rho(\rho)$ is given by

---

[7]W. K. H. Panofsky and M. Phillips, *Classical Electricity and Magnetism*, Reading, MA: Addison-Wesley, Chap. 6.

$$E_\rho(\rho) = \frac{2}{\rho} Q(\rho)$$

where $Q(\rho)$ is the total charge per unit length inside the radius ρ.

1.3. Consider a large, rectangular charged plane, with uniform surface charge density σ, lying at $z = 0$. Near the center of the plane and near $z = 0$, assume by symmetry that the field is in the z direction, $\mathbf{E} = \hat{e}_z E_z(z)$. Then show that

$$E_z = 2\pi\sigma\epsilon(z)$$

where $\epsilon(z) = 1$, $z > 0$, $\epsilon(z) = -1$, $z < 0$.

1.4. Now let the charged plane at $z = 0$ extend from $x = -l_1$ to $x = l_2$ and from $y = -l$ to $y = l$. Investigate the assumption that at $x = y = 0$, the fields E_x and E_y are zero, and E_z is independent of x and y. Give the conditions on l_1, l_2, l, and z under which the result of Problem 1.3 is approximately correct.

1.5. Do the same for the situation of Problem 1.2, where the cylindrical distribution runs from $z = -l_1$ to $z = l_2$. *Hint:* The final azimuthal integral can be done by going to the complex plane. Let $z = e^{i\varphi}$.

1.6. Prove that a conductor is an electrostatic shield. That is, show that the (static) field inside an empty hollow conductor is zero.

1.7. From Problem 1.6 it follows that a conductor inside a charged conducting shield will not become charged even if put in electrical contact with the outer conductor. This is independent of the shape of either conductor, but depends, via Gauss' law, etc., on the inverse square law. For a slightly different force law, the inner conductor will normally become charged when connected to a charged outer conductor. The calculation of that charge is difficult except for specially shaped conductors. Consider now that both conductors are thin spheres, concentric, with radii b and a, with $b > a$. The outer conductor b carries a charge Q and is electrically connected to the inner conductor.

(a) Calculate the charge δQ induced on the inner conductor if the force law is derived from a potential

$$V = \frac{q}{r} e^{-r/L}$$

where $L \gg b$. Note that this form of potential would result from a theory with massive photons.

(b) Same as (a) but with

$$V = \frac{q}{r}\left(\frac{L}{r}\right)^{\epsilon}$$

with $|\epsilon| \ll 1$.

(c) Same as (a) and (b) but with

$$V = \frac{q}{r}\left(1 + \epsilon\left(\frac{r_0}{r}\right)^{\beta}\right)$$

with $|\epsilon| \ll 1$ and $-1 < \beta < 1$.

Problems with images

1.8. Two conducting plates make an angle of 90° with each other. Let one plate be in the yz-plane, with $y > 0$; the other in the xz-plane, with $x > 0$; and place a point charge Q at $x = x_0 > 0$, $y = y_0 > 0$, $z = 0$. This configuration can be solved with images. Where and what are they?

1.9. Two conducting planes make an angle of 45° with each other. A point Q is placed at a distance R from the vertex, half-way between the plates. This configuration can be solved with images. Where and what are they?

1.10. A point charge Q is placed at a distance L from the center of a grounded conducting sphere of radius $R < L$. Show that the configuration is solved with an image charge Q' placed at a distance $R' = R^2/L$ from the center of the sphere. Of course, $R' < R$. Determine Q' as a function of Q, R, and L.

1.11. The same as Problem 1.10, except with a neutral isolated conducting circular cylinder, radius R, and a line charge λ a distance L from the center of the cylinder.

1.12. A point dipole is at a distance L and pointing away from the center of a grounded sphere of radius R. Using the method of images, find
   (a) The charge induced on the sphere.
   (b) The electric dipole moment (with respect to the center of the sphere) induced on the sphere.

1.13. Two isolated conducting spheres of radii R_1 and R_2 are placed a distance L apart, with $L \gg R_1$ and $L \gg R_2$. A charge Q is placed on sphere 1. This configuration can be solved by successive approximations involving images, images of images, etc.

(a) Calculate the lowest order (in $R_{1 \text{ or } 2}/L$) contribution to the force between the spheres. Carry the calculation one step farther, sufficient to identify the power of $R_{1 \text{ or } 2}/L$ of the next correction.

(b) If you calculated the force by setting it equal to the force between the image charges, your calculation was correct. Can you justify this statement? Although the answer "no" may be correct (i.e., you cannot justify this statement), it is not acceptable.

Problems involving spherical harmonics

1.14. An isolated conducting sphere is placed in a uniform electric field E_0 in the z direction, so that the applied potential is

$$\phi_0 = -E_0 z = -E_0 r \cos \theta = -E_0 r \, P_1(\cos \theta).$$

Find the potential $\delta\phi$ generated by the induced charge distribution on the surface of the sphere. From it, find the charge distribution itself and the dipole moment of the sphere. Assume that the sources of E_0 are far enough away so that they are unaffected by the introduction of the sphere.

1.15. (a) Suppose the applied field vanishes at the center of the sphere, with

$$E_z = E_0 \cdot \frac{z}{L}, \qquad E_x = -\frac{E_0 x}{2L}, \qquad E_y = -\frac{E_0 y}{2L}.$$

Find the potential $\phi_0(\mathbf{r})$ that corresponds to this field, the potential $\delta\phi$ generated by the induced charge distribution on the sphere, the induced charge distribution itself, and the total charge, dipole moment, and quadrupole moment of the sphere.

(b) How would you arrange a charge distribution to create this applied field near the sphere?

1.16. A nonconducting, very thin shell of radius b carries a surface charge distribution

$$\sigma = \sigma_0 (\cos \theta + \frac{1}{2} \sin \theta \cos \varphi).$$

Find the electrostatic potential for $r > b$ and $r < b$, assuming the field to be finite everywhere and no other charge to be present.

1.17. Solve the problem of a point charge and a grounded conducting sphere by expanding the Coulomb potential of the point charge in spherical harmonics with respect to the center of the sphere, solving for the potential for each l and resumming.

*1.18. Show that a single Maxwell quadrupole, characterized by two unit vectors $\hat{n}_1$, $\hat{n}_2$ and a magnitude Q, can produce the correct asymptotic potential of an arbitrarily confined $l = 2$ charge distribution. Clearly, the plane of $\hat{n}_1$ and $\hat{n}_2$ must contain two of the principle axes of the quadrupole. However, you must choose the right two.

Problems in two dimensions involving the use of analytic functions

(*Prerequisite*: Some knowledge of complex function theory.)

1.19. When the potential is independent of one rectangular coordinate, the Laplace equation for the potential ϕ becomes

$$\frac{\partial^2 \phi}{\partial x^2} + \frac{\partial^2 \phi}{\partial y^2} = 0.$$

The real and imaginary parts of any analytic function satisfy this equation, as we now show.

(a) An analytic function of a complex variable z

$$w = f(z),$$

has a derivative

$$\frac{dw}{dz} = \lim_{\Delta z \to 0} \left[\frac{f(z + \Delta z) - f(z)}{\Delta z} \right]$$

which is independent of the path by which $\Delta z \to 0$. With $z = x + iy$ and $w = u + iv$, show that this path independence implies the Cauchy–Riemann equations

$$\frac{\partial u}{\partial x} = \frac{\partial v}{\partial y} \quad \text{and} \quad \frac{\partial u}{\partial y} = -\frac{\partial v}{\partial x}.$$

(b) Show that the Cauchy–Riemann equations imply that u and v are harmonic.

(c) It is conventional to choose the imaginary part of w, v to be the potential.[8] With this choice, show that the lines of force are lines of constant u.

1.20. Given one or more two-dimensional conductors (i.e., three-dimensional cylinders), one looks for a function $w = f(z)$ such that the

---

[8]One could equally well choose u, the real part of w, as the potential.

cylinders are lines of constant v. The potential problem is thus automatically solved. Show then that the charge density on the conductor is $1/4\pi \; \partial u/\partial l$, where dl is the length variable along the conductor. To illustrate this technique, consider the function

$$u + iv = w = \phi_0 \sqrt{\frac{z}{b}} = \sqrt{\frac{x + iy}{b}} \; \phi_0$$

defined to be the positive square root for $x > 0$ when y approaches zero from above and analytic in the cut plane (cut from $y = 0$, $x = 0$ to $y = 0$, $x = \infty$).

(a) Find the equation for equipotential surfaces.

(b) The equipotential $v = 0$ is obtained by letting $y \to 0$ from above. For $0 < y \ll x$, give v as a function of x and y and find the charge density on the upper surface of the conductor as a function of x.

(c) The value of w for $y < 0$ is found by analytically continuing around the singularity. Use this procedure to calculate the charge density under the plate.

Note that although the charge density is singular as $x \to 0$, the integrated charge in a finite region of x is finite. Also, since the total charge is infinite (in three dimensions), there is no uniqueness here. In practice, the conditions that determine ϕ_0 and b will come from boundary conditions at ∞. (See Problems 1.22 and 1.23.)

1.21. The potential of a line charge λ is $2\lambda \log \rho$, where ρ is the cylindrical radius. If the line charge is at position $\boldsymbol{\rho}'$, the potential at $\boldsymbol{\rho}$ is

$$\phi(\boldsymbol{\rho}) = 2\lambda \log |\boldsymbol{\rho} - \boldsymbol{\rho}'|.$$

If the charge is distributed with a density $\sigma(\boldsymbol{\rho}')$, the resultant potential will be

$$\phi(\boldsymbol{\rho}) = 2 \int d\boldsymbol{\rho}' \; \sigma(\boldsymbol{\rho}') \log |\boldsymbol{\rho} - \boldsymbol{\rho}'|,$$

where the area element $d\boldsymbol{\rho}' = \rho' d\rho' d\varphi'$, and $|\boldsymbol{\rho} - \boldsymbol{\rho}'| = (\rho^2 + \rho'^2 - 2\rho\rho' \cos(\varphi - \varphi'))^{1/2}$. Show that the two-dimensional potential has a "multipole" expansion, for ρ outside of the charge distribution,

$$\phi(\boldsymbol{\rho}) = 2\lambda \log \frac{\rho}{\rho_0} - 2 \sum_{m=1}^{\infty} \frac{1}{m} \frac{1}{\rho^m} [c_m \cos m\varphi + d_m \sin m\varphi]$$

where ρ_0 is an arbitrary length scale, λ the total line charge,

$$\lambda = \int d\boldsymbol{\rho}' \, \sigma(\boldsymbol{\rho}'),$$

and

$$c_m = \int d\boldsymbol{\rho}' \, (\rho')^m \cos m\varphi' \, \sigma(\boldsymbol{\rho}'), \quad d_m = \int d\boldsymbol{\rho}' \, (\rho')^m \sin m\varphi' \, \sigma(\boldsymbol{\rho}').$$

You should make use of the identity

$$|z - z'| = |\rho^2 + \rho'^2 - 2\rho\rho' \cos(\varphi - \varphi')|^{1/2}$$

where z is the complex variable $z = \rho e^{i\varphi}$, and $\text{Re} \, \log(z - z') = \log|z - z'|$.

1.22. Consider the function

$$w = \phi_0 \sqrt{1 - \left(\frac{z}{b}\right)^2}$$

defined to be real and positive for $|x| < b$ and $y \to 0$ from above, and analytic in the cut plane from $y = 0$, $x = -b$ to $y = 0$, $x = b$.

(a) Find the value of w for $x = b + \epsilon$, $y = 0$ and $x = -b - \epsilon$, $y = 0$.

(b) Find the value of w as $x \to \pm\infty$. From this, find the physical system represented here.

1.23. Consider the analytic function $w(z)$ given implicitly by

$$z = w\pi + \frac{1}{2}(-1 + e^{-2\pi w})$$

(clearly, dimensional coordinates would scale z by a length unit and w by a potential unit). Let $w = u + iv$, with v as the potential.

(a) What is the line $v = 0$ in the x, y plane?

(b) What is the line $v = 1$ in the x, y plane?

(c) What is the electrostatic problem solved by this function?

(d) Sketch the equipotential lines with $v = 1 - \epsilon$ and $v = \epsilon$, when ϵ is very small.

(e) Give the potential and charge densities on both sides of the conducting surfaces as $x \to \infty$.

(f) Near $x = y = 0$, the relation between w and z is $w = \alpha \sqrt{z}$, where α is a constant. Relate the constant α to the potential difference and distance between the plates of this capacitor.

Problems involving dielectrics

1.24. Formulate and prove the uniqueness theorem for conductors and dielectrics with $\epsilon(x) > 0$.

1.25. Formulate and prove a variational principle for dielectrics and conductors analogous to (1.4.8).

1.26. A point charge q in vacuum is a distance $x = l$ from the plane surface of a dielectric extending to $x = \infty$. The dielectric constant is a constant, ϵ. The fields for this configuration of charges can be found using the method of images: The potential outside the dielectric is given by the Coulomb potential of the charge q plus the Coulomb potential of an image charge q' in the dielectric. The potential in the dielectric is given by Coulomb potential of an image charge q'' outside of the dielectric. Find the location and values of the two image charges. Calculate the force on the dielectric as the Coulomb force between q and q'; then verify that this is correct by integrating the stress tensor over an appropriate surface. Is there another reason for believing the result?

1.27. A dielectric sphere with dielectric constant ϵ is placed in a uniform electric field $\mathbf{E}_0$ (whose source is far enough away to be unaffected by the sphere). Find the electric field outside the sphere and the induced electric dipole moment of the sphere. Compare the result for large ϵ to that for a conducting sphere.

1.28. The problem of a point charge q a vector distance $\mathbf{b}$ from a neutral dielectric sphere of radius a cannot be solved with images; however, an integral for the correct potential at radius r can be obtained using the method of Problem 1.21. The answer for the potential outside the sphere $(r > a)$ is

$$\phi = \frac{q}{|\mathbf{r} - \mathbf{b}|} + \delta\phi$$

where

$$\delta\phi = -q \frac{a}{b} \frac{(\epsilon - 1)}{\epsilon + 1} I$$

with

$$I = \frac{1}{(1 + y^2 - 2y \cos \theta)^{1/2}} - \frac{1}{\gamma y^{1/\gamma}} \int_0^y dy' \frac{(y')^{1/\gamma - 1}}{(1 + y'^2 - 2y' \cos \theta)^{1/2}}$$

where $\gamma = 1 + \epsilon$ and $y = a^2/br$. Show that this answer is correct and give a similar answer for the potential inside the sphere ($r < a$).

Miscellaneous problems

1.29. Prove that the mean value of the electrostatic potential in vacuum averaged over the surface of a sphere is equal to the potential at the center of the sphere, provided there is no charge inside the sphere.

1.30. Find the equation for the lines of force in the xy plane around an electric dipole lying along the x-axis. The line of force points in the direction of the electric field, so its differential equation is

$$\frac{dy}{dx} = \frac{E_y}{E_x}.$$

Hint: Use polar coordinates in the plane to obtain an equation $r = r_0 f(\theta)$. Sketch a few lines.

1.31. Repeat Problem 1.30 for a quadrupole with a potential

$$\phi = \frac{1 - 3\cos^2\theta}{r^3}.$$

The answer is straightforward in the first quadrant ($0 \le \theta \le \pi/2$). What happens in the second quadrant ($\pi/2 \le \theta \le \pi$)?

1.32. (a) Calculate, from first principles and the definition of electric dipole moment, the force between two dipoles $\mathbf{p}_1$ and $\mathbf{p}_2$. Is $\mathbf{F}_{21} = -\mathbf{F}_{12}$?

(b) Calculate the torque $\boldsymbol{\tau}_{12}$ exerted by $\mathbf{p}_1$ on $\mathbf{p}_2$. Is $\boldsymbol{\tau}_{12} = -\boldsymbol{\tau}_{21}$? If so, fine. If not, explain what happened.

1.33. Prove directly, using a Green's type theorem, that the coefficients of potential are symmetric: $p_{ij} = p_{ji}$.

1.34. (a) Find the Green's function for a conducting insulated sphere.

(b) Find the Green's function for a conducting sphere held at fixed potential.

(c) Verify the symmetry $G(\mathbf{r}, \mathbf{r}_1) = G(\mathbf{r}_1, \mathbf{r})$ for both cases.

1.35. The simplest electrostatic variational calculation is quite complicated. Consider, for example, the problem of finding the field produced by a constant charge density

$$\rho = \rho_0, \qquad r < a$$

and

$$\rho = 0, \qquad r > a.$$

Try a variational function

$$\psi = \frac{A}{r^n}, \qquad r > a$$

$$\psi = Br^m + C, \qquad r < a.$$

From the boundary condition at $r = a$, find C as a function of A, B, and a. Calculate the variational integral I and minimize with respect to A and B. There remains a dimensionless function of n and m. Show that the minimum of I occurs for $n = 1$ and $m = 2$, the correct values.

1.36. There is a very successful model that takes into account the interaction of atomic dipoles with each other in calculating the dielectric constant. This model leads to a formula known as the Clausius–Mossotti relation. The argument starts by noting [as shown in (1.6.23)] that the macroscopic (i.e., average) field $\mathbf{E}$ is generated by external sources and by the average dipole moment per unit volume $\mathbf{P}$ of the dielectric. Thus,

$$\mathbf{E}(\mathbf{r}) = \mathbf{E}\,(\text{from external sources}) - \nabla \int d\mathbf{r}' \mathbf{P}(\mathbf{r}') \cdot \frac{\mathbf{r} - \mathbf{r}'}{|\mathbf{r} - \mathbf{r}'|^3}.$$

Presumably, the part of the above $d\mathbf{r}'$ integration coming from large values of $|\mathbf{r} - \mathbf{r}'|$ gives a good approximation to the contribution of distant dipoles to the local field at an atom. However, the contribution of nearby dipoles must be explicitly summed. A remarkable slight of hand follows: the integral over $d\mathbf{r}'$ coming from a very small, but macroscopic, sphere surrounding the atom at $\mathbf{r}$ is subtracted from the above formula and replaced by the sum of the fields of the point dipoles in that neighborhood. Thus,

$$\mathbf{E}_{\text{local}}(\mathbf{r}) = \mathbf{E} + \nabla \int_{\substack{\text{inside} \\ \text{sphere}}} d\mathbf{r}' \mathbf{P}(\mathbf{r}') \cdot \frac{(\mathbf{r} - \mathbf{r}')}{|\mathbf{r} - \mathbf{r}'|^3} - \nabla \sum_{\substack{i\,\text{inside} \\ \text{sphere}}} \mathbf{p}i \cdot \frac{\mathbf{r} - \mathbf{r}_i}{|\mathbf{r} - \mathbf{r}_i|^3}.$$

Now comes the slight of hand: For the two extremes of a crystal with cubic symmetry, on the one hand, and for randomly placed dipoles, on the other, the sum over i vanishes. Also, for a small enough sphere, $\mathbf{P}(\mathbf{r})$ will be approximately constant over the sphere. With these assumptions, show

(a) $$\mathbf{E}_{\text{local}} = \mathbf{E} + \frac{4\pi\mathbf{P}}{3}$$

and from this

(b)
$$\frac{\epsilon - 1}{\epsilon + 2} = \frac{4\pi}{3} n\alpha$$

where α is the atomic polarizability and n the number of atoms per unit volume. This formula predicts that the measurable quantity

$$\frac{(\epsilon + 2)n}{\epsilon - 1}$$

for a given substance should be approximately independent of external parameters, such as pressure and temperature. Note that weak coupling between the atoms corresponds to small $n\alpha$, so that

$$\epsilon - 1 \cong 4\pi n\alpha .$$

*1.37. For an interesting problem employing standard electrostatic methods, see Problem 2.14 (magnetic levitation of a super-conducting sphere).

CHAPTER 2

Steady Currents and Magnetostatics

2.1. STEADY CURRENTS

$\mathbf{W}$e describe the flow of currents in a medium by a vector current density $\mathbf{j}(\mathbf{x})$, where $\mathbf{j}(\mathbf{x}) \cdot d\mathbf{S}$ is the charge crossing the surface element $d\mathbf{S}$ in unit time, that is, the current through $d\mathbf{S}$. Total charge is conserved. Thus, with $\rho(\mathbf{x}, t)$ the charge density, the decrease of charge in a volume V must equal the flow of current through the boundary surface S:

$$-\frac{d}{dt} \int_V d\mathbf{r}\,\rho = \int_S d\mathbf{S} \cdot \mathbf{j}$$

or

$$-\int_V d\mathbf{r}\,\frac{\partial \rho}{\partial t} = \int_V d\mathbf{r}\,\nabla \cdot \mathbf{j}, \qquad (2.1.1)$$

for any volume V; hence,

$$\frac{\partial \rho}{\partial t} + \nabla \cdot \mathbf{j} = 0. \qquad (2.1.2)$$

Equation (2.1.2) is called the continuity equation. The conservation of total charge is the global reflection of the local law. [Almost all the known conserved quantities in physics have local densities that satisfy a local equation like (2.1.2). We shall see many of them.] For a static situation $\partial \rho / \partial t = 0$ and $\nabla \cdot \mathbf{j} = 0$. That is the case we will be considering in this section.

For a large class of media and sufficiently small electric fields, macroscopic currents are generated according to Ohm's law:

$$\mathbf{j} = \sigma \mathbf{E} \tag{2.1.3}$$

or more generally,

$$j_i = \sigma_{ij} E_j. \tag{2.1.4}$$

The tensor σ_{ij} is called the conductivity. The rate at which work is done on currents, producing heat, is

$$\frac{dW}{dt} = \int d\mathbf{r}\, E_i j_i = \int d\mathbf{r}\, E_i \sigma_{ij} E_j \tag{2.1.5}$$

so that the *symmetric* part of σ_{ij} must be positive. The antisymmetric part of σ_{ij} does not contribute to dW/dt. An example of an antisymmetric conductivity tensor can be found in the Hall effect, where σ_{ij} has an antisymmetric component

$$\sigma_{ij}^A = \text{constant } \epsilon_{ijk} B_k \tag{2.1.6}$$

with B_j the magnetic field. (See Problem 3.5.)

The equations that govern the conducting medium are, with (2.1.4),

$$\nabla \times \mathbf{E} = 0, \tag{2.1.7}$$

$$j_i = \sigma_{ij} E_j,$$

and

$$\nabla \cdot \mathbf{j} = 0, \tag{2.1.8}$$

identical to the equations for a dielectric medium, with the conductivity replacing the dielectric constant and $\mathbf{j}$ replacing $\mathbf{D}$. Equations (2.1.3), (2.1.7), and (2.1.8) imply Ohm's law for the relation between the potential difference $\Delta\phi$ along a conductor and the current I flowing through it:

$$\Delta\phi = RI \tag{2.1.9}$$

where R is called the resistance of the conductor. R is simply related to the conductivity for a cylindrical conductor:

$$R = \frac{L}{A\sigma},$$ (2.1.10)

where L is the length and A the cross-sectional area of the cylinder.

A current in conductors can be generated by electromagnetic induction (to be discussed later in this section) or by a chemical or thermal source of energy (for example a voltaic cell). We give a brief discussion of the latter here.

A voltaic cell is usually an arrangement of two electrodes of different material immersed in an ionized fluid, or electrolyte, such that it is energetically profitable to transport positive ions to one electrode (the positive terminal) and negative ions to the other electrode (the negative terminal). The electrochemical energy per unit charge transferred is called the electromotive force $\mathscr{E}$ of the cell. Charge will build up on the electrodes until the opposing potential difference of the electrodes $\Delta\phi = \mathscr{E}$. Thus, the open circuit voltage of a cell is equal to $\mathscr{E}$.

When the external circuit is closed, positive charge will flow in the external circuit from $+$ to $-$, and inside the cell from $-$ to $+$. What happens to the external potential $\Delta\phi$? At $I = 0$, $\Delta\phi = \mathscr{E}$. It is reasonable to expand $\Delta\phi$ about $\mathscr{E}$ in a power series in I and to keep the first two terms:

$$\Delta\phi = \mathscr{E} - rI + \cdots$$ (2.1.11)

We call r the internal resistance of the cell. r must be positive since when current I flows, the electrochemical power produced is $\mathscr{E}I$. This power cannot be smaller than the power $\Delta\phi I$ transformed into heat in the external resistance. Thus,

$$\mathscr{E}I \geq \Delta\phi I$$ (2.1.12)

or

$$r \geq 0.$$ (2.1.13)

For the equality to hold, there would have to be no dissipation in the cell itself, and the cell would be a perfect conductor. Normally, $r > 0$ and

$$I = \frac{\mathscr{E}}{R + r}.$$ (2.1.14)

2.2. MAGNETIC FIELDS

Units

We work in mixed Gaussian units: e.s.u. for electric field and charge, e.m.u. for magnetic fields (gauss or oersted). Current is fixed by units of charge and time: I is in e.s.u/sec. Note that 1 Ampere = 1 Coulomb/sec = 3×10^9 e.s.u./sec. Magnetic units start with the force between magnets, treated as dipoles. Thus, as in electrostatics, two unit magnetic poles situated 1 cm apart exert a force of 1 dyne on each other. Since magnetic poles do not appear in nature, one calculates (in principle) the force between two small, widely separated magnetic dipoles:

$$\mathbf{F}_{1 \text{ on } 2} = (\mathbf{m}_2 \cdot \nabla_2) \, \mathbf{B}(\mathbf{r}_2) = (\mathbf{m}_2 \cdot \nabla_2) \left(- \nabla_2 \frac{\mathbf{m}_1 \cdot (\mathbf{r}_2 - \mathbf{r}_1)}{r_{12}^3} \right) \quad (2.2.1)$$

or, with $\mathbf{r}_{12} = \mathbf{r}_2 - \mathbf{r}_1$,

$$\mathbf{F}_{1 \text{ on } 2} = \frac{3\mathbf{r}_{12}}{r_{12}^5} \mathbf{m}_2 \cdot \mathbf{m}_1 + \frac{3[\mathbf{m}_2 \mathbf{m}_1 \cdot \mathbf{r}_{12} + \mathbf{m}_1 \mathbf{m}_2 \cdot \mathbf{r}_{12}]}{r_{12}^5}$$

$$- 15\mathbf{r}_{12} \frac{\mathbf{m}_2 \cdot \mathbf{r}_{12} \mathbf{m}_1 \cdot \mathbf{r}_{12}}{r_{12}^7}. \quad (2.2.2)$$

With $\mathbf{F}$ in dynes and r_{12} in centimeters, (2.2.2) determines magnetic moment in e.m.u. The magnetic field $\mathbf{B}$ produced by a moment $\mathbf{m}$ is then given in gauss by

$$\mathbf{B} = - \nabla \frac{\mathbf{m} \cdot \mathbf{r}}{r^3} = - \frac{\mathbf{m}}{r^3} + \frac{3\mathbf{r}(\mathbf{m} \cdot \mathbf{r})}{r^5}. \quad (2.2.3)$$

The magnetostatics of permanent magnets had been clarified by the end of the eighteenth century. However, there was at that time no known connection between magnetic and electric phenomena. Then, in 1820, Oersted discovered the magnetic field surrounding a wire carrying a steady current. Within 12 years, Ampère, Biot, Savart, Faraday, and others had worked out the physics of magnetic fields and steady currents, culminating in Faraday's discovery of electromagnetic induction.

In working through this subject, we start from the discovery by Biot and Savart that the magnetic field at a point $\mathbf{r}$ due to a circuit carrying a current I could be calculated from the formula

$$\mathbf{B}(\mathbf{r}) = \frac{I}{c} \oint_C d\mathbf{l'} \times \frac{(\mathbf{r} - \mathbf{r'})}{|\mathbf{r} - \mathbf{r'}|^3}, \tag{2.2.4}$$

where $\oint d\mathbf{l'}$ signifies a line integral around the circuit and $\mathbf{r'}$ is the position vector of $d\mathbf{l'}$. The constant c has the dimension of velocity, $c \cong 3 \times 10^{10}$ cm/sec.

Equation (2.2.4) contains a strong hint as to the force of a magnetic field on a circuit. Let us use (2.2.4) to calculate the force of $\mathbf{B}$ on a hypothetical magnetic pole p. It is

$$\mathbf{F}_{C \text{ on } p} = p\mathbf{B} = -\frac{I}{c} \oint d\mathbf{l'} \times p \frac{(\mathbf{r'} - \mathbf{r})}{|\mathbf{r} - \mathbf{r'}|^3}. \tag{2.2.5}$$

We recognize

$$p \frac{(\mathbf{r'} - \mathbf{r})}{|\mathbf{r} - \mathbf{r'}|^3} = \mathbf{B}_p(\mathbf{r'}), \tag{2.2.6}$$

where $\mathbf{B}_p(\mathbf{r'})$ is the magnetic field that would be produced at $\mathbf{r'}$ by the hypothetical pole at $\mathbf{r}$. Thus,

$$\mathbf{F}_{C \text{ on } p} = -\frac{I}{c} \oint d\mathbf{l'} \times \mathbf{B}_p(\mathbf{r'}), \tag{2.2.7}$$

and if we assume that action is equal and opposite to reaction (a treacherous assumption here, as we shall see later, but correct for circuits carrying steady currents), we find that the force on a circuit is suggested to be

$$\mathbf{F} = \frac{I}{c} \oint d\mathbf{l'} \times \mathbf{B}(\mathbf{r'}). \tag{2.2.8}$$

The artificiality of the use of poles could be avoided—we could arrive at the same result (2.2.8) by considering a real dipole instead of a hypothetical pole. More important, the formula (2.2.8) is experimentally correct. Thus, one can calculate the magnetic field due to a current in a wire *as if* it were due to a sum of contributions from each circuit element $d\mathbf{l'}$, with

$$d\mathbf{B} = \frac{I}{c} d\mathbf{l'} \times \frac{(\mathbf{r} - \mathbf{r'})}{|\mathbf{r} - \mathbf{r'}|^3} \tag{2.2.9}$$

and the force on a circuit *as if* it were the sum of forces on each circuit element $d\mathbf{l}'$ of

$$dF = \frac{I}{c}d\mathbf{l}' \times \mathbf{B}.\qquad(2.2.10)$$

We might try to guess the field of a single moving charge from (2.2.9) and the magnetic force on a single moving charge from (2.2.10). To do this, we note that with n charges per unit volume, each of magnitude q and velocity $\mathbf{v}$,

$$I = nqvA\qquad(2.2.11)$$

where A is the cross-sectional area of the wire. Multiplying by dl/dl converts v to $\mathbf{v}$, and multiplying A by dl produces the volume of the current element dl. Thus, $Id\mathbf{l} = Nq\mathbf{v}$, where N is the number of charges in dl. With one charge, we would guess from (2.2.9)

$$\mathbf{B}_q = q\frac{\mathbf{v}}{c} \times \frac{\mathbf{r} - \mathbf{r}'}{|\mathbf{r} - \mathbf{r}'|^3}\qquad(2.2.12)$$

and from (2.2.10)

$$\mathbf{F}_q = q\frac{\mathbf{v}}{c} \times \mathbf{B},\qquad(2.2.13)$$

the Lorentz force law.

Remarkably, it turns out that (2.2.13) is exactly right (if $\mathbf{F}$ is properly interpreted) and (2.2.12) is approximately right for low frequencies and velocities $\ll c$.

One disconcerting discovery is that (2.2.12) and (2.2.13) do *not* satisfy Newton's third law. Thus,

$$\mathbf{F}_{2\text{ on }1} = q_1\frac{\mathbf{v}_1}{c} \times \mathbf{B}_2(\mathbf{r}_1)$$

$$= q_1\frac{\mathbf{v}_1}{c} \times \left[q_2\frac{\mathbf{v}_2}{c} \times \frac{(\mathbf{r}_1 - \mathbf{r}_2)}{|\mathbf{r}_1 - \mathbf{r}_2|^3}\right]\qquad(2.2.14)$$

or

$$\mathbf{F}_{2\text{ on }1} = \frac{q_1 q_2}{c^2 r_{12}^3}\mathbf{v}_1 \times \left(\mathbf{v}_2 \times (\mathbf{r}_1 - \mathbf{r}_2)\right)\qquad(2.2.15)$$

and

$$\mathbf{F}_{1\text{ on }2} = \frac{q_2 q_1}{c^2 r_{12}^3}\mathbf{v}_2 \times \left(\mathbf{v}_1 \times (\mathbf{r}_2 - \mathbf{r}_1)\right).\qquad(2.2.16)$$

But

$$\mathbf{v}_1 \times (\mathbf{v}_2 \times \mathbf{r}) = \mathbf{v}_1 \cdot \mathbf{r}\mathbf{v}_2 - \mathbf{v}_1 \cdot \mathbf{v}_2\mathbf{r} \tag{2.2.17}$$

and

$$\mathbf{v}_2 \times (\mathbf{v}_1 \times (-\mathbf{r})) = -\mathbf{v}_2 \cdot \mathbf{r}\mathbf{v}_1 + \mathbf{v}_1 \cdot \mathbf{v}_2\mathbf{r}; \tag{2.2.18}$$

the two expressions are only opposite when

$$\mathbf{v}_1 \cdot \mathbf{r}\mathbf{v}_2 - \mathbf{v}_2 \cdot \mathbf{r}\mathbf{v}_1 = 0 \qquad \text{or} \qquad \mathbf{r} \times (\mathbf{v}_2 \times \mathbf{v}_1) = 0,$$

that is, either $\mathbf{v}_2 \| \mathbf{v}_1$, or $\mathbf{r} \perp$ to the plane of $\mathbf{v}_1$ and $\mathbf{v}_2$. We shall see later that in spite of this inequality of action and reaction, momentum can still be defined for electromagnetic systems and is conserved; however, one must add to the particle momentum a field momentum that gives overall balance.

We should, however, check for action and reaction in the case of circuits carrying steady currents. Thus, for the force of circuit 2 on circuit 1, we have

$$\mathbf{F}_{2 \text{ on } 1} = \frac{I_1}{c} \oint_{C_1} d\mathbf{l}_1 \times \mathbf{B}_2(\mathbf{r}_1)$$

$$= \frac{I_1 I_2}{c^2} \oint_{C_1} d\mathbf{l}_1 \times \oint_{C_2} \left(\frac{d\mathbf{l}_2 \times (\mathbf{r}_1 - \mathbf{r}_2)}{r_{12}^3} \right) \tag{2.2.19}$$

or

$$\mathbf{F}_{2 \text{ on } 1} = \frac{I_1 I_2}{c^2} \left[\oint_{C_2} d\mathbf{l}_2 \oint_{C_1} d\mathbf{l}_1 \cdot \frac{(\mathbf{r}_1 - \mathbf{r}_2)}{|\mathbf{r}_1 - \mathbf{r}_2|^3} - \int \int_{C_1 C_2} \frac{d\mathbf{l}_1 \cdot d\mathbf{l}_2}{r_{12}^3} (\mathbf{r}_1 - \mathbf{r}_2) \right]. \tag{2.2.20}$$

The first integral is zero, since $(\mathbf{r}_1 - \mathbf{r}_2)/r_{12}^3 = -\nabla_1 1/r_{12}$ and $\int_{C_1} d\mathbf{l}_1 \cdot \nabla_1 1/r_{12} = 0$ for every $\mathbf{r}_2$. Thus,

$$\mathbf{F}_{2 \text{ on } 1} = -\frac{I_1 I_2}{c^2} \oint_{C_1} \oint_{C_2} \frac{d\mathbf{l}_1 \cdot d\mathbf{l}_2}{r_{12}^3} (\mathbf{r}_1 - \mathbf{r}_2) \tag{2.2.21}$$

which evidently satisfies the law of action and reaction. We can also read off the sign of the force for I_1 and I_2 in the same sense: It is attractive.

We turn now to different ways of expressing the basic formula (2.2.4). As we pointed out earlier, $I d\mathbf{l} = nqv d^3\mathbf{r}$, or

$$I d\mathbf{l} = \rho \mathbf{v} d^3\mathbf{r} = \mathbf{j} d^3\mathbf{r} \qquad (2.2.22)$$

where $d^3\mathbf{r}$ is the volume element of the wire, ρ the charge density of the moving charge, $\mathbf{v}$ its velocity, and $\mathbf{j}$ the current density. More directly,

$$I d\mathbf{l} = \mathbf{j} A dl = \mathbf{j} d^3\mathbf{r}. \qquad (2.2.23)$$

Therefore, a continuous distribution of current—imagine many wires lined up together—generates a field

$$\mathbf{B}(\mathbf{r}) = \int d\mathbf{r}' \frac{\mathbf{j}(\mathbf{r}')}{c} \times \frac{\mathbf{r} - \mathbf{r}'}{|\mathbf{r} - \mathbf{r}'|^3}, \qquad (2.2.24)$$

and a field $\mathbf{B}$ exerts a volume force on a current $\mathbf{j}$ of

$$\mathbf{f} = \frac{\mathbf{j}}{c} \times \mathbf{B}. \qquad (2.2.25)$$

"Volume force" $\mathbf{f}$ means that the actual force on an element of current occupying a volume V is

$$\mathbf{F} = \int_V d\mathbf{r}\,\mathbf{f}. \qquad (2.2.26)$$

The formulae we have derived are sufficient to calculate the magnetic field $\mathbf{B}$ of a given steady current distribution. Since the current is steady, we must have, in the absence of an indefinite piling up of charge density, $\nabla \cdot \mathbf{j} = 0$.

We note that (2.2.24) is

$$\mathbf{B}(\mathbf{r}) = -\int d\mathbf{r} \frac{\mathbf{j}(\mathbf{r}')}{c} \times \nabla \frac{1}{|\mathbf{r} - \mathbf{r}'|} \qquad (2.2.27)$$

or

$$\mathbf{B}(\mathbf{r}) = \nabla \times \frac{1}{c} \int d\mathbf{r}' \frac{\mathbf{j}(\mathbf{r}')}{|\mathbf{r} - \mathbf{r}'|}. \qquad (2.2.28)$$

From (2.2.28) we can derive the differential equations satisfied by $\mathbf{B}$ [analogous to (1.1.17) and (1.1.36) for the electric field].

First,

$$\nabla \cdot \mathbf{B} = 0 \tag{2.2.29}$$

since $\mathbf{B}$ is a curl. Second,

$$\begin{aligned}
\nabla \times \mathbf{B} &= \frac{1}{c} \nabla \times \left(\nabla \times \int d\mathbf{r}' \, \frac{\mathbf{j}(\mathbf{r}')}{|\mathbf{r} - \mathbf{r}'|} \right) \\
&= \frac{1}{c} \nabla \nabla \cdot \int d\mathbf{r}' \, \frac{\mathbf{j}(\mathbf{r}')}{|\mathbf{r} - \mathbf{r}'|} - \frac{1}{c} \nabla^2 \int d\mathbf{r}' \, \frac{\mathbf{j}(\mathbf{r}')}{|\mathbf{r} - \mathbf{r}'|}.
\end{aligned} \tag{2.2.30}$$

The first term vanishes, since $\nabla \cdot \mathbf{j} = 0$. To see this, note that

$$\begin{aligned}
\nabla \cdot \int d\mathbf{r}' \, \frac{\mathbf{j}(\mathbf{r}')}{|\mathbf{r} - \mathbf{r}'|} &= \int d\mathbf{r}' \nabla \frac{1}{|\mathbf{r} - \mathbf{r}'|} \cdot \mathbf{j}(\mathbf{r}') \\
&= - \int d\mathbf{r}' \nabla' \frac{1}{|\mathbf{r} - \mathbf{r}'|} \cdot \mathbf{j}(\mathbf{r}') \\
&= \int d\mathbf{r}' \frac{1}{|\mathbf{r} - \mathbf{r}'|} \nabla' \cdot \mathbf{j}(\mathbf{r}') = 0.
\end{aligned}$$

The second term in (2.2.30) is $(4\pi/c)\mathbf{j}(\mathbf{r})$. The source equation for $\mathbf{B}$ is therefore

$$\nabla \times \mathbf{B} = \frac{4\pi \mathbf{j}}{c}. \tag{2.2.31}$$

Ampère's circuital law follows directly from (2.2.31) by integration over any open surface bounded by a curve C:

$$\int_S \nabla \times \mathbf{B} \cdot d\mathbf{S} = \frac{4\pi}{c} \int_S d\mathbf{S} \cdot \mathbf{j} \tag{2.2.32}$$

or by Stokes' theorem,

$$\oint_C \mathbf{B} \cdot d\mathbf{l} = \frac{4\pi}{c} I_{\text{enclosed}}, \tag{2.2.33}$$

where I_{enclosed} is the current flowing through the surface S.

Finally, (2.2.28) invites us to define a vector potential

$$\mathbf{A} = \frac{1}{c} \int d\mathbf{r}' \, \frac{\mathbf{j}(\mathbf{r}')}{|\mathbf{r} - \mathbf{r}'|} \qquad (2.2.34)$$

with

$$\mathbf{B} = \nabla \times \mathbf{A}. \qquad (2.2.35)$$

Note that (2.2.29) implies the possibility of introducing a vector potential, for which (2.2.34) supplied an explicit formula.

The notion of gauge invariance makes its first appearance here. We can always add the gradient of a scalar $\nabla \psi$ to any $\mathbf{A}$ without changing the field $\mathbf{B}$, since $\nabla \times (\nabla \psi) = 0$. The choice of $\mathbf{A}$ among all these possibilities is called the choice of gauge. The transformation

$$\mathbf{A} \rightarrow \mathbf{A} + \nabla \psi \qquad (2.2.36)$$

is called a gauge transformation, and $\mathbf{B}$ is said to be gauge invariant. The gauge choice in (2.2.34) is evidently $\nabla \cdot \mathbf{A} = 0$. Note that (2.2.36) informs us that we can always find a ψ to make $\nabla \cdot \mathbf{A} = 0$. Suppose $\nabla \cdot \mathbf{A}_0 \neq 0$. Then let $\mathbf{A}_1 = \mathbf{A}_0 + \nabla \psi$. We can make $\nabla \cdot \mathbf{A}_1 = \nabla \cdot \mathbf{A}_0 + \nabla^2 \psi = 0$ by solving the equation $\nabla^2 \psi = -\nabla \cdot \mathbf{A}_0$, which is always possible.

The underlying theories that physicists work with today are all theories with gauge invariance under transformations similar to (2.2.36). Not surprisingly, they are called gauge theories. In particular, the so-called standard model of the strong, weak, and electromagnetic interactions has all interactions mediated by gauge fields: eight colored gluon fields for the strong interactions, the $W^{\pm}$ and Z_0 fields for the weak interactions, and of course the electromagnetic field.

2.3. MAGNETIC MULTIPOLES

Starting from (2.2.34) for the vector potential

$$\mathbf{A}(\mathbf{r}) = \frac{1}{c} \int d\mathbf{r}' \, \frac{\mathbf{j}(\mathbf{r}')}{|\mathbf{r} - \mathbf{r}'|}, \qquad (2.2.34)$$

with $\nabla \cdot \mathbf{j} = 0$, we can expand $\mathbf{A}(\mathbf{r})$ in a power series in r'/r (convergent for $r' < r$):

$$\mathbf{A}(\mathbf{r}) = \frac{1}{c} \sum_l \frac{(-1)^l}{l!} \int d\mathbf{r}' \mathbf{j}(\mathbf{r}') \left(\mathbf{r}' \cdot \nabla \right)^l \frac{1}{r}. \qquad (2.3.1)$$

We consider first the terms $l = 0$ and $l = 1$, and then go on to the general term. First, $l = 0$:

$$A^{(0)}(\mathbf{r}) = \frac{1}{cr} \int d\mathbf{r}' \mathbf{j}(\mathbf{r}'). \tag{2.3.2}$$

$A^{(0)}(\mathbf{r})$ can be seen to be zero from the identity

$$0 = \int d\mathbf{r}\, x_i \frac{\partial j_k}{\partial x_k} = - \int d\mathbf{r}\, \frac{\partial x_i}{\partial x_k} j_k(\mathbf{r}) = - \int d\mathbf{r}\, j_i(\mathbf{r}). \tag{2.3.3}$$

Thus, there is no vector potential of order $1/r$ for large r and, hence, no magnetic field of order $1/r^2$. This conclusion has nothing to do with the nonexistence of magnetic poles. If magnetic poles existed, there, of course, would be magnetostatic fields going like $1/r^2$, where r would be the distance from the pole. There still would be no $1/r^2$ field generated by steady electric currents. Note that a single moving point charge does not constitute a steady current and, hence, *will* give rise to a $1/r^2$ field.
    The term $l = 1$:

$$A^{(1)}(\mathbf{r}) = -\frac{1}{c} \int d\mathbf{r}'\, \mathbf{j}(\mathbf{r}')\, \mathbf{r}' \cdot \nabla \frac{1}{r}. \tag{2.3.4}$$

It is useful here to change to tensor notation:

$$A_i^{(1)}(\mathbf{r}) = -\frac{1}{c} \int d\mathbf{r}'\, j_i(\mathbf{r}')\, x_k' \frac{\partial}{\partial x_k} \frac{1}{r}. \tag{2.3.5}$$

We proceed by decomposing the term $j_i x_k'$ into symmetric and antisymmetric parts:

$$j_i x_k' = \frac{1}{2}\left(j_i x_k' + j_k x_i'\right) + \frac{1}{2}\left(j_i x_k' - j_k x_i'\right). \tag{2.3.6}$$

The first term in (2.3.6) integrates to zero, as in (2.3.3):

$$0 = \int d\mathbf{r}\, x_l x_k \frac{\partial}{\partial x_i} j_i = - \int d\mathbf{r}(j_l x_k + j_k x_l). \tag{2.3.7}$$

The second term in (2.3.6), inserted into (2.3.5), gives

$$A_i^{(1)}(\mathbf{r}) = -\frac{1}{c}\int d\mathbf{r}'\left(\frac{j_i(\mathbf{r}')\,x_k' - j_k(\mathbf{r}')\,x_i'}{2}\right)\frac{\partial}{\partial x_k}\frac{1}{r}. \qquad (2.3.8)$$

We define the magnetic moment density $\mathbf{M}$ by

$$\mathbf{M}(\mathbf{r}') = \frac{\mathbf{r}' \times \mathbf{j}(\mathbf{r}')}{2c}, \qquad (2.3.9)$$

so that, since

$$\frac{1}{2}(\mathbf{r}' \times \mathbf{j}) \times \nabla = \frac{1}{2}[-\mathbf{r}'\mathbf{j}\cdot\nabla + \mathbf{j}(\mathbf{r}'\cdot\nabla)],$$

$$\mathbf{A}^{(1)}(\mathbf{r}) = -\int d\mathbf{r}'\,\mathbf{M}(\mathbf{r}') \times \nabla\frac{1}{r} \qquad (2.3.10)$$

$$= -\mathbf{m} \times \nabla\frac{1}{r} \qquad (2.3.11)$$

where $\mathbf{m}$ is the magnetic moment,

$$\mathbf{m} = \int \mathbf{M}(\mathbf{r}')d\mathbf{r}'. \qquad (2.3.12)$$

The magnetic field of the moment $\mathbf{m}$ is

$$\mathbf{B}^{(1)} = \nabla \times \mathbf{A}^{(1)} = -\nabla \times \left(\mathbf{m} \times \nabla\frac{1}{r}\right) \qquad (2.3.13)$$

or, since $\nabla^2(1/r) = 0$ for $r \neq 0$,

$$\mathbf{B}^{(1)} = (\mathbf{m}\cdot\nabla)\,\nabla\frac{1}{r}$$

$$= \nabla(\mathbf{m}\cdot\nabla)\frac{1}{r}$$

$$= -\nabla\frac{\mathbf{m}\cdot\mathbf{r}}{r^3} \qquad (2.3.14)$$

corresponding to a magnetic "potential"

$$\phi^*(\mathbf{r}) = \frac{\mathbf{m} \cdot \mathbf{r}}{r^3}. \tag{2.3.15}$$

In general, of course, one cannot express $\mathbf{B}$ as the gradient of a potential, since $\nabla \times \mathbf{B} \neq 0$. However, in a region where $\nabla \times \mathbf{B} = 0$, that is, outside of a current-carrying region, one can define a potential ϕ^*, called the magnetic pseudopotential, such that

$$\mathbf{B} = -\nabla \phi^*. \tag{2.3.16}$$

We discuss the magnetic dipole further by considering two special cases:

1. A circuit carrying current I:

$$\mathbf{m} = \frac{1}{2c} \int_C d\mathbf{r}\, \mathbf{r} \times \mathbf{j} = \frac{I}{2c} \oint_C \mathbf{r} \times d\mathbf{l} \tag{2.3.17}$$

where $d\mathbf{l}$ is in the direction of the current, or by Stokes' theorem,

$$\mathbf{m} = \frac{I}{c} \mathbf{A}, \tag{2.3.18}$$

where $\mathbf{A}$ is the "area" of any surface bounded by the circuit C:

$$\mathbf{A} \equiv \int d\mathbf{S},$$

with the direction of $d\mathbf{S}$ determined by the right-hand rule applied to the circuit C.

2. A point particle of charge q and velocity $\mathbf{v}$ in orbit, radius vector $\mathbf{r}_p$:

$$\mathbf{m} = \frac{1}{2c} \int d\mathbf{r}(\mathbf{r} \times \mathbf{j}) = \frac{1}{2c} \int d\mathbf{r}\, q\delta(\mathbf{r} - \mathbf{r}_p)\, \mathbf{r} \times \mathbf{v}_p$$

or

$$\mathbf{m} = \frac{q\mathbf{r}_p \times \mathbf{v}_p}{2c} = \frac{q\mathbf{l}_p}{2mc} \tag{2.3.19}$$

where $\mathbf{l}_p$ is the angular momentum of the particle. The factor $q/2mc$ is called the gyromagnetic ratio of the particle. Note that in this example $j(\mathbf{r}, t) = q\mathbf{v}_p\delta(\mathbf{r} - \mathbf{r}_p(t))$ is not a constant. Nevertheless, the formula for the magnetic moment turns out to be correct if one first time-averages $\mathbf{m}$ over the fast orbital motion.

The general term, (2.3.1), can be rewritten in tensor form:

$$A_i^l(\mathbf{r}) = \frac{(-1)^l}{c}\frac{1}{l!}\int d\mathbf{r}'\, j_i(\mathbf{r}')\, x_{i_1}'\cdots x_{i_l}'\frac{\partial}{\partial x_{i_1}}\cdots\frac{\partial}{\partial x_{i_l}}\frac{1}{r}. \quad (2.3.20)$$

The above expression for $A_i^l(\mathbf{r})$ permits the construction of a pseudo-potential that we give without proof (but see Problem 2.12 for the case $l = 2$):

$$\phi^{*l}(\mathbf{r}) = -\frac{2l}{(l+1)!}(-1)^l\int d\mathbf{r}'\,(\mathbf{r}'\cdot\nabla)^{l-1}\frac{\mathbf{M}(\mathbf{r}')\cdot\mathbf{r}}{r^3}. \quad (2.3.21)$$

Recall the analogous electrostatic formula

$$\phi^l(\mathbf{r}) = \frac{(-1)^l}{l!}\int d\mathbf{r}'\,\rho(\mathbf{r}')\,(\mathbf{r}'\cdot\nabla)^l\frac{1}{r}, \quad (1.3.1)$$

which showed that the potential of an arbitrary charge distribution can be written as a sum of multipoles of order l, each multipole itself being an integral over $\mathbf{r}'$ of Maxwell multipoles with the tensor structure

$$Q_{i_1\ldots i_l}^{\text{electric}} = \hat{r}_{i_1}'\ldots\hat{r}_{i_l}'$$

and magnitude

$$Q_l^{\text{electric}}(\mathbf{r}') = d\mathbf{r}'\, r'^l\rho(\mathbf{r}')\frac{(-1)^l}{l!}. \quad (2.3.22)$$

Similarly, $\phi^*(\mathbf{r})$ is given as a sum of multipoles of order l, each multipole itself being an integral over $\mathbf{r}'$ of a Maxwell multipole of the tensor structure

$$Q_{i_1\ldots i_l}^{\text{magnetic}} = \hat{r}_{i_2}'\ldots\hat{r}_{i_l}'\hat{M}_{i_1} \quad (2.3.23)$$

and magnitude

$$Q_l^{\text{magnetic}} = \frac{2l(-1)^{l+1}}{(l+1)!}r'^{l-1}M(\mathbf{r}'). \quad (2.3.24)$$

As usual, only the symmetric, trace-free part of Q_l^{magnetic} contributes to the integral (2.3.21) for ϕ^*. Since there is no magnetic monopole, the Maxwell multipoles can be thought of as coming from successive displacement of a dipole, rather than a charge.

2.4. MAGNETIC FIELDS IN MATTER

We deal only briefly here with the averaging process since the essential issues are very similar to the electrostatic case. As in that case, we define an average field

$$\mathbf{B} = \int d\mathbf{r}' \, f(\mathbf{r} - \mathbf{r}')\mathbf{b}(\mathbf{r}') \tag{2.4.1}$$

where $\mathbf{b}$ is the microscopic field. The differential equations for $\mathbf{b}$,

$$\nabla \times \mathbf{b} = \frac{4\pi}{c} \mathbf{j}_m \tag{2.2.31}$$

and

$$\nabla \cdot \mathbf{b} = 0 \tag{2.2.29}$$

become, on averaging,

$$\nabla \cdot \mathbf{B} = 0 \tag{2.4.2}$$

and

$$\nabla \times \mathbf{B} = \frac{4\pi}{c} \bar{\mathbf{j}}_m. \tag{2.4.3}$$

It remains to calculate $\bar{\mathbf{j}}_m$. An applied magnetic field will induce a magnetic dipole density, the dipole moment per unit volume $\mathbf{M}$. The vector potential due to $\mathbf{M}$ will be given by (2.3.10)

$$\mathbf{A}_M = \int d\mathbf{r}' \, \nabla \frac{1}{|\mathbf{r} - \mathbf{r}'|} \times \mathbf{M}(\mathbf{r}') \tag{2.4.4}$$

$$= \int \frac{d\mathbf{r}'}{|\mathbf{r} - \mathbf{r}'|} \nabla' \times \mathbf{M}(\mathbf{r}'). \tag{2.4.5}$$

Evidently, $\nabla' \times \mathbf{M}(\mathbf{r}')$ plays the same role with respect to $\bar{\mathbf{j}}/c$ as $-\nabla \cdot \mathbf{P}$ does with respect to $\bar{\rho}_b$. In integrating (2.4.4) by parts to arrive at (2.4.5), we have as usual dropped surface terms, understanding that they will emerge from the behavior of $\nabla \times \mathbf{M}$ at an (approximate) discontinuity. Thus, we see that

$$\bar{\mathbf{j}}_m = c\nabla \times \mathbf{M} + \bar{\mathbf{j}}_f \qquad (2.4.6)$$

where $\bar{\mathbf{j}}_f$ is the conduction current of moving charges. Calling $\bar{\mathbf{j}}_f = \mathbf{j}$, we have from (2.4.2), (2.4.3), and (2.4.6), the field equations

$$\nabla \cdot \mathbf{B} = 0 \qquad (2.4.2)$$

and

$$\nabla \times \mathbf{B} = \frac{4\pi\mathbf{j}}{c} + 4\pi\nabla \times \mathbf{M}. \qquad (2.4.7)$$

Analogous to our definition of $\mathbf{D}$ in electrostatics, we define

$$\mathbf{H} = \mathbf{B} - 4\pi\mathbf{M}, \qquad (2.4.8)$$

leading to the final form for (2.4.7):

$$\nabla \times \mathbf{H} = \frac{4\pi\mathbf{j}}{c}. \qquad (2.4.9)$$

Together with the relation between $\mathbf{B}$ and $\mathbf{H}$, (2.4.9) and (2.4.2) determine the magnetic field.

A word on nomenclature: Before the electrical origin of magnetic fields was known, the electrical analogue to $\mathbf{B}$ seemed to be $\mathbf{D}$, and the analogue to $\mathbf{H}$ seemed to be $\mathbf{E}$, the differences in the right-hand sides reflecting the absence of magnetic poles and currents. Consequently, the historic name given to the vector $\mathbf{H}$ is magnetic field, that given to $\mathbf{B}$ is magnetic induction, one conventionally measured in oersteds, the other in gauss. We do not differentiate these units from each other.

The boundary conditions on $\mathbf{B}$ and $\mathbf{H}$ at a material discontinuity follow as usual from (2.4.2) and (2.4.9). A Gaussian pillbox applied to (2.4.2) tells us that B_{normal} is continuous; a Stokesian rectangle applied to (2.4.9) tells us that

$$\Delta H_{\text{tangential}} = \frac{4\pi}{c} K$$

where K is the surface current, that is, the current going through the infinitesimal Stokesian rectangle per unit length along the tangent under consideration. A surface current requires an infinite current density and hence infinite conductivity. For finite conductivity, we shall see that the "surface" current is finite near the surface, and hence there is no $\Delta H_{\text{tangential}}$, even for a sharply bounded surface.

If **B** is a linear function of **H**, $B_i = \mu_{ij}H_j$, where the tensor μ_{ij} is called the permeability. As usual, the isotropic case has $\mu_{ij} = \delta_{ij}\mu$.

We discuss briefly the relation between **B** and **H**. First, let us consider permanent magnets. These are endowed with a fixed dipole moment $\mathbf{M}(\mathbf{r})$ per unit volume, giving rise to an effective pole volume density $-\nabla \cdot \mathbf{M}$ and pole surface density M_{normal}. (We do not discuss the atomic physics of permanent magnetization.) The equations determining the field configuration are

$$\nabla \times \mathbf{H} = \frac{4\pi\mathbf{j}}{c} = 0 \qquad (2.4.9)$$

and

$$\nabla \cdot \mathbf{B} = 0. \qquad (2.4.2)$$

Thus, since $\mathbf{j} = 0$,

$$\mathbf{H} = -\nabla\phi^* \qquad (2.4.10)$$

$$\nabla \cdot \mathbf{H} = -4\pi\nabla \cdot \mathbf{M}, \qquad (2.4.11)$$

and so

$$\nabla^2\phi^* = 4\pi\nabla \cdot \mathbf{M} \qquad (2.4.12)$$

plus boundary conditions determines ϕ^* and, hence, **H** and **B**.

Second, let us consider paramagnetic substances: $\mu > 1$. These consist of atoms possessing permanent magnetic dipole moments, which in the normal state are randomly oriented, cancel out, and average to zero. The presence of a magnetic field will polarize the moments and tend to align them with itself. A simple classical calculation shows this effect. Consider a gas of atoms, each having a permanent magnetic moment $\boldsymbol{\mu}$, in an externally applied field $\mathbf{B}_0$. The energy of the magnetic moment in the field is

$$W = -\boldsymbol{\mu} \cdot \mathbf{B}_0$$

and the Boltzmann distribution function for $\boldsymbol{\mu}$ is

$$F = \frac{e^{\boldsymbol{\mu}\cdot\mathbf{B}_0/kT}}{\displaystyle\int e^{\boldsymbol{\mu}\cdot\mathbf{B}_0/kT}\,d\Omega}. \qquad (2.4.13)$$

The mean value of $\boldsymbol{\mu}$ will be in the direction of $\mathbf{B}_0$, so $\bar{\boldsymbol{\mu}} = \bar{\mu}\mathbf{B}_0/B_0$

and

$$\bar{\mu} = \frac{\mathbf{B}_0 \cdot \bar{\boldsymbol{\mu}}}{B_0} = \frac{1}{B_0} \frac{\int d\Omega \, \boldsymbol{\mu} \cdot \mathbf{B}_0 \, e^{\boldsymbol{\mu} \cdot \mathbf{B}_0 / kT}}{\int d\Omega \, e^{\boldsymbol{\mu} \cdot \mathbf{B}_0 / kT}}. \tag{2.4.14}$$

To obtain some notion of the order of magnitude of $\mu B_0 / kT$, we take μ to be one electron Bohr magneton ($e\hbar/2m_e c$) and measure B_0 in tesla (10^4 gauss) and T in degrees absolute. We then find

$$\frac{\mu B_0}{kT} \cong \frac{1}{2} \frac{B_0}{T}, \tag{2.4.15}$$

so that even for high fields and low temperatures $\mu B_0 / kT$ is quite small. The calculation of $\bar{\mu}$ is very simple for $\mu B_0 / kT \ll 1$. From (2.2.14)

$$\bar{\mu} = \frac{\mu \int_{-1}^{1} w e^{w\mu B_0/kT} dw}{\int_{-1}^{1} e^{w\mu B_0/kT} dw} \approx \frac{\mu^2 B_0}{kT} \frac{\int_{-1}^{1} w^2 \, dw}{\int_{-1}^{1} dw} = \frac{\mu^2 B_0}{3kT} \tag{2.4.16}$$

and the atomic paramagnetic polarizability is[1]

$$\alpha_P = \frac{\mu^2}{3kT}. \tag{2.4.17}$$

Let us see what this volume is. With $\mu = e\hbar/2mc$, where e and m are the charge and mass, respectively, of an electron,

$$\alpha_P = \frac{e^2}{3kT} \left(\frac{\hbar}{2mc} \right)^2$$

$$= \frac{e^2}{3a_B kT} \cdot \frac{a_B^3}{4} \left(\frac{e^2}{\hbar c} \right)^2 \tag{2.4.18}$$

$$\approx \frac{1}{T} a_B^3 \tag{2.4.19}$$

---

[1]The formula (2.4.17) is called the Langevin–Debye equation. It evidently holds equally well for a collection of freely rotating electric dipoles.

with T in degrees absolute. Here, a_B is the Bohr radius, $a_B \cong \frac{1}{2} \times 10^{-8}$ cm.

Thus, except for very low temperatures, the paramagnetic susceptibilities are small compared to the electric ones. Note that the high T limit calculated here is necessary for the classical calculation to be valid; that is, kT must be larger than the alignment energy μB_0, since otherwise the integral over angles, $d\Omega$, would have to be replaced by a sum over discrete levels.

Third, we consider diamagnetic materials. Diamagnetism is present in all matter; however, it is dominated by paramagnetism when the latter is present.

Diamagnetism comes from a basic property of magnetic interactions (called Lenz's law): They oppose any change in the magnetic field. The mechanism used is electromagnetic induction, which will be addressed in Section 2.7.[2] A careful calculation of the polarizability of an electron orbit must wait until then. However, we can make a rough estimate of the order of magnitude. The magnetic moment $\boldsymbol{\mu}$ of an orbiting electron in an atom, we have seen in (2.3.19), is

$$\boldsymbol{\mu} = \frac{e\mathbf{L}}{2mc}, \tag{2.4.20}$$

where $\mathbf{L}$ is the angular momentum of the electron. $\mathbf{L}$, in turn, will have a component that is proportional to $\mathbf{B}$ and will not average to zero over many orbits. Dimensionally,

$$L \propto mr^2\omega \tag{2.4.21}$$

where r is the radius of the orbit and ω a frequency of rotation *caused by the magnetic field*. This characteristic frequency[3] associated with the magnetic field is

$$\omega_c \sim \frac{eB}{mc}, \tag{2.4.22}$$

---

[2]In fact, one is treading on dangerous ground in attempting a quantitative classical calculation of diamagnetism, since it is a famous theorem of classical statistical mechanics that a temporally constant magnetic field can have *no* effect on a thermodynamic system at equilibrium: Paramagnetic and diamagnetic effects cancel. Therefore, some consequences of quantum mechanics must be added to the calculation. For example, the derivation given above of (2.4.17) assumes the existence of a permanent unique magnetic moment—not possible in classical mechanics. The derivation to be given in Section 2.7 assumes the existence of unique orbits and time scales for the field free motion of the system, which is again not possible in classical mechanics.

[3]Think of the equation $m\ddot{\mathbf{r}} = \mathbf{F} + (e/c)\dot{\mathbf{r}} \times \mathbf{B}$, and note the characteristic frequency ω_c contained in $\dot{\mathbf{r}}/\mathbf{r}$. $\mathbf{F}$ in this equation stands for the sum of the electric forces on the electron.

giving

$$\mu \sim \frac{e}{2mc} \cdot mr^2 \cdot \frac{eB}{mc} = \frac{e^2}{mc^2} \cdot r^2 B; \qquad (2.4.23)$$

Thus, the diamagnetic polarizability of an atom is very roughly

$$\alpha_D \sim \frac{e^2}{mc^2} a_B^2, \qquad (2.4.24)$$

which is smaller than the electric polarizability by a factor $(e^2/mc^2)/a_B \cong (1/137)^2$. It is also smaller than the paramagnetic polarizability by a factor $T/(137)^2$, so that diamagnetism is normally observed only in materials for which the paramagnetic susceptibility vanishes. This occurs when the intrinsic atomic or molecular moment μ [as in (2.4.17)] vanishes identically for reasons of symmetry.

2.5. MOTIONAL ELECTROMOTIVE FORCE AND ELECTROMAGNETIC INDUCTION

If we move a conductor through a magnetic field, the $ev/c \times \mathbf{B}$ force will act on electrons, giving an effective electric field $\mathbf{E}_{ef} = \mathbf{v}/c \times \mathbf{B}$. Thus, in general, if a circuit in a magnetic field is displaced, we can expect an effective electric field to be generated in the conducting wire, and we will find an effective electromotive force

$$\mathscr{E}_{ef} = \oint_C \mathbf{E}_{ef} \cdot d\mathbf{l} = \oint_C \frac{\mathbf{v}}{c} \times \mathbf{B} \cdot d\mathbf{l}. \qquad (2.5.1)$$

Note here that $\mathbf{v}$ is the velocity of displacement of the circuit element $d\mathbf{l}$.

We define the magnetic flux through the circuit

$$\Phi = \int_S \mathbf{B} \cdot d\mathbf{S}, \qquad (2.5.2)$$

where S is any surface bounded by C, and the normal $d\mathbf{S}$ and the direction of circulation $d\mathbf{l}$ are connected by the right-hand rule. Note that since $\nabla \cdot \mathbf{B} = 0$, Φ is independent of the surface S.

As we displace and deform the circuit, the flux Φ will normally change. We now show that that rate of change determines the effective electromotive force via the equation

$$\mathcal{E}_{ef} = -\frac{1}{c}\frac{d\Phi}{dt}.$$ (2.5.3)

Note that (2.5.3) is *not* Faraday's law, since the magnetic field is not changing with time.

With **B** given by a vector potential **A**, the flux Φ is

$$\Phi = \int_S \mathbf{B} \cdot d\mathbf{S} = \oint_C \mathbf{A} \cdot d\mathbf{l},$$ (2.5.4)

and, as we change from contour C_1 to contour C_2, we have

$$\delta\Phi = \oint_{C_2} \mathbf{A} \cdot d\mathbf{l} - \oint_{C_1} \mathbf{A} \cdot d\mathbf{l}.$$ (2.5.5)

We parametrize the path of the contour with a parameter τ such that

$$x_i = x_i(\tau), \qquad 0 \le \tau \le 1, \qquad x_i(1) = x_i(0).$$ (2.5.6)

Then

$$\delta\Phi = \int_0^1 d\tau A_i(\mathbf{x} + \delta\mathbf{x})\left(\frac{dx_i}{d\tau} + d\frac{\delta x_i}{d\tau}\right) - \int d\tau A_i(\mathbf{x})\frac{dx_i}{d\tau}$$ (2.5.7)

$$= \int_0^1 d\tau\left[A_i(\mathbf{x})d\frac{\delta x_i}{d\tau} + \frac{\partial A_i}{\partial x_j}\delta x_j\frac{dx_i}{d\tau}\right].$$ (2.5.8)

We integrate the first term in (2.5.8) by parts; the integrated term is zero, because $\delta\mathbf{x}$ and $\mathbf{x}$ have the same values at $\tau = 1$ and at $\tau = 0$. There results

$$\delta\Phi = \int_0^1 d\tau\left[\frac{\partial A_i}{\partial x_j} - \frac{\partial A_j}{\partial x_i}\right]\frac{dx_i}{d\tau}\delta x_j.$$ (2.5.9)

We recognize that $(dx_i/d\tau)\, d\tau = dx_i$ (i.e., the dl_i of a line integral) and

$$\left(\frac{\partial A_i}{\partial x_j} - \frac{\partial A_j}{\partial x_i}\right)dx_i\,\delta x_j = \mathbf{B} \cdot \delta\mathbf{x} \times d\mathbf{l}.$$

Equation (2.5.9) then becomes

$$\delta\Phi = -\int \delta\mathbf{x} \times \mathbf{B} \cdot d\mathbf{l} \tag{2.5.10}$$

and

$$-\frac{1}{c}\frac{d\Phi}{dt} = \int \frac{\mathbf{v}}{c} \times \mathbf{B} \cdot d\mathbf{l} = \mathcal{E}_{ef}, \tag{2.5.11}$$

which is the equation we set out to prove.

Faraday's law of induction was, of course, a great experimental discovery. Nevertheless, it is interesting to observe that it follows from (2.5.11) and the assumption of Galilean (or of Lorentz) invariance. Imagine a magnet and a circuit. Consider two operations. First, move the circuit between the pole pieces of the magnet. An electromotive force $\mathcal{E}_{ef}$ given by (2.5.11) will appear in the circuit, a corresponding current $I = \mathcal{E}_{ef}/R$ will flow, and the total charge transferred (e.g., deposited from an electrolyte or measured by a ballistic galvanometer) will be

$$Q = \int dt \, I = \frac{1}{R}\int dt \, \mathcal{E}_{ef} = -\frac{1}{Rc}\Delta\Phi, \tag{2.5.12}$$

all this *no matter how slowly* the circuit is moved. Second, move the magnet past the circuit with equal and opposite velocity. (In the rest frame of the magnet, this operation looks the same as the first one.) If we are not to be able to tell one reference system from another, the second operation must induce an electromotive force in the circuit having the same value as $\mathcal{E}_{ef}$. In this case, however, the electromotive force is genuinely induced, as discovered by Faraday. The law is the same as (2.5.11), of course:

$$\mathcal{E} = -\frac{1}{c}\frac{d\Phi}{dt}, \tag{2.5.13}$$

where now $\mathcal{E}$ is a true electromotive force and $d\Phi/dt$ the rate of change of flux through the circuit. Obviously, we could have a third operation, where both the circuit and magnet move in opposite directions. The result is again the same equation (2.5.13), now with $\mathcal{E}$ = total electromotive force—true and motional—and $d\Phi/dt$ the total change of flux through C, whether from the motion of C, or a change of magnetic field, or both.

We deduce the differential equation corresponding to the integral relation (2.5.13) by holding the circuit fixed. Then (2.5.13) is equivalent to

$$-\frac{1}{c}\int_S d\mathbf{S} \cdot \frac{\partial \mathbf{B}}{\partial t} = \oint_C \mathbf{E} \cdot d\mathbf{l} = \int_S d\mathbf{S} \cdot \nabla \times \mathbf{E} \qquad (2.5.14)$$

for any surface S; hence,

$$\nabla \times \mathbf{E} = -\frac{1}{c}\frac{\partial \mathbf{B}}{\partial t}. \qquad (2.5.15)$$

This is the differential form of Faraday's law. Note that $\mathbf{E}$ has now acquired a curl.

2.6. MAGNETIC ENERGY AND FORCE

Analogously to our procedure in electrostatics, we calculate the rate at which the induced electric field acting on the current causes a loss of magnetic field energy. It is

$$-\frac{dW}{dt} = \int d\mathbf{r}\, \mathbf{E} \cdot \mathbf{j}. \qquad (2.6.1)$$

We transform this expression as follows:

$$-\frac{dW}{dt} = \frac{c}{4\pi}\int d\mathbf{r}\, \mathbf{E} \cdot \nabla \times \mathbf{H}$$

$$= \frac{c}{4\pi}\int d\mathbf{r}\, \mathbf{E} \times \nabla \cdot \mathbf{H} \qquad (2.6.2)$$

and, after dropping a surface term that vanishes, provided $\mathbf{E} \times \mathbf{H}$ goes to zero faster than $1/r^2$, we obtain

$$-\frac{dW}{dt} = -\frac{c}{4\pi}\int d\mathbf{r}\, \mathbf{E} \times \overleftarrow{\nabla} \cdot \mathbf{H}, \qquad (2.6.2)$$

where the symbol $\overleftarrow{\nabla}$ means the gradient differentiates $\mathbf{E}$, but is algebraically to the right of $\mathbf{E}$. We remedy this position by putting ∇ algebraically to the left of $\mathbf{E}$ and changing the sign. Equation (2.6.2) then becomes

$$-\frac{dW}{dt} = \frac{c}{4\pi}\int d\mathbf{r}\, \nabla \times \mathbf{E} \cdot \mathbf{H} = -\frac{1}{4\pi}\int d\mathbf{r}\, \mathbf{H} \cdot \frac{\partial \mathbf{B}}{\partial t} \qquad (2.6.3)$$

for the loss of energy by the field. As in the electric case, we can integrate (2.6.3), provided the medium is linear and the permeability tensor symmetric. We find

$$W = \frac{1}{8\pi} \int d\mathbf{r} \, \mathbf{H} \cdot \mathbf{B}. \tag{2.6.4}$$

We can discuss systems of current circuits, analogous to conductors in the electric case. We limit ourselves here to nonmagnetic media. The analogues to charge and potential are flux and current. To see this, we introduce the vector potential $\mathbf{A}$:

$$W = \frac{1}{8\pi} \int d\mathbf{r} \, \mathbf{B} \cdot \mathbf{H}$$

$$= \frac{1}{8\pi} \int d\mathbf{r} \, \nabla \times \mathbf{A} \cdot \mathbf{H}$$

$$= \frac{1}{8\pi} \int d\mathbf{r} \, \mathbf{A} \cdot \nabla \times \mathbf{H}$$

$$= \frac{1}{2c} \int d\mathbf{r} \, \mathbf{A} \cdot \mathbf{j}$$

which for a system of circuits becomes

$$W = \frac{1}{2c} \sum_i I_i \oint_{C_i} \mathbf{A} \cdot d\mathbf{l}_i \tag{2.6.5}$$

or, by Stokes' theorem,

$$W = \frac{1}{2c} \sum_i I_i \Phi_i \tag{2.6.6}$$

where I_i is the current in and Φ_i the flux through the ith circuit.

As with charge and potential, the fluxes and currents are linear functions of each other:

$$\Phi_i = c \sum_j L_{ij} I_j. \tag{2.6.7}$$

The L_{ij}'s are called coefficients of induction. Note that L_{ij} is defined so that the electromotive force in C_i induced by a changing current in C_j is

$$\mathcal{E}_i = -L_{ij}\frac{dI_j}{dt} \quad \text{(no sum)}. \tag{2.6.8}$$

We list several sometimes useful expressions for the energy W of a system of current carrying circuits:

$$W = \frac{1}{2c}\sum_i I_i\Phi_i$$

$$= \frac{1}{2}\sum_{i,j} I_i L_{ij} I_j \tag{2.6.9}$$

and defining a matrix G_{ij} which is the inverse of the matrix L_{ij}, we have

$$W = \frac{1}{2c^2}\sum_{i,j} \Phi_i G_{ij}\Phi_j. \tag{2.6.10}$$

For an extended current $\mathbf{j(r)}$, in the absence of a magnetic medium $(\mu = 1)$,

$$W = \frac{1}{2c}\int d\mathbf{r}\, \mathbf{A}\cdot\mathbf{j}$$

$$= \frac{1}{2c^2}\int d\mathbf{r}\, d\mathbf{r}'\, \mathbf{j}(\mathbf{r}')\cdot\frac{1}{|\mathbf{r}-\mathbf{r}'|}\mathbf{j}(\mathbf{r}). \tag{2.6.11}$$

The contribution to W in (2.6.11) from two separated circuits, 1 and 2, is

$$W_{12} = \oint\oint \frac{d\mathbf{l}_1\cdot d\mathbf{l}_2}{|\mathbf{r}_1-\mathbf{r}_2|}\frac{I_1 I_2}{c^2}$$

so that

$$L_{12} = \frac{1}{c^2}\oint\oint \frac{d\mathbf{l}_1\cdot d\mathbf{l}_2}{|\mathbf{r}_1-\mathbf{r}_2|}. \tag{2.6.12}$$

The idealization of line currents in (2.6.12) cannot be made for the diagonal element of the inductance matrix. (See Problem 2.16.)

The generalized force on a circuit is given, with $\delta\xi_k$ a generalized displacement of the circuit, by

$$F_k = -\frac{\delta W}{\delta \xi_k}\bigg|_{\text{constant } \Phi} = \frac{\delta W}{\delta \xi_k}\bigg|_{\text{constant } I}$$

(as we have seen in the electrostatic case). So

$$F_k = \frac{1}{2c^2}\sum_{ij} I_i I_j \frac{\delta L_{ij}}{\delta \xi_k} \qquad (2.6.13)$$

in agreement with (2.2.21).

As in the electric case, the calculation of forces on material bodies is difficult. As in the electric case, however, we can find the total force on all the current (free or bound) in a region surrounded by vacuum as follows. The total force on currents inside the volume V is

$$\mathbf{F} = \frac{1}{c}\int_V d\mathbf{r}\,\mathbf{j} \times \mathbf{B}, \qquad (2.6.14)$$

which we transform into

$$\mathbf{F} = \frac{1}{4\pi}\int d\mathbf{r}(\nabla \times \mathbf{B}) \times \mathbf{B}.$$

In tensor notation,

$$F_i = \frac{1}{4\pi}\int d\mathbf{r}(B_k \partial_k B_i - B_k \partial_i B_k); \qquad (2.6.15)$$

since $\partial_k B_k = 0$, we rewrite (2.6.15) as

$$F_i = \frac{1}{4\pi}\int d\mathbf{r}\,\partial_k\left(B_k B_i - \frac{\delta_{ik}}{2}\mathbf{B}^2\right), \qquad (2.6.16)$$

leading to a magnetic stress tensor

$$T_{ik}^M = \frac{1}{4\pi}\left(B_k B_i - \delta_{ik}\frac{\mathbf{B}^2}{2}\right) \qquad (2.6.17)$$

and a total force on the currents inside V

$$F_i = \int_S dS_k \, T^M_{ik} \tag{2.6.18}$$

where S is the surface enclosing V. dS_k is, of course, the outward normal. Since the surface S in (2.6.18) is in vacuum, the microscopic and macroscopic **B** are equal.

2.7. DIAMAGNETISM

We return now to the question of diamagnetism, which was discussed qualitatively in Section 2.4.

We consider an isolated atom, containing n electrons and a heavy nucleus. Imagine applying a constant magnetic field $\mathbf{B}_0$ to the atom. A famous result, due to Larmor, is that, to first order in $\mathbf{B}_0$, the system looks just like the same system with $\mathbf{B}_0 = 0$, but rotating gently with a rotational angular velocity (called Larmor precession)

$$\boldsymbol{\omega}_L = -\frac{e\mathbf{B}_0}{2m_e c}, \tag{2.7.1}$$

where e is the charge of the electron (negative!) and m_e its mass. We prove the result by considering the equation of motion for each electron e:

$$m_e \frac{d^2\mathbf{r}_e}{dt^2} = \mathbf{F}_e + e\frac{\mathbf{v}_e}{c} \times \mathbf{B}_0 \tag{2.7.2}$$

where $\mathbf{F}_e$ is the total remaining force on the eth electron, assumed velocity-independent and angular-momentum conserving, thus, for example, electrostatic.

What happens when we rotate the system about a point with a fixed angular velocity $\boldsymbol{\omega}$? We specify a vector **A** as

$$\mathbf{A} = \hat{\mathbf{i}}A_x + \hat{\mathbf{j}}A_y + \hat{\mathbf{k}}A_z \tag{2.7.3}$$

where A_x, A_y, A_z are the components of **A** along the rotating axes. The rotation of the coordinate axes is described as

$$\frac{d\hat{\mathbf{i}}}{dt} = \boldsymbol{\omega} \times \hat{\mathbf{i}}, \qquad \frac{d\hat{\mathbf{j}}}{dt} = \boldsymbol{\omega} \times \hat{\mathbf{j}}, \qquad \frac{d\hat{\mathbf{k}}}{dt} = \boldsymbol{\omega} \times \hat{\mathbf{k}} \tag{2.7.4}$$

so that

$$\frac{d\mathbf{A}}{dt} = \boldsymbol{\omega} \times \mathbf{A} + \frac{d\mathbf{A}}{d[t]} \tag{2.7.5}$$

where

$$\frac{d\mathbf{A}}{d[t]} = \hat{\mathbf{i}} \frac{dA_x}{dt} + \hat{\mathbf{j}} \frac{dA_y}{dt} + \hat{\mathbf{k}} \frac{dA_z}{dt}. \tag{2.7.6}$$

That is, $d\mathbf{A}/d[t]$ is the rate of change of $\mathbf{A}$ seen by an observer in the rotating coordinate system. Applying (2.7.5) to (2.7.2), we have

$$m_e \left(\frac{d^2 \mathbf{r}_e}{d[t]^2} + 2\boldsymbol{\omega} \times \frac{d\mathbf{r}_e}{d[t]} + \boldsymbol{\omega} \times (\boldsymbol{\omega} \times \mathbf{r}_e) \right) = \mathbf{F}_e + \frac{e}{c} \left[\frac{d\mathbf{r}_e}{d[t]} + \boldsymbol{\omega} \times \mathbf{r}_e \right] \times \mathbf{B}_0. \tag{2.7.7}$$

If we set $\boldsymbol{\omega} = \boldsymbol{\omega}_L = -e\mathbf{B}_0/2m_ec$ and neglect quadratic terms in $\boldsymbol{\omega}$ or $\mathbf{B}$, we return to the **B**-less equation

$$m_e \frac{d^2 \mathbf{r}_e}{d[t]^2} = \mathbf{F}_e, \tag{2.7.8}$$

as stated earlier. Thus, the solution of (2.7.8) for the components of the **r**'s are *as if* there were no field and no rotation. The neglect of quadratic terms is justified if $\omega_L \mathbf{r} \ll d\mathbf{r}/d[t]$ and $\omega_L d\mathbf{r}/d[t] \ll d^2\mathbf{r}/d[t]^2$, which for ordinary atoms and magnetic fields is true.

The velocity of each electron is now

$$\frac{d\mathbf{r}_e}{dt} = \frac{d\mathbf{r}_e}{d[t]} + \boldsymbol{\omega}_L \times \mathbf{r}_e; \tag{2.7.9}$$

the angular momentum of each electron is

$$\mathbf{l}_e = m_e \mathbf{r}_e \times \frac{d\mathbf{r}_e}{dt} = m_e \mathbf{r}_e \times \frac{d\mathbf{r}_e}{d[t]} + m_e \mathbf{r}_e \times (\boldsymbol{\omega}_L \times \mathbf{r}_e) \tag{2.7.10}$$

and the magnetic moment of each electron orbit is

$$\mathbf{m} = \frac{e}{2m_ec} \mathbf{l}_e = \frac{e}{2c} \mathbf{r}_e \times \frac{d\mathbf{r}_e}{d[t]} + \frac{e}{2c} \mathbf{r}_e \times (\boldsymbol{\omega}_L \times \mathbf{r}_e). \tag{2.7.11}$$

Of course, we have taken the origin of $\mathbf{r}$ (i.e., the center of rotation) at the center of the atom. Adding up all the electrons in the atom and averaging over atoms [assuming that the first term on the right-hand side of (2.7.11) averages to zero], we have for the average dipole moment per atom $\bar{\mathbf{m}}$

$$\bar{\mathbf{m}} = -\frac{e^2}{4m_ec^2}\sum_e \left(r_e^2 \mathbf{B}_0 - (\mathbf{r}_e \cdot \mathbf{B}_0)\mathbf{r}_e\right) \tag{2.7.12}$$

$$= -\frac{e^2}{6m_ec^2}\left(\sum_e \overline{r_e^2}\right)\mathbf{B}_0 \tag{2.7.13}$$

for a diamagnetic polarizability

$$\alpha_D = -\frac{e^2}{6m_ec^2}\sum_e \overline{r_e^2} \tag{2.7.14}$$

in agreement with our earlier estimate.

It remains to be shown that as the magnetic field is turned on, starting from zero, the induced electric field converts the original motion of the system to the Larmor precessing motion.

We proceed by deriving a differential equation for the time dependence of the magnetic moment of an atom, averaged over many atoms and many cycles of the fast atomic motion.

The equation governing the motion of the electrons in an atom is

$$m\frac{d^2\mathbf{r}_e}{dt^2} = \mathbf{F}_e + e\left(\mathbf{v}_e \times \frac{\mathbf{B}_e}{c} + \mathbf{E}_e\right) \tag{2.7.15}$$

where $\mathbf{F}_e$ is the force of the nucleus and other electrons on e, $\mathbf{B}_e$ the magnetic field acting on e, and $\mathbf{E}_e$ the induced electric field acting on e. We neglect nuclear motion and take $\mathbf{r}_e$ to be the radius vector from the nucleus to e.

We take the cross product of $\mathbf{r}_e$ with (2.7.15) and sum over e. Since the force $\mathbf{F}_e$ is angular-momentum conserving, it drops out and we have the equation

$$\frac{d\mathbf{L}}{dt} = m\sum_e \mathbf{r}_e \times \frac{d^2\mathbf{r}_e}{dt^2} = \sum_e \mathbf{r}_e \times e\left(\frac{\mathbf{v}_e}{c} \times \mathbf{B}_e + \mathbf{E}_e\right) \tag{2.7.16}$$

where $\mathbf{L}$ is the total angular momentum of the electrons about the nucleus.

The triple vector product $\mathbf{r}_e \times (\mathbf{v}_e \times \mathbf{B}_e)$ is

$$
\begin{aligned}
\mathbf{r}_e \times (\mathbf{v}_e \times \mathbf{B}_e) &= \mathbf{r}_e \cdot \mathbf{B}_e \mathbf{v}_e - \mathbf{r}_e \cdot \mathbf{v}_e \mathbf{B}_e \\
&= \frac{\mathbf{r}_e \cdot \mathbf{B}_e \mathbf{v}_e + \mathbf{B}_e \cdot \mathbf{v}_e \mathbf{r}_e}{2} + \frac{\mathbf{r}_e \cdot \mathbf{B}_e \mathbf{v}_e - \mathbf{B}_e \cdot \mathbf{v}_e \mathbf{r}_e}{2} - \mathbf{r}_e \cdot \mathbf{v}_e \mathbf{B}_e \\
&= \frac{\mathbf{B}_e \times (\mathbf{v}_e \times \mathbf{r}_e)}{2} + \frac{1}{2}\frac{d}{dt}(\mathbf{r}_e \mathbf{r}_e \cdot \mathbf{B}_e - \mathbf{r}_e^2 \mathbf{B}_e) \\
&\quad - \frac{1}{2}\left(\mathbf{r}_e \mathbf{r}_e \cdot \frac{d\mathbf{B}_e}{dt} - \mathbf{r}_e^2 \frac{d\mathbf{B}_e}{dt}\right) \tag{2.7.17}
\end{aligned}
$$

so that

$$
\begin{aligned}
\frac{d\mathbf{L}}{dt} &= \frac{e}{2}\sum_e \mathbf{B}_e \times \left(\frac{\mathbf{v}_e}{c} \times \mathbf{r}_e\right) + \frac{e}{2c}\frac{d}{dt}\sum_e (\mathbf{r}_e \mathbf{r}_e \cdot \mathbf{B}_e - \mathbf{r}_e^2 \mathbf{B}_e) \\
&\quad + e\sum_e \left(\mathbf{r}_e \times \mathbf{E}_e - \frac{1}{2c}\left(\mathbf{r}_e \mathbf{r}_e \cdot \frac{d\mathbf{B}_e}{dt} - \mathbf{r}_e^2 \frac{d\mathbf{B}_e}{dt}\right)\right). \tag{2.7.18}
\end{aligned}
$$

We simplify (2.7.18) by assuming that the spatial variation of $\mathbf{B}$ is small over the atom, so that $\mathbf{B}_e$ can be replaced by $\mathbf{B}$, and $d\mathbf{B}_e/dt$ by $\partial\mathbf{B}/\partial t$. We cannot do the same for $\mathbf{E}_e$, since the coefficient $\Sigma \mathbf{r}_e$ will average to zero over many cycles of the fast motion and over many atoms. We therefore expand $\mathbf{E}_e$ about the nucleus of the atom:

$$
\mathbf{E}_e = \mathbf{E}_{\text{nucleus}} + \frac{1}{2c}\mathbf{r}_e \times \frac{\partial\mathbf{B}}{\partial t} + \delta\mathbf{E} \tag{2.7.19}
$$

where $\delta\mathbf{E}$ is linear in $\mathbf{r}_e$, but such that

$$
\sum_e \mathbf{r}_e \times \delta\mathbf{E}
$$

averages to zero (see Problem 2.17). The result is

$$
\begin{aligned}
\frac{d\mathbf{L}}{dt} &= \frac{e}{2mc}\mathbf{L} \times \mathbf{B} + \frac{d}{dt}\frac{e}{2c}\sum_e (\mathbf{r}_e \mathbf{r}_e \cdot \mathbf{B} - \mathbf{r}_e^2 \mathbf{B}) + \sum_e \mathbf{r}_e \times \delta\mathbf{E} \\
&\quad + e\sum_e \left(\mathbf{r}_e \times \frac{1}{2c}\left(\mathbf{r}_e \times \frac{\partial\mathbf{B}}{\partial t}\right)\right) - \frac{1}{2c}\left(\mathbf{r}_e \mathbf{r}_e \cdot \frac{\partial\mathbf{B}}{\partial t} - \mathbf{r}_e^2 \frac{\partial\mathbf{B}}{\partial t}\right) \tag{2.7.20}
\end{aligned}
$$

or

$$\frac{d\mathbf{M}}{dt} = \boldsymbol{\omega}_L \times \mathbf{M} + \frac{d}{dt}\frac{e^2}{4mc^2}\sum_e (\mathbf{r}_e\mathbf{r}_e \cdot \mathbf{B} - r_e^2\mathbf{B}) + e\sum_e \mathbf{r}_e \times \delta\mathbf{E}. \tag{2.7.21}$$

We average over atoms and find

$$\frac{d\mathbf{M}}{dt} = \boldsymbol{\omega}_L \times \mathbf{M} - \frac{d}{dt}\left(\frac{e^2}{6mc^2}\sum_e \overline{r_e^2}\right)\mathbf{B}, \tag{2.7.22}$$

where $\mathbf{M} = \Sigma_e \mathbf{m}_e$ is the total magnetic dipole moment, $\mathbf{M} = e\mathbf{L}/2mc$.

We can solve (2.7.22) easily for a field $\mathbf{B}$ in one direction, say, z. In that case, starting with M_z and B that are both zero, we find

$$M_z = -\frac{e^2}{6mc^2}\sum_e \overline{r_e^2}B_z \tag{2.7.23}$$

confirming our earlier guess, (2.7.14). Evidently, the component of $\mathbf{M}$ in the x, y plane precesses around the field with the precession frequency ω_L, as expected.

CHAPTER 2 PROBLEMS

2.1. Show that the functions $\rho(\mathbf{x}, t) = qf(\mathbf{x} - \mathbf{x}_p(t))$ and $\mathbf{j}(\mathbf{x}, t) = q(d\mathbf{x}_p/dt)f(\mathbf{x} - \mathbf{x}_p(t))$ with $\mathbf{x}_p(t)$ an arbitrary function of t satisfy the continuity equation

$$\frac{\partial\rho}{\partial t} + \nabla \cdot \mathbf{j} = 0$$

and are therefore possible candidates for charge and current density. For a point particle, the function $f(\mathbf{x} - \mathbf{x}_p(t))$ would go over to $\delta^3(\mathbf{x} - \mathbf{x}_p(t))$.

2.2. Two closed metal surfaces are immersed in a conducting fluid. Construct and prove a uniqueness theorem for the current and field.

2.3. Two metal spheres, one very small, the other of radius b, are immersed in a conducting medium of conductivity σ. The centers of the spheres are separated by a distance L. The small sphere has a current I_0 flowing out of it; the sphere of radius b is maintained at a potential V with respect to a very large conductor containing the system. What is the total current I flowing into the large sphere?

2.4. Current I_0 enters an "infinite" thin conducting plane (of conductivity

σ) at a point and departs at infinity. Assuming the flow is uniform across the width Δ of the thin plane, give a formula for the electrostatic potential at a point ρ, φ in the plane.

2.5. A circular hole of radius a is cut from the plane of Problem 2.4. Its center is a distance $b > a$ from the entrance point of the current. Using the method of Problem 1.21, find a formula for the electrostatic potential at ρ, φ.

2.6. A current I of uniform current density flows down a circular cylindrical wire of radius b. Using Ampère's circuital law, find the magnetic field at a distance ρ from the center of the wire, for $\rho < b$ and $\rho > b$.

2.7. Imagine a uniform current I_0 flowing in the z direction for $|\boldsymbol{\rho} - \boldsymbol{\rho}_1| < b$ and in the $-z$ direction for $|\boldsymbol{\rho} - \boldsymbol{\rho}_2| < b$, where $|\boldsymbol{\rho}_1 - \boldsymbol{\rho}_2| < b$. Draw a picture of the resultant current distribution and give a formula for the (uniform) magnetic field in the overlap area. Now let $|\boldsymbol{\rho}_1 - \boldsymbol{\rho}_2| \to 0$ and $I_0 \to \infty$ so that $I_0|\boldsymbol{\rho}_1 - \boldsymbol{\rho}_2|$ remains finite. Give a formula for the resultant current distribution as a function of ρ and φ.

2.8. Calculate the field **B** inside and outside of a perfect solenoid, that is, an infinite thin cylinder of radius a carrying a uniform circulating current density $\mathbf{j} = \hat{k} \times \hat{\rho} J \delta(\rho - \rho_a)$, where $\hat{\rho}$ is the unit radial vector from the axis of the cylinder to the point $\boldsymbol{\rho}$. $\hat{k}$ is a unit vector parallel to the axis of the cylinder. Check that your answer is consistent with Ampère's circuital law (2.2.33).

2.9. Calculate the **B** and **H** fields of a magnetized spherical shell of radius b with a constant dipole moment per unit volume **M**.

2.10. Calculate the **B** and **H** fields on the axis of a circular cylindrical magnet of radius a and length h with a constant dipole moment per unit volume **M**. Give the fields both inside and outside the magnet.

2.11. A spherical shell of radius R carries a uniform charge distribution of surface charge density σ. The shell is rotated about an axis with constant angular velocity $\boldsymbol{\omega}$. Find the magnetic field **B** inside and outside the sphere.

Suppose you wish to wind a current-carrying wire around a sphere in such a way that the field inside the sphere is uniform. How should the wire be wound?

2.12. Study the next approximation in r'/r to the magnetic field of a confined current. That is, from the general formula

$$\mathbf{A}(\mathbf{r}) = \frac{1}{c} \int d\mathbf{r}' \, \frac{\mathbf{j}(\mathbf{r}')}{|\mathbf{r} - \mathbf{r}'|}$$

find the term following the dipole field. In particular, find the magnetic scalar potential and show that it is a quadrupole field.

2.13. A current distribution $\mathbf{J}(\mathbf{r})$ exists in vacuum in the space $z > 0$, all x and y. The space $z < 0$ is filled with a medium of permeability μ.

Show that a $\mathbf{B}$ field satisfying the field equation and boundary conditions is given by the following:

(a) For $z > 0$, the unperturbed $\mathbf{B}$ field of the current $\mathbf{J}(\mathbf{r})$, to which must be added the $\mathbf{B}$ field of an image current $\mathbf{L}(\mathbf{r})$ with

$$L_x(x, y, z) = (\mu - 1)/(\mu + 1)J_x(x, y, -z),$$

$$L_y(x, y, z) = (\mu - 1)/(\mu + 1)J_y(x, y, -z),$$

and

$$L_z(x, y, z) = -(\mu - 1)/(\mu + 1)J_z(x, y, -z).$$

(b) The $\mathbf{B}$ field for $z < 0$ is given by the unperturbed field of an image current $2\mu/(1 + \mu)\mathbf{J}(\mathbf{r})$.

(c) The equations do not appear at first sight to lead to a unique solution. What makes the solution unique?

*2.14. A superconductor behaves like a perfect diamagnet, that is, a material with $\mu = 0$. An interesting property of a superconducting sphere is that it is repelled by a magnetic field and, therefore, can be balanced above a magnet.

(a) Simulate the magnetic field as resulting from a very long circular cylindrical magnet of radius b and uniform magnetization M, with the $-M$ pole far enough away to neglect. The superconducting sphere, with radius a, is placed at a vertical height h above the center of the positive face of the magnet. Calculate the force on the sphere. Take $a \ll h$ and carry out a multipole expansion of the B field generated by the superconducting sphere in the presence of the magnet. Then integrate the stress tensor around the sphere to get the force. You will have to keep the dipole and quadrupole terms to obtain a nonvanishing magnetic force.

(b) Study the vertical stability of the suspension.

(c) Study the horizontal stability of the suspension. (*Answer*: Stable for $b/h > \sqrt{24}$.)

2.15. Assume that force $\mathbf{F}$ is invariant (up to and including terms of order v/c) to a change of coordinate system going from one observer to another (the primed observer) with relative velocity $\mathbf{v}$. From this, assuming $\mathbf{E}' = \mathbf{E} + 0(v/c)$, $\mathbf{B}' = \mathbf{B} + 0(v/c)$, and the Lorentz force law, show that

$$\mathbf{E}' = \mathbf{E} + \frac{\mathbf{v}}{c} \times \mathbf{B} + 0\left(\frac{v^2}{c^2}\right).$$

2.16. A current is carried by a wire of radius r. The wire is in the form of a circular loop, of radius a, with $a \gg r$.

    (a) Calculate the self-inductance L of the loop accurate to order $1/\log(a/r)$.

    (b) From this result, estimate the tension in the wire when carrying a current I. Make sure your answer is a tension.

2.17. Complete the argument leading from (2.7.18) to (2.7.20) by showing that $\Sigma \, \mathbf{r}_e \times \delta \mathbf{E}$, where $\delta \mathbf{E}$ is linear in $\mathbf{r}_e$, averages to zero over many cycles of fast motion and over many atoms.

2.18. Solve Eq. (2.7.21) for the $\mathbf{M}_\perp$, the components of $\mathbf{M}$ perpendicular to $\boldsymbol{\omega}_L$, given the initial value of $\mathbf{M}_\perp$. Assume $\boldsymbol{\omega}_L$ is in a fixed direction, but has arbitrary time dependence.

CHAPTER 3

Time-Dependent Fields and Currents

3.1. MAXWELL'S EQUATIONS

The static and quasistatic (including electromagnetic induction) equations for **E** are

$$\nabla \times \mathbf{E} = -\frac{1}{c}\frac{\partial \mathbf{B}}{\partial t} \qquad (3.1.1)$$

and

$$\nabla \cdot \mathbf{E} = 4\pi\rho; \qquad (3.1.2)$$

for **B** they are

$$\nabla \cdot \mathbf{B} = 0 \qquad (3.1.3)$$

and

$$\nabla \times \mathbf{B} = \frac{4\pi\mathbf{j}}{c}. \qquad (3.1.4)$$

Equations (3.1.3) and (3.1.1) are clearly consistent; however, if ρ is time-dependent, the continuity equation, (2.1.2), tells us that $\nabla \cdot \mathbf{j}$ will not vanish, and (3.1.4) will be inconsistent. We repair this inconsistency in the simplest possible way. We add a term $4\pi\mathbf{j}/c$ to the right-hand side of

81

(3.1.4):

$$\nabla \times \mathbf{B} = \frac{4\pi}{c}(\mathbf{j} + \mathbf{j}').$$ (3.1.5)

Consistency requires

$$\nabla \cdot (\mathbf{j} + \mathbf{j}') = 0$$

or, from the continuity equation,

$$\nabla \cdot \mathbf{j}' - \frac{\partial \rho}{\partial t} = 0.$$ (3.1.6)

Assuming (again, the simplest assumption) that (3.1.2) still holds, we find

$$\nabla \cdot \mathbf{j}' = \frac{1}{4\pi}\frac{\partial}{\partial t}\nabla \cdot \mathbf{E}$$

$$= \nabla \cdot \frac{1}{4\pi}\frac{\partial \mathbf{E}}{\partial t}$$ (3.1.7)

from which

$$\mathbf{j} = \frac{1}{4\pi}\frac{\partial \mathbf{E}}{\partial t} + \nabla \times \mathbf{Q}$$ (3.1.8)

where $\mathbf{Q}$ can be any vector. Again, the simplest assumption is $\mathbf{Q} = 0$. This yields a new, consistent equation to replace (3.1.4):

$$\nabla \times \mathbf{B} = \frac{1}{c}\frac{\partial \mathbf{E}}{\partial t} + \frac{4\pi\mathbf{j}}{c}.$$ (3.1.9)

Maxwell called the new current $\mathbf{j}'$ the displacement current. Together with $\mathbf{j}$, the "total" current is "conserved," in the sense that no flux of $\mathbf{j} + \mathbf{j}'$ emerges from a closed surface, that is, $\nabla \cdot (\mathbf{j} + \mathbf{j}') = 0$.

Maxwell's full equations require us to look in a new way at the causal relationships between the fields.

First, observe that two of the equations involve time derivatives:

$$\frac{\partial \mathbf{E}}{\partial t} = c\nabla \times \mathbf{B} - 4\pi\mathbf{j}$$ (3.1.10)

and

$$\frac{\partial \mathbf{B}}{\partial t} = -c\nabla \times \mathbf{E}. \qquad (3.1.11)$$

It follows that the right-hand sides of (3.1.10) and (3.1.11) tell us how $\mathbf{E}$ and $\mathbf{B}$ change. This is not the way the static equations led us to think: Current was viewed as *causing* a magnetic field and changing magnetic field as *causing* an electromotive force. When we come to the study of radiation, we will see how our normal way of thinking about causality can be restored. Nevertheless, as a mathematical physics boundary value problem, one must think of (3.1.10) and (3.1.11) as calculating $\partial \mathbf{B}/\partial t$ and $\partial \mathbf{E}/\partial t$, given $\mathbf{B}$, $\mathbf{E}$, and $\mathbf{j}$ as functions of position at an initial time.

Second, observe that taking the divergence of (3.1.10) and (3.1.11) leads to the time derivatives of the divergence equations (3.1.2) and (3.1.3). Thus, from (3.1.10),

$$\nabla \cdot \frac{\partial \mathbf{E}}{\partial t} = -4\pi \nabla \cdot \mathbf{j} = 4\pi \frac{\partial \rho}{\partial t} \qquad \text{or} \qquad \frac{\partial}{\partial t}(\nabla \cdot \mathbf{E} - 4\pi \rho) = 0, \qquad (3.1.12)$$

and from (3.1.11),

$$\nabla \cdot \frac{\partial \mathbf{B}}{\partial t} = 0 = \frac{\partial}{\partial t}\nabla \cdot \mathbf{B} \qquad (3.1.13)$$

so that provided the divergence equations hold at one time, the equations for $\partial \mathbf{E}/\partial t$ and $\partial \mathbf{B}/\partial t$ will guarantee that they hold for all time. Clearly, therefore, the divergence equations should be viewed as enforcing certain boundary conditions on the $\mathbf{E}$ and $\mathbf{B}$ fields, whereas the time derivative equations are the dynamical equations. Just as in most physical theories, one formulates the time-dependent problem as the prediction of the future (or past) state from the present. For example, in mechanics, we specify $\mathbf{r}$ and $d\mathbf{r}/dt$, and then predict the future course of $\mathbf{r}$ and $d\mathbf{r}/dt$ via Newton's second law. Similarly, in quantum theory, we give the wave function at one time and use the Schrödinger equation to predict its value at another time. We see that the analogous problem in electrodynamics is to give $\mathbf{E}$ and $\mathbf{B}$ at one time, subject to the constraints $\nabla \cdot \mathbf{E} = 4\pi \rho$ and $\nabla \cdot \mathbf{B} = 0$. Maxwell's equations then predict the future (or past) values of $\mathbf{E}$ and $\mathbf{B}$.

This assumes, of course, that ρ and $\mathbf{j}$ are given functions of space and time, satisfying the continuity equation $\nabla \cdot \mathbf{j} + (\partial \rho/\partial t) = 0$. When we consider the interaction of particles and fields as complete dynamical systems, ρ and $\mathbf{j}$ can no longer be considered as given functions. Their time dependence must be calculated as well. We will, of course, come back to this. For the moment, however, we suppose that the sources ρ and $\mathbf{j}$ are composed of heavy objects (magnets, large capacitors, atomic nuclei, etc.) on which the reaction of the fields can be ignored.

3.2. ELECTROMAGNETIC FIELDS IN MATTER

In order to obtain equations for the macroscopic fields, we proceed to average the microscopic fields $\mathbf{e}$ and $\mathbf{b}$, as we did in Sections 1.6 and 2.4. In our earlier discussions of electric and magnetic fields in matter, we relied heavily on the explicit form of the integrals giving the static fields as functions of the charge and current densities. Since those integrals no longer hold in the time-dependent regime we are now considering, we must proceed differently. Instead of working with the solutions, we work directly with the differential equations for the microscopic fields. These are first the homogeneous equations

$$\nabla \cdot \mathbf{b} = 0 \tag{3.2.1}$$

and

$$\nabla \times \mathbf{e} = -\frac{1}{c}\frac{\partial \mathbf{b}}{\partial t}. \tag{3.2.2}$$

The spatially averaged fields $\mathbf{B}$ and $\mathbf{E}$ clearly satisfy the same equations:

$$\nabla \cdot \mathbf{B} = 0 \tag{3.2.3}$$

and

$$\nabla \times \mathbf{E} = -\frac{1}{c}\frac{\partial \mathbf{B}}{\partial t}. \tag{3.2.4}$$

Note that we average our fields over space, but not time. The space average is necessary to smooth the fluctuations of the microscopic fields in going from atom to atom. Therefore, the averaging function must extend over a volume that contains many atoms. On the other hand, we need the time resolution to be fine enough to describe light emitted by atoms—that is, to be finer than a characteristic atomic time. In dealing with normal atomic phenomena, there is, therefore, no need for, and nothing to be gained by, a time average. As emphasized in Section 1.6, it is necessary that the averaging volume, although large enough to contain many atoms, must be small enough to resolve the distances we wish to study. These requirements are easily compatible for visible light and a gas at normal temperature and pressure, as we show below, following (3.2.34).

The inhomogeneous microscopic equations

$$\nabla \cdot \mathbf{e} = 4\pi\rho_m \tag{3.2.5}$$

and

$$\nabla \times \mathbf{b} = \frac{1}{c}\frac{\partial \mathbf{e}}{\partial t} + \frac{4\pi}{c}\mathbf{j}_m \tag{3.2.6}$$

are averaged to give

$$\nabla \cdot \mathbf{E} = 4\pi\overline{\rho}_m \tag{3.2.7}$$

and

$$\nabla \times \mathbf{B} = \frac{1}{c}\frac{\partial \mathbf{E}}{\partial t} + \frac{4\pi}{c}\overline{\mathbf{j}}_m. \tag{3.2.8}$$

In order to carry out the required averages, we again separate ρ_m and $\mathbf{j}_m$ into bound and free components:

$$\rho_m = \rho_f + \rho_b \tag{3.2.9}$$

and

$$\mathbf{j}_m = \mathbf{j}_f + \mathbf{j}_b : \tag{3.2.10}$$

$\overline{\rho}_f$ and $\overline{\mathbf{j}}_f$ will be the macroscopically observed charge and current densities:

$$\overline{\rho}_f = \rho \tag{3.2.11}$$

and

$$\overline{\mathbf{j}}_f = \mathbf{j}. \tag{3.2.12}$$

We continue with our model of Section 1.6 in which ρ_b and j_b are assumed to come from neutral atoms or molecules. Therefore, we can write

$$\rho_b = \sum_n \rho_n(\mathbf{r} - \mathbf{r}_n, t) \tag{3.2.13}$$

where $\mathbf{r}_n$ is the location of the nth neutral atom or molecule. $\rho_n(\mathbf{x})$ falls rapidly to zero for x larger than a_B, an atomic radius, and $\int d\mathbf{x}\rho_n(\mathbf{x}) = 0$. Similarly,

$$\mathbf{j}_b = \sum_n \mathbf{j}_n(\mathbf{r} - \mathbf{r}_n, t). \tag{3.2.14}$$

However, here, as we shall see, we cannot [as we did in (2.3.3)] require $\int \mathbf{j}(\mathbf{x}) \, d\mathbf{x} = 0$.

We now average:

$$\bar{\rho}_b = \sum_n \int d\mathbf{r}' f(\mathbf{r} - \mathbf{r}') \rho_n(\mathbf{r}' - \mathbf{r}_n)$$

or, with $\mathbf{r}' - \mathbf{r}_n = \mathbf{x}$,

$$\bar{\rho}_b = \sum_n \int d\mathbf{x} f(\mathbf{r} - \mathbf{r}_n - \mathbf{x}) \rho_n(\mathbf{x}). \qquad (3.2.15)$$

Since x is restricted by ρ_n to be $\lesssim a_B$, and f, the averaging function, varies on a scale that includes many atoms, we expand f in powers of $\mathbf{x}$, up to and including the linear term. The first term vanishes by the assumed neutrality of ρ_n. The second term gives

$$\bar{\rho}_b = -\sum_n \int d\mathbf{x}\, \mathbf{x} \rho_n(\mathbf{x}) \cdot \nabla_r f(\mathbf{r} - \mathbf{r}_n) \qquad (3.2.16)$$

or

$$\bar{\rho}_b = -\nabla \cdot \mathbf{P}(\mathbf{r}) \qquad (3.2.17)$$

where

$$\mathbf{P}(\mathbf{r}) = \sum_n f(\mathbf{r} - \mathbf{r}_n) \mathbf{p}_n. \qquad (3.2.18)$$

$\mathbf{p}_n$ is the dipole moment of the nth atom and $\mathbf{P}$ the average dipole moment per unit volume.

We may neglect the next term in the expansion, since we will be considering electric and magnetic fields that are weak enough that a linear theory suffices. Thus, the next term has a contribution that is independent of the electric field and does not contribute to our linear equation, plus a contribution which is linear in the electric field and does but is negligible compared to (3.2.16) for the reasons given in Section 1.6.

For $\bar{\mathbf{j}}_b$, we carry out a similar expansion:

$$\bar{\mathbf{j}}_b = \sum_n \int d\mathbf{x}\, f(\mathbf{r} - \mathbf{r}_n - \mathbf{x}) \mathbf{j}_n(\mathbf{x}) \qquad (3.2.19)$$

$$= \sum_n f(\mathbf{r} - \mathbf{r}_n) \int d\mathbf{x}\, \mathbf{j}_n(\mathbf{x}) - \int d\mathbf{x}\, \mathbf{x} \cdot \nabla_r \sum_n f(\mathbf{r} - \mathbf{r}_n) \mathbf{j}_n(\mathbf{x}). \qquad (3.2.20)$$

The first term in (3.2.20) involves

$$\int d\mathbf{x}\, \mathbf{j}_n(\mathbf{x}) = -\int d\mathbf{x}\, \mathbf{x} \nabla \cdot \mathbf{j}_n = \int d\mathbf{x}\, \frac{\partial \rho_n}{\partial t}\mathbf{x} = \frac{d\mathbf{p}_n}{dt} \tag{3.2.21}$$

so that the first term becomes

$$\overline{\mathbf{j}}_{b_1} = \sum_n f(\mathbf{r} - \mathbf{r}_n)\frac{d\mathbf{p}_n}{dt} = \frac{\partial}{\partial t}\mathbf{P}. \tag{3.2.22}$$

The second term we write using a tensorial notation. Its ith component is

$$\overline{j}_{b_{2i}} = -\int x_k \frac{\partial}{\partial r_k}\sum_n f(\mathbf{r} - \mathbf{r}_n)j_{ni}(\mathbf{x})\, d\mathbf{x}. \tag{3.2.23}$$

We decompose $x_k j_{ni}$ into a symmetric and antisymmetric part. We neglect the symmetric part for the same reasons as those given above following (3.2.18). The antisymmetric part will, of course, give our usual magnetic dipole density. Thus, (3.2.23), antisymmetrized, becomes

$$\overline{j}_{b_{2i}} = -\frac{1}{2}\sum_n \frac{\partial}{\partial r_k}f(\mathbf{r} - \mathbf{r}_n)\int d\mathbf{x}(x_k j_{ni} - x_i j_{nk}) \tag{3.2.24}$$

$$\overline{\mathbf{j}}_{b_2} = -\frac{1}{2}\sum_n \int d\mathbf{x}[\mathbf{j}_n(\mathbf{x}\cdot\nabla_r f) - \mathbf{x}(\mathbf{j}_n\cdot\nabla_r f)]\, d\mathbf{x}$$

$$= \frac{1}{2}\sum_n (\nabla_r f_n) \times \int d\mathbf{x}(\mathbf{x}\times\mathbf{j}_n) \tag{3.2.25}$$

or

$$\overline{\mathbf{j}}_{b_2} = c\sum_n \nabla_r f(\mathbf{r} - \mathbf{r}_n)\times\mathbf{m}_n \tag{3.2.26}$$

where $\mathbf{m}_n$ is the magnetic moment of the nth atom. Equation (3.2.26) then leads to

$$\overline{\mathbf{j}}_{b_2} = c\nabla\times\sum_n f(\mathbf{r} - \mathbf{r}_n)\mathbf{m}_n = c\nabla\times\mathbf{M} \tag{3.2.27}$$

where $\mathbf{M}$ is the magnetic moment per unit volume. Thus, we have

$$\overline{\rho}_b = -\nabla\cdot\mathbf{P} \quad\text{and}\quad \overline{\mathbf{j}}_b = \frac{\partial\mathbf{P}}{\partial t} + c\nabla\times\mathbf{M}. \tag{3.2.28}$$

The averaged inhomogeneous equations are therefore

$$\nabla \cdot \mathbf{E} = 4\pi\rho - 4\pi\nabla \cdot \mathbf{P} \tag{3.2.29}$$

and

$$\nabla \times \mathbf{B} = \frac{1}{c}\frac{\partial \mathbf{E}}{\partial t} + \frac{4\pi}{c}\frac{\partial \mathbf{P}}{\partial t} + 4\pi\nabla \times \mathbf{M} + \frac{4\pi\mathbf{j}}{c} \tag{3.2.30}$$

so, introducing the $\mathbf{D}$ and $\mathbf{H}$ fields, we have

$$\nabla \cdot \mathbf{D} = 4\pi\rho \tag{3.2.31}$$

and

$$\nabla \times \mathbf{H} = \frac{1}{c}\frac{\partial \mathbf{D}}{\partial t} + \frac{4\pi\mathbf{j}}{c}. \tag{3.2.32}$$

Equations (3.2.31) and (3.2.32), together with the homogeneous equations (3.2.3) and (3.2.4), determine the boundary conditions to be imposed at a material discontinuity:

$$\begin{aligned} \Delta D_{\text{normal}} &= 4\pi\sigma \quad (\sigma = \text{surface charge}) \\ \Delta B_{\text{normal}} &= 0 \\ \Delta E_{\text{tangential}} &= 0 \end{aligned} \tag{3.2.33}$$

and

$$\Delta H_{\text{tangential}} = \frac{4\pi K}{c} \quad (K = \text{surface current}).$$

Together with the constitutive relations between $\mathbf{E}$ and $\mathbf{D}$ on the one hand, and $\mathbf{H}$ and $\mathbf{B}$ on the other, these equations and boundary conditions determine the time dependence of the fields. However, unlike the matter-free case, the fields at one time in the presence of matter do not determine the fields at all later times. This is because there is a finite time and space lag between the imposition of an electric field and the appearance of a nonvanishing polarization. Consequently, the relation between the macroscopic electric field and the polarization will be

$$\mathbf{P}_i(\mathbf{r}, t) = \int_{-\infty}^{t} dt' \chi_{ij}(\mathbf{r}, t - t') E_j(\mathbf{r}, t'), \tag{3.2.34}$$

where χ_{ij} is the susceptibility tensor at position $\mathbf{r}$. The relationship between $\mathbf{E}$ and $\mathbf{P}$ will be linear, as indicated, for fields that are weak compared with the internal fields in matter ($e/a_B^2 \sim 5 \times 10^9$ V/cm). The space lag whose scale would be set by atomic sizes, $\sim 10^{-8}$ cm, is left out of (3.2.34), because our fields are averaged over a volume containing many atoms, so that whatever space lag existed would be washed out.

That χ depends only on $t - t'$ assumes that the affect of the field $\mathbf{E}$ at time t' on the polarization at time t only depends on the difference $t - t'$, in a way that is independent of the time t'. We say (3.2.34) is time-translation-invariant. Equation (3.2.34) shows that specifying $\mathbf{E}$ at one time is not, in general, enough information to determine $\mathbf{E}$ at later times, since the integral over t' requires $\mathbf{E}$ to be known at all previous times. There are still physically obvious sets of consistent initial conditions, but they depend explicitly on the geometry under consideration. In addition, they require specifying the entire previous history of the system. A simple example is discussed in Section 3.5 and Problem 3.14.

The $t - t'$ dependence of χ is governed by the characteristic frequencies of atomic motion. Since these are the same frequencies that atoms radiate, it is necessary to keep track of the temporal nonlocality of χ. Note that if we wish to describe the spatial variation of the fields, consistency requires that our averaging process allow us to resolve the length scale of wavelengths, which, in turn, must be much larger than atomic lengths to justify our assumption of spatial locality. This is the case for atomic radiation. With f the radiation frequency and λ its wavelength, we have, very approximately, $\lambda \sim c/f \sim 4\pi c/(e^2/a_B\hbar) \approx 1000$ Å. On the other hand, a 1000 Å cube of gas contains about 10^5 atoms at normal temperature and pressure.

Equation (3.2.34) invites a Fourier transform. With

$$E_j(t) = \frac{1}{\sqrt{2\pi}} \int_{-\infty}^{\infty} d\omega \, e^{-i\omega t} \tilde{E}_j(\omega) \tag{3.2.35}$$

and

$$P_j(t) = \frac{1}{\sqrt{2\pi}} \int_{-\infty}^{\infty} d\omega \, e^{-i\omega t} \tilde{P}_j(\omega) \tag{3.2.36}$$

we have

$$\tilde{P}_i(\mathbf{r}, \omega) = \tilde{\chi}_{ij}(\mathbf{r}, \omega) \, \tilde{E}_j(\mathbf{r}, \omega) \tag{3.2.37}$$

where

$$\tilde{\chi}_{ij}(\mathbf{r}, \omega) = \int_{-\infty}^{\infty} dt \, \chi_{ij}(\mathbf{r}, t) \, e^{i\omega t}. \tag{3.2.38}$$

It follows that

$$\tilde{D}_i(\mathbf{r}, \omega) = \epsilon_{ij}(\mathbf{r}, \omega)\, \tilde{E}_i(\mathbf{r}, \omega) \qquad (3.2.39)$$

where ϵ_{ij}, the dielectric tensor, is given by

$$\epsilon_{ij} = \delta_{ij} + 4\pi \chi_{ij} \qquad (3.2.40)$$

and, of course,

$$\mathbf{D}(\mathbf{r}, t) = \mathbf{E}(\mathbf{r}, t) + 4\pi \mathbf{P}(\mathbf{r}, t).$$

Since the boundary conditions (3.2.33) must hold at all times, they must also hold for the Fourier transformed fields.

The rules for complex conjugation of the Fourier transformed fields and dielectric constants follow directly from the reality of the fields themselves:

$$\tilde{E}(\omega)^* = \tilde{E}(-\omega), \quad \text{etc.} \qquad (3.2.41)$$

and

$$\epsilon_{ij}(\omega)^* = \epsilon_{ij}(-\omega). \qquad (3.2.42)$$

An important property of the dielectric constant follows from the presumption that if the field $\mathbf{E}$ is zero before $t = 0$, the polarization $\mathbf{P}$ and displacement $\mathbf{D}(t)$ should also be zero before $t = 0$. If $\mathbf{E}(t) = 0$ for $t < 0$, then

$$\tilde{E}(\omega) = \frac{1}{\sqrt{2\pi}} \int_0^\infty dt'\, e^{i\omega t'} \mathbf{E})t') \qquad (3.2.43)$$

and so $\tilde{E}(\omega)$ is analytic in the upper half ω plane.[1] Conversely, if $\tilde{E}(\omega)$ is analytic in the upper half ω plane, then

$$\tilde{E}(t) = \frac{1}{\sqrt{2\pi}} \int_{-\infty}^\infty d\omega\, e^{-i\omega t} \tilde{E}(\omega) \qquad (3.2.44)$$

vanishes for $t < 0$. This can be seen from (3.2.44) by closing the ω contour

---

[1] Strictly speaking, $\tilde{E}(\omega)$ for real ω is the boundary as ω approaches the real axis of a function that is analytic in the upper half ω plane. We see here for the first time the importance of analytic functions to notions of causality.

in the upper half-plane, presuming $\tilde{\mathbf{E}}(\omega)$ to go to zero sufficiently rapidly as $\omega \to \infty$.

Similarly,

$$\tilde{\mathbf{D}}(\omega) = \frac{1}{\sqrt{2\pi}} \int_0^\infty dt'\, e^{i\omega t}\mathbf{D}(t') \tag{3.2.45}$$

so that $\tilde{\mathbf{D}}(\omega)$ must also be analytic in the upper half ω plane. Finally, since

$$\tilde{D}_i(\omega) = \epsilon_{ij}(\omega)\, E_j(\omega), \tag{3.2.46}$$

$\epsilon_{ij}(\omega)$ must also be analytic in the upper half-plane. One can rule out upper half-plane poles of ϵ, since they would have to be compensated by zeroes of E. However, $E(\omega)$ is essentially arbitrary (except for its analyticity properties) and cannot be required to have zeroes at a predetermined value of ω. Remarkably, the upper half-plane analyticity of $\epsilon(\omega)$ is almost sufficient to guarantee causal propagation in a material medium (i.e., signal propagation with velocity limited by light velocity c). It must be supplemented only by the requirement that field energy can be lost, but not gained from the medium. This is shown in Problems 3.13 and 3.14. We turn in the next section to a discussion of field energy in a dielectric.

3.3. MOMENTUM AND ENERGY

We consider the force on charges and currents in the absence of dielectric matter. The total force on charges and currents inside a volume V is

$$F_i = \int_V d\mathbf{r}\left[\rho E_i + \frac{1}{c}(\mathbf{j} \times \mathbf{B})_i\right]. \tag{3.3.1}$$

We substitute $\nabla \cdot (\mathbf{E}/4\pi)$ for ρ and $c(\nabla \times \mathbf{B}/4\pi) - 1/4\pi\, \partial\mathbf{E}/\partial t$ for $\mathbf{j}$ to obtain

$$F_i = \int_V \frac{d\mathbf{r}}{4\pi}\left[E_i\partial_k E_k - \frac{1}{c}\left(\frac{\partial\mathbf{E}}{\partial t} \times \mathbf{B}\right)_i - \frac{1}{4\pi}[\mathbf{B} \times (\nabla \times \mathbf{B})]_i\right]. \tag{3.3.2}$$

We follow a by now familiar path:

$$E_i\partial_k E_k = \partial_k(E_i E_k) - E_k(\partial_k E_i - \partial_i E_k) - E_k\partial_i E_k, \tag{3.3.3}$$

$$[\mathbf{B} \times (\nabla \times \mathbf{B})]_i = B_k \partial_i B_k - \mathbf{B} \cdot \nabla B_i = \partial_i \frac{\mathbf{B}^2}{2} - \partial_k (B_k B_i), \qquad (3.3.4)$$

and

$$E_k (\partial_k E_i - \partial_i E_k) = -[\mathbf{E} \times (\nabla \times \mathbf{E})]_i = \frac{1}{c} \left(\mathbf{E} \times \frac{\partial \mathbf{B}}{\partial t} \right)_i. \qquad (3.3.5)$$

Putting it all together, we obtain

$$F_i = - \int_V \frac{d\mathbf{r}}{4\pi c} \left(\frac{\partial \mathbf{E}}{\partial t} \times \mathbf{B} + \mathbf{E} \times \frac{\partial \mathbf{B}}{\partial t} \right)_i + \int d\mathbf{r} \, \partial_k T_{ik} \qquad (3.3.6)$$

where T_{ik} is the sum of the familiar electric and magnetic stress tensors (1.3.19) and (2.6.17). The new term on the right of (3.3.6) is the time derivative of a vector

$$-\mathbf{p}_{em} = - \int_V d\mathbf{r} \, \frac{\mathbf{E} \times \mathbf{B}}{4\pi c}. \qquad (3.3.7)$$

Transpose $-d\mathbf{p}_{em}/dt$ to the left-hand side of (3.3.6) and recognize that $\mathbf{F} = d\mathbf{p}_m/dt$, where $\mathbf{p}_m$ is the material momentum of the charges and currents acted on by the fields.[2] Thus, we have

$$\frac{d}{dt} (p_{m_i} + p_{em_i}) = \int_S dS_k \, T_{ik} \qquad (3.3.8)$$

which clearly identifies $\mathbf{p}_{em}$ as the electromagnetic momentum contained in the volume V.

We turn next to energy. In this case, we prefer to keep the possibility of describing dielectric media. The dynamical equations are then

$$\nabla \times \mathbf{E} = - \frac{1}{c} \frac{\partial \mathbf{B}}{\partial t} \qquad (3.3.9)$$

and

---

[2]We know that this statement holds for nonrelativistic systems. For relativistic systems, the final equation (3.3.8) is correct, although the force defined as $d\mathbf{p}_m/dt$ does not have simple relativistic properties. We will come back to this issue when we discuss relativity in Chapter 6.

$$\nabla \times \mathbf{H} = \frac{1}{c}\frac{\partial \mathbf{D}}{\partial t} + \frac{4\pi \mathbf{j}}{c}. \tag{3.3.10}$$

Now dot $\mathbf{H}$ into (3.3.9), $\mathbf{E}$ into (3.3.10), and subtract. There results

$$\frac{1}{c}\left(\mathbf{E}\cdot\frac{\partial \mathbf{D}}{\partial t} + \mathbf{H}\cdot\frac{\partial \mathbf{B}}{dt}\right) = -4\pi\frac{\mathbf{j}}{c}\cdot\mathbf{E} + (\mathbf{E}\cdot\nabla\times\mathbf{H} - \mathbf{H}\cdot\nabla\times\mathbf{E}) \tag{3.3.11}$$

or, if we integrate over a volume V,

$$-\frac{1}{4\pi}\int_V d\mathbf{r}\left(E\cdot\frac{\partial \mathbf{D}}{\partial t} + \mathbf{H}\cdot\frac{\partial \mathbf{B}}{\partial t}\right) = \int_V d\mathbf{r}\,\mathbf{j}\cdot\mathbf{E} + \frac{c}{4\pi}\int_V d\mathbf{r}\,\nabla\cdot(\mathbf{E}\times\mathbf{H}) \tag{3.3.12}$$

$$= \int_V d\mathbf{r}\,\mathbf{j}\cdot\mathbf{E} + \int_S \mathcal{P}\cdot d\mathbf{S} \tag{3.3.13}$$

where $\mathcal{P}$, the Poynting vector, is

$$\mathcal{P} = \frac{c}{4\pi}\mathbf{E}\times\mathbf{H}. \tag{3.3.14}$$

Equation (3.3.13) clearly is an energy balance equation: On the left is the rate at which field energy is lost, on the right the two loss mechanisms, doing work on charges and escaping through the surface surrounding V. The increase in electrical field energy (including possible absorption by matter) is

$$\delta W = \frac{1}{4\pi}\int d\mathbf{r}\,\mathbf{E}\cdot\delta\mathbf{D} \tag{3.3.15}$$

with an analogous term for magnetic energy. For a material with a symmetric, time-independent susceptibility tensor

$$\delta D_i = \epsilon_{ij}\delta E_j \tag{3.3.16}$$

and

$$\delta W = \frac{1}{4\pi}E_i\,\delta E_j\,\epsilon_{ij} = \delta\frac{E_iE_j\epsilon_{ij}}{8\pi} \tag{3.3.17}$$

so that (3.3.15) can be integrated to give

$$W = \frac{1}{8\pi} \int d\mathbf{r}\, E_i E_j \epsilon_{ij}. \tag{3.3.18}$$

To study the general case, we consider a situation in which the field is cycled from zero to zero, and integrate δW to see if energy is absorbed in the process. The total absorption will be

$$\Delta W = \frac{1}{4\pi} \int\limits_{-\infty}^{\infty} dt \int d\mathbf{r}\, \mathbf{E} \cdot \frac{\partial \mathbf{D}}{\partial t} \tag{3.3.19}$$

and, for passive matter, must be nonnegative. We evaluate (3.3.19) by inserting Fourier transforms for $\mathbf{E}$ and $\mathbf{D}$:

$$\Delta W = \frac{1}{4\pi} \int\limits_{-\infty}^{\infty} dt \int d\mathbf{r} \int \frac{d\omega}{\sqrt{2\pi}} \tilde{E}_j(\omega)\, e^{-i\omega t} \frac{\partial}{\partial t} \int \frac{d\omega'}{\sqrt{2\pi}} \tilde{D}_j(\omega')\, e^{-i\omega' t}$$

$$\tag{3.3.20}$$

or

$$\Delta W = \frac{1}{4\pi} \int d\mathbf{r} \int\limits_{-\infty}^{\infty} d\omega\, i\omega \tilde{E}_j(\omega)\, \epsilon_{jk}(-\omega)\, \tilde{E}_k(-\omega). \tag{3.3.21}$$

We introduce the real and imaginary parts of E_j and ϵ_{jk}:

$$\tilde{E}_j(\omega) = R_j(\omega) + iI_j(\omega) \tag{3.3.22}$$

and

$$\epsilon_{jk} = \epsilon_{jk}^R + i\epsilon_{jk}^I. \tag{3.3.23}$$

It follows from (3.2.41) and (3.2.42) that R_j and ϵ_{jk}^R are even functions of ω and I_j and ϵ_{jk}^I are odd functions of ω. Thus,

$$\Delta W = \frac{1}{2\pi} \int d\mathbf{r} \int\limits_{0}^{\infty} \omega\, d\omega [(R_j R_k + I_j I_k)\, \epsilon_{jk}^I - (I_j R_k - I_k R_j)\, \epsilon_{jk}^R]. \tag{3.3.24}$$

We see that absorption comes (for $\omega > 0$) from the anti-hermitian part of ϵ_{jk}:

$$\epsilon_{ajk} = \left(\frac{\epsilon - \epsilon^\dagger}{2i}\right)_{jk} = \frac{\epsilon_{jk} - \epsilon^*_{kj}}{2i} \tag{3.3.25}$$

which must be a nonnegative matrix, that is,

$$(f_j^*(\epsilon_a)_{jk}f_k) \geq 0 \tag{3.3.26}$$

for any vector f.

For the special case of a scalar dielectric constant ϵ, (3.3.26) becomes

$$\text{Im } \epsilon(\omega) > 0 \tag{3.3.27}$$

for $\omega > 0$.

Upper half-plane analyticity and imaginary part positivity on the real axis further restrict the properties of ϵ.

First, they imply that $\epsilon(\omega)$ has no zeros in the upper half ω plane. Second, they relate the signs of the real part and the imaginary part of $\sqrt{\epsilon}$ on the real axis: They must be equal. (See Problems 3.13 and 3.14.)

We shall see in the next section an example in which the upper half-plane analyticity of $\epsilon(\omega)$ is closely related to the positivity of ϵ_a. This is not surprising since both these properties are related to the passivity of matter. Clearly, if there is power being put into matter, then the absorption ΔW can be negative, and the polarization density in the medium can precede the application of the electric field.

3.4. POLARIZABILITY AND ABSORPTION BY ATOMIC SYSTEMS

We consider an atom in a uniform but time-dependent electric field, and calculate the Schrödinger wave function $\psi(\mathbf{r}_1, \ldots, \mathbf{r}_n, t)$ for the atom, where $\mathbf{r}_1, \ldots, \mathbf{r}_n$ are the coordinates of the electrons in the atom. The neglect of the motion of the nucleus is a very good approximation and does not affect the conclusions we shall draw.

The applied field is

$$\mathbf{E}(t) = \int \frac{d\omega}{\sqrt{2\pi}} e^{-i\omega t} \tilde{\mathbf{E}}(\omega)$$

and the Hamiltonian is

$$H = H_0 + H_1 \tag{3.4.1}$$

where $H_1 = -\mathbf{E} \cdot \mathbf{X}$ and $\mathbf{X}$ is the electric dipole operator of the atom:

$$\mathbf{X} = e \sum_n \mathbf{r}_n. \tag{3.4.2}$$

We try to solve the Schrödinger equation

$$-\frac{\hbar}{i} \frac{\partial \psi}{\partial t} = H\psi \tag{3.4.3}$$

in the presence of $\mathbf{E}$. Since we are looking for linear effects, we will use first-order perturbation theory to calculate ψ and then use ψ to calculate the expected value of the dipole operator $\mathbf{X}$:

$$\langle \mathbf{X} \rangle = \int \psi^*(\mathbf{r}_1, \ldots, \mathbf{r}_n) \mathbf{X} \psi(\mathbf{r}_1, \ldots, \mathbf{r}_n) \, d\mathbf{r}_1, \ldots, d\mathbf{r}_n. \tag{3.4.4}$$

We solve (3.4.3) by assuming[3] $\psi = \psi_0 + \psi_1$, where ψ_1 is first-order in $\mathbf{E}$. ψ_0 satisfies the equation

$$H_0 \psi_0 = -\frac{\hbar}{i} \frac{\partial \psi_0}{\partial t} \tag{3.4.5}$$

and will be taken here as the ground state wave function, $u_0 e^{-iW_0 t/\hbar}$, for which

$$H_0 u_0 = W_0 u_0. \tag{3.4.6}$$

Of course, this is appropriate only for a system at a temperature T such that no significant excitation is present.

We rewrite (3.4.3) as

$$-\frac{\hbar}{i} \frac{\partial \psi_0}{\partial t} - \frac{\hbar}{i} \frac{\partial \psi_1}{\partial t} = (H_0 + H_1)(\psi_0 + \psi_1). \tag{3.4.7}$$

The zeroth-order term is satisfied by (3.4.5); the first-order equation is

---

[3] We take the ground state to be nondegenerate.

$$-\frac{\hbar}{i}\frac{\partial \psi_1}{\partial t} = H_0\psi_1 + H_1\psi_0. \tag{3.4.8}$$

To solve (3.4.8), we expand ψ_1 in the complete set of eigenfunctions of H_0:

$$\psi_1 = \sum_n c_n(t)\, u_n. \tag{3.4.9}$$

Since $H_0 u_n = W_n u_n$, our equation (3.4.8) becomes

$$\sum_n \left(-\frac{\hbar}{i}\frac{\partial c_n}{\partial t} - c_n W_n\right) u_n = -\mathbf{X}\cdot\int\frac{d\omega}{\sqrt{2\pi}}\,e^{-i\omega t}\tilde{\mathbf{E}}(\omega)\,u_0\,e^{-i(W_0/\hbar)t} \tag{3.4.10}$$

We look for solutions

$$c_n = \int_{-\infty}^{\infty} d\omega\, e^{-i\omega t - i(W_0/\hbar)t} d_n(\omega) \tag{3.4.11}$$

from which

$$\sum_n (W_0 - W_n + \hbar\omega) d_n(\omega) u_n = -\mathbf{X}\cdot\frac{\tilde{\mathbf{E}}(\omega)}{\sqrt{2\pi}}u_0 \tag{3.4.12}$$

and hence

$$d_n(\omega) = -\frac{1}{W_0 - W_n + \hbar\omega}\int d\mathbf{r}\, u_n^*\mathbf{X}u_0 \cdot \frac{\tilde{\mathbf{E}}(\omega)}{\sqrt{2\pi}}$$

and

$$\psi_1 = -\int_{-\infty}^{\infty} d\omega\, e^{-i\omega t - i(W_0/\hbar)t}\sum_n\frac{u_n}{W_0 - W_n + \hbar\omega}\int d\mathbf{r}\, u_n^*\mathbf{X}u_0 \cdot \frac{\tilde{\mathbf{E}}(\omega)}{\sqrt{2\pi}}, \tag{3.4.13}$$

where $\int d\mathbf{r}$ stands for $\int d\mathbf{r}_1\, d\mathbf{r}_2 \ldots d\mathbf{r}_n$. Note that we have calculated the driven part of ψ_1. One can always add to ψ_1 a solution of the homogeneous equation, $c_n = c_n(0)\, e^{-iW_n t/\hbar}$, with $c_n(0)$ completely arbitrary. However, in the real world such an added term with $n \neq 0$ would damp out rapidly by radiation. An added term with $n = 0$ is simply a change of normalization that must be canceled in the integral (3.4.4).

Continuing from (3.4.13), we calculate the electric dipole moment:

$$\langle \mathbf{X} \rangle = \int (\psi_0^* + \psi_1^*) \, \mathbf{X} \, (\psi_0 + \psi_1) \, d\mathbf{r}. \tag{3.4.14}$$

The zeroth-order term is zero, since reflection invariance of H_0 requires a nondegenerate u_0 to be either even or odd under the transformation $\mathbf{r} \to -\mathbf{r}$. The first-order term is

$$\langle \mathbf{X} \rangle = \int \psi_0^* \mathbf{X} \psi_1 \, d\mathbf{r} + \text{complex conjugate} \tag{3.4.15}$$

or

$$\langle \mathbf{X} \rangle = - \int_{-\infty}^{\infty} \frac{d\omega}{\sqrt{2\pi}} e^{-i\omega t} \sum_n \frac{\mathbf{X}_{0n} \mathbf{X}_{n0} \cdot \tilde{\mathbf{E}}(\omega)}{W_0 - W_n + \hbar\omega} + \text{c.c.} \tag{3.4.16}$$

where $\mathbf{X}_{n0}$ is the electric dipole matrix element from u_0 to u_n:

$$\mathbf{X}_{n0} = \int u_n^* \mathbf{X} u_0 \, d\mathbf{r}. \tag{3.4.17}$$

We explicitly add the complex conjugate to obtain

$$\langle \mathbf{X} \rangle = - \int_{-\infty}^{\infty} \frac{d\omega}{\sqrt{2\pi}} e^{-i\omega t} \sum_n \frac{\mathbf{X}_{0n} \mathbf{X}_{n0} \cdot \tilde{\mathbf{E}}(\omega)}{W_0 - W_n + \hbar\omega}$$

$$- \int_{-\infty}^{\infty} \frac{d\omega}{\sqrt{2\pi}} e^{i\omega t} \sum_n \frac{\mathbf{X}_{0n}^* \mathbf{X}_{n0}^* \cdot \tilde{\mathbf{E}}^*(\omega)}{W_0 - W_n + \hbar\omega}. \tag{3.4.18}$$

Next, change ω to $-\omega$ in the second term of (3.4.18):

$$\langle X_i \rangle = - \int_{-\infty}^{\infty} \frac{d\omega}{\sqrt{2\pi}} e^{-i\omega t} \tilde{E}_j(\omega) \sum_n \left\{ \frac{X_{i0n} X_{jn0}}{W_0 - W_n + \hbar\omega} + \frac{X_{in0} X_{j0n}}{W_0 - W_n - \hbar\omega} \right\} \tag{3.4.19}$$

where we have used the hermiticity of the operator $\mathbf{X}$, that is, $X_{0n}^* = X_{n0}$, and the reflection property of $\tilde{\mathbf{E}}$, that is, $\tilde{\mathbf{E}}^*(-\omega) = \tilde{\mathbf{E}}(\omega)$. We then have a final formula for polarizability by an applied field:

$$\alpha_{ij}(\omega) = - \sum_n \left\{ \frac{X_{i0n}X_{jn0}}{W_0 - W_n + \hbar\omega} + \frac{X_{j0n}X_{in0}}{W_0 - W_n - \hbar\omega} \right\}. \qquad (3.4.20)$$

There are some important properties of α_{ij} to be noted. First, the numerators for each n are Hermitian matrices in i, j space. That is,

$$(X_{i0n}X_{jn0})^\dagger = (X_{j0n}X_{in0})^* = X_{i0n}X_{jn0}. \qquad (3.4.21)$$

Therefore, any anti-Hermitian part must come in some way from the denominators.

Second, the numerator matrices in (3.4.20) are nonnegative. That is, since $(X_i)_{0n}(X_j)_{n0} = (X_i)_{0n}(X_j)_{0n}^*$, any vector V_i will make

$$V_i^* X_{i0n} X_{jn0} V_j = V_i^* X_{i0n} X_{j0n}^* V_j = |V_i^* X_{i0n}|^2 \geq 0.$$

Third, if the Hamiltonian H_0 is real—not only self-adjoint, but real—then the wave functions u_n can also be chosen to be real, and $X_{i0n} = X_{in0}$, so that α_{ij} is automatically symmetric. The Hamiltonian H_0 can always be made real if time reversal holds. Thus, time reversal invariance produces a symmetric α_{ij}.[4] Note that a *fixed* magnetic field $\mathbf{B}_0$ will violate time reversal invariance since the $\mathbf{v} \times \mathbf{B}/c$ force depends on the sign of the velocity, which reverses under time reversal. (See Problem 3.5.)

Fourth, when $\hbar\omega$ is equal to $W_n - W_0$ for some n, the first term of (3.4.20) for $\alpha_{ij}(\omega)$ becomes infinite. This comes about because we have ignored the fact that the excited atoms can radiate. Classically, this radiation limits the amplitude of oscillation that can be produced by the applied field and therefore keeps $\langle X_i \rangle$ finite. (See Problem 4.8 and Section 5.9.) Quantum mechanically, the possibility of radiation gives the energy level W_n a width, which appears as a negative imaginary part

$$W_n \rightarrow W_n - i\frac{\Gamma_n}{2} \qquad (3.4.22)$$

and prevents the pole from appearing at real values of ω, or, in this approximation, anywhere in the upper half-plane.

Fifth, the polarizability $\alpha_{jk}(\omega) \rightarrow 0$ as $\omega \rightarrow \infty$, provided the sum $\sum_n |\mathbf{X}_{0n}|^2$ converges. This is presumably a general property of material systems, where the convergence of the sum reflects the inability of matter to follow an infinitely rapid oscillation. It then follows that the dielectric tensor approaches δ_{jk}: $\lim_{\omega \rightarrow \infty} \epsilon_{jk}(\omega) = \delta_{jk}$.

---

[4]Time reversal invariance is discussed briefly in Section 6.1.

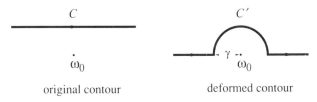

Figure 3.1.

Finally, we turn to the question of absorption. Recall that $\alpha_{ij}(\omega)$ will be causal if

$$\langle X_j \rangle = \int\limits_{-\infty}^{\infty} \frac{d\omega}{\sqrt{2\pi}} e^{-i\omega t} \alpha_{jk}(\omega) \tilde{\mathbf{E}}_k(\omega) \tag{3.4.23}$$

vanishes for $t < 0$, provided $\mathbf{E}(t)$ vanishes for $t < 0$—that is, provided $\tilde{\mathbf{E}}(\omega)$ is analytic in the upper half-plane. For that to be the case, we must for $t < 0$ be able to close the contour in (3.4.23) in the upper half ω plane. We will be able to do so [since $\alpha_{ij}(\omega)$ is an analytic function of ω, except on the real axis] if we define the integral (3.4.23) to be calculated with ω slightly above the real axis. This prescription guarantees that our causal condition will hold. At the same time, it gives a finite definition to the integral (3.4.23), as well as to the sum (3.4.20) for α_{ij} in the case where $W_0 + \hbar\omega$ is in a continuum region of the W_n's.

Consider as an example the integral

$$f(t) = \int\limits_{C} e^{-i\omega t} \frac{d\omega\, g(\omega)}{\omega_0 - \omega} \tag{3.4.24}$$

where ω_0 is real, $g(\omega)$ is analytic in the upper half-plane and on the real axis, and the contour C is above the real axis. Then for $t < 0$, we can close the contour in the upper half ω plane and find $f(t) = 0$. For $t > 0$, we proceed by deforming the contour (as shown in Figure 3.1) onto the real axis, except for a semicircle of radius γ, which we will eventually let shrink to zero. $f(t)$ is then given by

$$f(t) = \lim_{\gamma \to 0} \left[\int\limits_{-\infty}^{\omega_0 - \gamma} d\omega\, e^{-i\omega t} \frac{g(\omega)}{\omega_0 - \omega} + \int\limits_{\omega_0 + \gamma}^{\infty} d\omega\, e^{-i\omega t} \frac{g(\omega)}{\omega_0 - \omega} \right.$$
$$\left. + \int\limits_{C'} d\omega\, e^{-i\omega t} \frac{g(\omega)}{\omega_0 - \omega} \right]. \tag{3.4.25}$$

The first two integrals in (3.4.25) define the Cauchy principle value P. In the C' integral, since g is analytic on the real axis, we can expand ω about ω_0:

$$\omega = \omega_0 + \gamma e^{i\theta} \tag{3.4.26}$$

and keep the only term that fails to vanish as $\gamma \to 0$:

$$\lim_{\gamma \to 0} \int_{C'} \frac{d\omega\, e^{-i\omega t} g(\omega)}{\omega_0 - \omega} = \gamma i \int_{\pi}^{0} \frac{d\theta\, e^{i\theta}\, e^{-i\omega_0 t}}{-\gamma e^{i\theta}} g(\omega_0) \tag{3.4.27}$$

so that

$$f(t) = P \int \frac{d\omega\, e^{-i\omega t}}{\omega_0 - \omega} g(\omega) + \pi i\, e^{-i\omega_0 t} g(\omega_0), \tag{3.4.28}$$

A suitable mnemonic for $1/(\omega_0 - \omega)$ is thus

$$\lim_{\gamma \to 0} \frac{1}{\omega_0 - \omega - i\gamma} = P \frac{1}{\omega_0 - \omega} + i\pi \delta(\omega_0 - \omega) \tag{3.4.29}$$

where the first term is real, the second imaginary.

Note that if the Fourier transform of a differential equation (in our case, the Schrödinger equation) leads to an algebraic equation

$$(\omega_0 - \omega)f = g, \tag{3.4.30}$$

then the solution f can contain a term $A\delta(\omega_0 - \omega)$ with arbitrary A, since $x\delta(x) = 0$. The result for $f(t)$, (3.4.28), shows that causality requires the coefficient of $\delta(\omega_0 - \omega)$ to be $i\pi g(\omega_0)$.

Returning to (3.4.20), we see that for $\omega > 0$, α_{jk} has an anti-Hermitian part

$$\left(\frac{\alpha - \alpha^+}{2i} \right)_{jk} = \pi \sum_n X_{j0n} X_{kn0} \delta(W_n - W_0 - \hbar\omega), \tag{3.4.31}$$

corresponding to absorption of energy from the external field by the atomic target.

3.5. FREE FIELDS IN ISOTROPIC MATERIALS

We look here for monochromatic free field solutions of Maxwell's equations in isotropic materials with scalar dielectric constant $\epsilon(\omega)$, permeability $\mu(\omega)$, and conductivity $\sigma(\omega)$. (We will take up anisotropic media in Section 3.7.) The equations are, with all quantities depending on time like $e^{-i\omega t}$,

$$\nabla \times \mathbf{H} = \frac{1}{c}(-i\omega\mathbf{D} + 4\pi\mathbf{j}) \qquad (3.5.1)$$

and

$$\nabla \times \mathbf{E} = i\frac{\omega}{c}\mathbf{B}. \qquad (3.5.2)$$

The constraint (divergence) equations are automatically satisfied for $\omega \neq 0$.

Set

$$\mathbf{j} = \sigma\mathbf{E}, \qquad \mathbf{D} = \epsilon\mathbf{E} \quad \text{and} \quad \mathbf{H} = \frac{\mathbf{B}}{\mu},$$

Equation (3.5.1) becomes

$$\nabla \times \frac{\mathbf{B}}{\mu} = \frac{1}{c}(-i\omega\epsilon + 4\pi\sigma)\mathbf{E}$$

$$= -i\frac{\omega}{c}\left(\epsilon + \frac{4\pi i\sigma}{\omega}\right)\mathbf{E}. \qquad (3.5.3)$$

We see that conductivity is equivalent to an imaginary part of ϵ, with a pole at $\omega = 0$ representing finite static conductivity. We may therefore assume that $4\pi i\sigma/\omega$ is included in ϵ, giving the simpler equation

$$\nabla \times \frac{\mathbf{B}}{\mu} = -i\frac{\omega\epsilon}{c}\mathbf{E}. \qquad (3.5.4)$$

We can eliminate either $\mathbf{E}$ or $\mathbf{B}$, and find propagation equations for $\mathbf{E}$ and $\mathbf{B}$ alone. Thus, from (3.5.2), $\mathbf{B} = (c/i\omega)\nabla \times \mathbf{E}$, substituted in (3.5.4), gives

$$\nabla \times \left(\frac{\nabla \times \mathbf{E}}{\mu} \right) = \frac{\omega^2}{c^2} \epsilon \mathbf{E}, \tag{3.5.5}$$

whereas substituting $\mathbf{E}$ from (3.5.4) into (3.5.2) yields

$$\nabla \times \frac{1}{\epsilon} \left(\nabla \times \frac{\mathbf{B}}{\mu} \right) = \frac{\omega^2}{c^2} \mathbf{B}. \tag{3.5.6}$$

Equations (3.5.5) and (3.5.6) are equivalent, in that if $\mathbf{E}$ is a solution of (3.5.5), then $\mathbf{B} = \text{const.} \; \nabla \times \mathbf{E}$ satisfies (3.5.6); if $\mathbf{B}$ satisfies (3.5.6), then $\mathbf{E} = \text{const.} \; 1/\epsilon(\nabla \times (\mathbf{B}/\mu))$ satisfies (3.5.5). Thus, either equation describes the propagation of a monochromatic electromagnetic signal in a linear isotropic material medium.

Of course, there are many solutions for a given ϵ, μ, and ω; the choice between them requires a specification of spatial boundary conditions. We will return to this point later. For now we confine ourselves to homogeneous media, so that ϵ and μ are spatial constants. In that case, (3.5.5) and (3.5.6) become

$$-\nabla^2 \mathbf{E} = \frac{\omega^2}{c^2} \epsilon \mu \mathbf{E} \tag{3.5.7}$$

with

$$\nabla \cdot \mathbf{E} = 0$$

and

$$-\nabla^2 \mathbf{B} = \frac{\omega^2}{c^2} \epsilon \mu \mathbf{B} \tag{3.5.8}$$

with

$$\nabla \cdot \mathbf{B} = 0$$

and

$$\mathbf{B} = \frac{c}{i\omega} \nabla \times \mathbf{E}, \qquad \mathbf{E} = -\frac{c}{i\omega} \frac{\nabla \times \mathbf{B}}{\epsilon \mu}. \tag{3.5.9}$$

The solutions of (3.5.9) with definite wave number $\mathbf{k}$ are, with $\omega^2 = c^2 k^2 / \epsilon \mu$,

$$\mathbf{E} = \mathbf{e} \, e^{i\mathbf{k} \cdot \mathbf{x}} \tag{3.5.10}$$

and

$$\mathbf{B} = \mathbf{b}\, e^{i\mathbf{k}\cdot\mathbf{x}} \tag{3.5.11}$$

with

$$\mathbf{b} = \frac{c\mathbf{k}}{\omega} \times \mathbf{e} \tag{3.5.12}$$

and

$$\mathbf{e} = -\frac{c}{\epsilon\mu\omega}\,\mathbf{k} \times \mathbf{b}. \tag{3.5.13}$$

Thus, the solution, for given $\mathbf{k}$ and ω, with $c^2 k^2/\epsilon\mu = \omega^2$, is determined by the vector $\mathbf{e}$, which must be transverse ($\mathbf{e} \cdot \mathbf{k} = 0$). For each $\mathbf{k}$ and ω, there are two linearly independent directions of polarization, $\mathbf{e}_\lambda$. For example, with $\mathbf{k}$ in the z direction, these modes could be $\mathbf{e}_{\lambda_1} = \hat{\mathbf{e}}_x$ and $\mathbf{e}_{\lambda_2} = \hat{\mathbf{e}}_y$.

The most general complex vector $\mathbf{e}_\lambda$ represents a state of elliptic polarization. Recall that the real fields $\mathbf{E}_r$ and $\mathbf{B}_r$ are calculated by taking the real part of the complex vectors $\mathbf{E}$ and $\mathbf{B}$. With $\mathbf{e}_\lambda = \mathbf{e}_1 + i\mathbf{e}_2$, orthogonal to $\mathbf{k}$, we have, at any one point $\mathbf{x}$, with $\phi = \mathbf{k} \cdot \mathbf{x}$,

$$\mathbf{E}_r = \text{Re}(\mathbf{e}_1 + i\mathbf{e}_2)e^{-i\omega t + i\phi} \quad \text{or} \quad \mathbf{E}_r = \mathbf{e}_1 \cos(\omega t - \phi) + \mathbf{e}_2 \sin(\omega t - \phi) \tag{3.5.14}$$

which describes an ellipse in the $\mathbf{e}_1$, $\mathbf{e}_2$-plane. There are two special cases: $\mathbf{e}_1$ parallel to $\mathbf{e}_2$, in which $\mathbf{E}_r$ varies without changing direction. This is called plane, or linearly polarized. Second, $\mathbf{e}_1$ is equal in magnitude and perpendicular to $\mathbf{e}_2$. This is circularly polarized: $\mathbf{E}_r$ moves along a circle of radius $|\mathbf{e}_1| = |\mathbf{e}_2|$.

The handedness of the polarization is defined by the screw sense of the rotation of the electric field with respect to the direction of propagation. Let the direction of propagation be z, with x, y, z being a right-handed system, and $\mathbf{e} = \hat{\mathbf{e}}_x + i\hat{\mathbf{e}}_y$ so that $\mathbf{E}_r = \hat{\mathbf{e}}_x \cos \omega t + \hat{\mathbf{e}}_y \sin \omega t$ moves like a right-handed screw. Thus, $\mathbf{e}_\pm = \hat{\mathbf{e}}_x \pm i\hat{\mathbf{e}}_y$ is right/left circular polarized.[5]

The most general solutions of the propagation equations in the isotropic medium are, with $\hat{\mathbf{e}}_{\lambda_i}$ chosen once and for all for each $\mathbf{k}$,

---

[5]There seems to be some disagreement in the literature on the definition of left and right circular polarization. Our choice makes right circular polarization coincide with positive photon helicity. See Section 3.8 for a discussion of helicity.

$$\mathbf{E}(\mathbf{r}, t) = \int d\omega \int_{k^2 = \epsilon\mu\omega^2/c^2} d\mathbf{k} \sum_{\lambda_i} \mathbf{e}_{\lambda_i} e^{i(\mathbf{k}\cdot\mathbf{x} - \omega t)} a(\mathbf{k}, \omega, \lambda_i) \quad (3.5.15)$$

and

$$\mathbf{B}(\mathbf{r}, t) = \int d\omega \int_{k^2 = \epsilon\mu\omega^2/c^2} d\mathbf{k} \, e^{i(\mathbf{k}\cdot\mathbf{x} - \omega t)} \frac{c}{\omega} \sum_{\lambda_i} \mathbf{k} \times \mathbf{e}_{\lambda_i} a(\mathbf{k}, \omega, \lambda_i) . \quad (3.5.16)$$

Before continuing, we distinguish between two situations. First, a signal that lasts for a finite time is described by a function $f(t)$ with a Fourier transform $\tilde{f}(\omega) = \int (f(t)/\sqrt{2\pi})e^{i\omega t} dt$, such that integrated fluxes, like $\int f(t) g(t) dt$, will be given by

$$\int_{-\infty}^{\infty} f(t) g(t) \, dt = \int_{-\infty}^{\infty} \tilde{f}^*(\omega) \, \tilde{g}(\omega) \, d\omega . \quad (3.5.17)$$

Second is the case of a monochromatic signal, or a sum of monochromatic signals for which

$$f(t) = \mathrm{Re}(f_0 e^{-i\omega t})$$

and

$$g(t) = \mathrm{Re}(g_0 e^{-i\omega t}) ;$$

the time-integrated flux will, of course, be infinite; the item of interest will usually be the time-averaged flux (or energy density). From

$$f = (\mathrm{Re}\, f_0) \cos \omega t + (\mathrm{Im}\, f_0) \sin \omega t$$
$$g = (\mathrm{Re}\, g_0) \cos \omega t + (\mathrm{Im}\, g_0) \sin \omega t \quad (3.5.18)$$

we have

$$\overline{fg} = \frac{1}{2} [\mathrm{Re}\, f_0 \, \mathrm{Re}\, g_0 + \mathrm{Im}\, f_0 \, \mathrm{Im}\, g_0]$$

$$= \frac{1}{2} \mathrm{Re}(f_0^* \, g_0) . \quad (3.5.19)$$

We can now calculate the average energy density and energy flux of a polarized monochromatic plane wave (with real ϵ and μ):

$$\mathbf{E}_r = \text{Re } \mathbf{e}_\lambda \, e^{i(\mathbf{k} \cdot \mathbf{x} - \omega t)} \qquad (3.5.20)$$

$$\mathbf{B}_r = \text{Re } \frac{c\mathbf{k}}{\omega} \times \mathbf{e}_\lambda \, e^{i(\mathbf{k} \cdot \mathbf{x} - \omega t)}. \qquad (3.5.21)$$

The energy density is, from (3.3.13),

$$u = \epsilon \frac{\mathbf{E}_r^2}{8\pi} + \frac{\mathbf{B}_r^2}{8\pi\mu},$$

whose time average, from (3.5.19), is

$$\bar{u} = \frac{1}{2} \cdot \frac{1}{8\pi} \left\{ \epsilon \mathbf{e}_\lambda^* \cdot \mathbf{e}_\lambda + \frac{c^2 k^2}{\omega^2 \mu} \mathbf{e}_\lambda^* \cdot \mathbf{e}_\lambda \right\} \qquad (3.5.22)$$

$$= \frac{\epsilon}{8\pi} \mathbf{e}_\lambda^* \cdot \mathbf{e}_\lambda. \qquad (3.5.23)$$

Note that the magnetic and electric contributions to $\bar{u}$ are equal.
   The time-averaged flux of energy is, from (3.3.14),

$$\overline{\mathcal{P}} = \frac{c}{4\pi} \overline{\mathbf{E} \times \mathbf{H}} = \frac{1}{2} \cdot \frac{c^2}{4\pi} \mathbf{e}_\lambda^* \cdot \mathbf{e}_\lambda \frac{\mathbf{k}}{\omega\mu}$$

$$= \frac{\epsilon}{8\pi} \mathbf{e}_\lambda^* \cdot \mathbf{e}_\lambda \cdot \frac{c^2 \mathbf{k}}{\epsilon\mu\omega} = \bar{u} \cdot \frac{c}{\sqrt{\epsilon\mu}} \hat{k} \qquad (3.5.24)$$

corresponding to a velocity $c/\sqrt{\epsilon\mu}$.
   How do we take into account the imaginary part of $\epsilon\mu$ in the equation

$$\omega^2 = \frac{c^2 k^2}{\epsilon\mu}? \qquad (3.5.25)$$

We can understand most clearly what happens here by observing that for $\sqrt{\epsilon\mu}$ complex, either $\mathbf{k}$ or ω (or both) must be complex. Therefore, our considerations cannot apply to a medium occupying all of space for all time, since complex ω implies an exponentially growing field in time (either future or past), complex $\mathbf{k}$ an exponentially growing field in space. Imposing consistent initial conditions requires that the dielectric be of limited spatial extent (including possibly semi-infinite). The incoming field may then be specified in free space outside the dielectric for all previous times, with the field in the dielectric zero for all previous times.

We shall see next how a sensible formulation of boundary conditions and time development presents itself naturally with a semi-infinite medium. For this purpose, we consider the simplest possible situation: a semi-infinite dielectric to the right of $x = 0$, and a plane polarized electromagnetic wave incident normally from the left.

The incident electric field is $\mathbf{E} = \hat{e}_y E_{\text{inc}}$ with

$$E_{\text{inc}} = \int_{-\infty}^{\infty} d\omega \, f(\omega) \, e^{i\omega[(x/c)-t-(b/c)]} \tag{3.5.26}$$

where $b < 0$.

We chose $f(\omega)$ to be analytic in the upper half ω plane. This ensures that $\mathbf{E} = 0$ for $(x/c) - t - (b/c) > 0$; in particular, at $t = 0$, $\mathbf{E}$ and $\partial \mathbf{E}/\partial t$ vanish for $x > b$. It also makes $\mathbf{E}$ independent of the path of the ω integral in the upper half-plane, provided $f(\omega) \to 0$ sufficiently rapidly as $\omega \to \infty$.

The incident magnetic field is $\mathbf{B} = \hat{\mathbf{e}}_z B_{\text{inc}}$ with

$$B_{\text{inc}} = \int_{-\infty}^{\infty} d\omega \, f(\omega) \, e^{i\omega[(x/c)-t-(b/c)]} . \tag{3.5.27}$$

Maxwell's equations become, for this simple geometry,

$$\frac{\partial E}{\partial x} = -\frac{1}{c} \frac{\partial B}{\partial t} \tag{3.5.28}$$

and

$$\frac{\partial B}{\partial x} = -\frac{1}{c} \frac{\partial D}{\partial t} \tag{3.5.29}$$

where

$$\mathbf{D} = \hat{e}_y D \qquad \text{and} \qquad \mathbf{B} = \mathbf{H}.$$

The divergence equations are satisfied identically.

The boundary conditions at $x = 0$ follow directly from (3.5.28) and (3.5.29). They are E and B continuous.

The appropriate solution of (3.5.28) and (3.5.29), with $\mu = 1$ and ϵ a function of ω, is

$$E = \int d\omega \, e^{-i\omega t - (i\omega b/c)} \left[e^{i\omega x/c} - R(\omega) e^{-i\omega x/c} \right] f(\omega), \qquad x < 0 \qquad (3.5.30)$$

and

$$E = \int d\omega \, e^{-i\omega t - (i\omega b/c)} T(\omega) e^{ik'x} f(\omega), \qquad x > 0 \qquad (3.5.31)$$

where $R(\omega)$ is the reflected amplitude and $T(\omega)$ the transmitted amplitude. The wave number in the dielectric is $k' = \omega \sqrt{\epsilon}/c$. In order for $\mathbf{E}$ not to grow exponentially, we must choose $\operatorname{Im} k' > 0$.

The magnetic field is given by

$$B = \int d\omega \, e^{-i\omega t - (i\omega b/c)} \left[e^{i\omega x/c} + R(\omega) e^{-i\omega x/c} \right] f(\omega), \qquad x < 0 \qquad (3.5.32)$$

and

$$B = \sqrt{\epsilon} \int d\omega \, e^{-i\omega t - (i\omega b/c)} T(\omega) e^{ik'x} f(\omega), \qquad x > 0.$$

The boundary conditions at $x = 0$ are

$$1 - R = T \qquad (3.5.33)$$

and

$$1 + R = \sqrt{\epsilon} T \qquad (3.5.34)$$

so that

$$T = \frac{2}{1 + \sqrt{\epsilon}} \qquad (3.5.35)$$

and

$$R = \frac{\sqrt{\epsilon} - 1}{\sqrt{\epsilon} + 1}. \qquad (3.5.36)$$

This solution gives E and B the desired properties: for $t < 0$,

$$E = E_{\text{inc}}, \qquad (3.5.37)$$

$$\frac{\partial E}{\partial t} = \frac{\partial E_{\text{inc}}}{\partial t}, \tag{3.5.38}$$

$$B = B_{\text{inc}} \tag{3.5.39}$$

and

$$\frac{\partial B}{\partial t} = \frac{\partial B_{\text{inc}}}{\partial t}. \tag{3.5.40}$$

This follows from the known properties of $\epsilon(\omega)$: $\epsilon(\omega)$ is analytic and nonzero in the upper half-plane, and $\sqrt{\epsilon} \to 1$ as $\omega \to \infty$ since we have chosen Im $k' > 0$. Therefore, T, R, and k' are analytic in the upper half-plane, $k' \to \omega/c$ as $\omega \to \infty$, and the reflected and transmitted waves vanish for $t < 0$. Therefore, we have correctly incorporated the initial condition $E = E_{\text{inc}}$ and $B = B_{\text{inc}}$ into our solution. Given these analytic properties, one can also show that the wavefronts propagate causally, that is, that no transmitted or reflected wave shows up before transmission at velocity c would permit it to do so. (See Problems 3.13 and 3.14.)

3.6. REFLECTION AND REFRACTION

We consider now a plane wave of polarization $\hat{e}$, wave number $\mathbf{k}$, and frequency $\omega = ck$ incident in the x–y plane from the left in air (or vacuum) on the plane surface of a dielectric medium with an index of refraction $n = \sqrt{\epsilon}$ extending from $x = 0$ to the right. We take $\mu = 1$. The angle of incidence is θ, the angle of refraction θ'. We take the polarization (which we define as the $\mathbf{E}$ direction) in the plane of incidence. This is illustrated in Figure 3.2. (See also Problem 3.7.)

The boundary conditions at $x = 0$ must hold for all y and z, so that k_y and k_z must be continuous across the boundary (taken here at $x = 0$). Thus, with $\omega^2 = c^2 k^2 = c^2 k'^2/\epsilon$, we have $k'^2 = k^2 \cdot \epsilon$ or, since $k_z = 0$,

$$k_x'^2 + k_y^2 = (k_x^2 + k_y^2)\epsilon. \tag{3.6.1}$$

Consider first real $\epsilon\mu$. Then we have $k_y^2 = k^2 \sin^2\theta$ and $k_y^2 = k'^2 \sin^2\theta'$, so

$$\frac{\sin\theta'}{\sin\theta} = \frac{k}{k'} = \frac{1}{\sqrt{\epsilon}} \tag{3.6.2}$$

which is Snell's law.

For complex ϵ, we return to (3.6.1). We consider first the case of

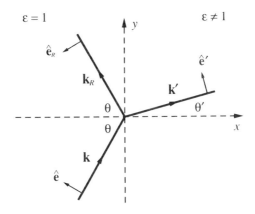

Figure 3.2.

high conductivity, $4\pi\sigma/\omega \gg 1$, for which

$$\epsilon \sim \frac{4\pi i\sigma}{\omega},$$

$$k_x'^2 \sim k^2 \frac{4\pi i\sigma}{\omega},$$

and

$$k_x' \sim \frac{1+i}{\sqrt{2}} k \cdot \sqrt{\frac{4\pi\sigma}{\omega}} \tag{3.6.3}$$

so that all fields in the conductor damp out like $e^{-kx\sqrt{2\pi\sigma/\omega}}$; beyond the skin depth

$$\delta = \frac{1}{k} \sqrt{\frac{\omega}{2\pi\sigma}} \tag{3.6.4}$$

the fields go rapidly to zero.

If ϵ has only a small imaginary part, the real part of k_x' will be determined, as usual, as $\text{Re } k_x' \cong k' \cos \theta'$ with θ' given by Snell's law. The imaginary part, $\text{Im } k_x'$, will be given by the equation

$$(\text{Re } k_x' + i \text{Im } k_x')^2 + k_y^2 = k^2\epsilon \tag{3.6.5}$$

and

$$\text{Im } k'_x \cong \frac{k^2}{2 \text{ Re } k'_x} \text{ Im } \epsilon$$

$$= \frac{k}{2} \frac{\sin \theta'}{\cos \theta' \sin \theta} \text{ Im } \epsilon , \tag{3.6.6}$$

neglecting $(\text{Im } k'_x)^2$. Thus to a good approximation (for small $\text{Im } \epsilon$), the transmitted wave behaves as if ϵ is real, except, as we move into the medium, for a damping given by

$$|f(x, y, z)| \cong e^{-\text{Im } k'_x x}|f(0, y, z)| \tag{3.6.7}$$

with $\text{Im } k'_x$ given by (3.6.6). The direction of energy flow is given by the real part of the wave number, $\text{Re } \mathbf{k}' = \hat{e}_x \text{ Re } k'_x + \hat{e}_y k_y$.

In going from (3.6.5) to (3.6.6), we have chosen the positive root of (3.6.5) for k'_x. For a medium extending to ∞, this choice is evidently required and, as we have seen in the previous section, is consistent with the causality requirement that $\sqrt{\epsilon(\omega)} \to 1$ as $\omega \to \infty$. If the medium extends only a finite distance in the x-direction, there will be a second set of boundary conditions at the second surface. In order to ensure that our solution corresponds to the physical input (incident wave on left, reflected wave on left, transmitted wave on right with no incident wave on right), we will have to use both roots of (3.6.5) in our solution. However, if the wave has substantially decayed by the time it hits the second boundary, the numerical effect of this change is small. In the following we assume that the second surface is far enough away to be neglected.

We return now to the first surface and construct the solution that satisfies the boundary conditions.

The electric field is given, for $x < 0$, by

$$\mathbf{E} = e^{i(k_x x + k_y y - \omega t)} \mathbf{e} - e^{i(-k_x x + k_y y - \omega t)} \mathbf{e}_R \tag{3.6.8}$$

(where $\mathbf{e}_R$ is the reflected amplitude) and, for $x > 0$, by

$$\mathbf{E} = e^{i(k'_x x + k_y y - \omega t)} \mathbf{e}_T \tag{3.6.9}$$

The minus sign preceding $\mathbf{e}_R$ in (3.6.8) is chosen for convenience.

Thus, with

$$\mathbf{B} = \mathbf{b} \, e^{i(k_x x + k_y y - \omega t)} - \mathbf{b}_R \, e^{i(-k_x x + k_y y - \omega t)} \tag{3.6.10}$$

for $x < 0$, and

$$\mathbf{B} = \mathbf{b}_T \, e^{i(k'_x x + k_y y - \omega t)} \tag{3.6.11}$$

for $x > 0$, we must have

$$\mathbf{b} = \frac{c\mathbf{k}}{\omega} \times \mathbf{e}, \tag{3.6.12}$$

$$\mathbf{b}_R = \frac{c\mathbf{k}_R}{\omega} \times \mathbf{e}_R, \tag{3.6.13}$$

and

$$\mathbf{b}_T = \frac{c\mathbf{k}'}{\omega} \times \mathbf{e}_T. \tag{3.6.14}$$

The condition $\nabla \cdot \mathbf{E} = 0$ determines the vectors $\mathbf{e}$, $\mathbf{e}_R$, and $\mathbf{e}_T$ to be

$$e_x = -\sin\theta \tag{3.6.15}$$
$$e_y = \cos\theta \tag{3.6.16}$$
$$e_{Rx} = R\sin\theta \tag{3.6.17}$$
$$e_{Ry} = R\cos\theta \tag{3.6.18}$$
$$e'_x = -T\sin\theta' \tag{3.6.19}$$

and

$$e'_y = T\cos\theta' \tag{3.6.20}$$

where R and T, the scalar reflection and transmission amplitudes, are to be determined from the boundary conditions: $\Delta E_y = 0$ and $\Delta B_z = 0$. From $\Delta E_y = 0$, we find

$$\cos\theta(1 - R) = \cos\theta' T, \tag{3.6.21}$$

and from $\Delta B_z = 0$, we find

$$\frac{k}{\omega c}(1 + R) = \frac{k'}{\omega c} T. \tag{3.6.22}$$

The solution for R and T is

$$R = \frac{k' \cos \theta - k \cos \theta'}{k' \cos \theta + k \cos \theta'} \qquad (3.6.23)$$

and

$$T = \frac{2k \cos \theta}{k' \cos \theta + k \cos \theta'}. \qquad (3.6.24)$$

(See Problem 3.4 for a discussion of energy balance).

There are two interesting limiting cases to consider. First, high conductivity, where k' will dominate in (3.6.23) and make $R \cong 1$. Second, Im $\epsilon = 0$. Then k'_x and k'^2 are real, and (3.6.23) shows that R can vanish. This happens at θ_B (Brewster's angle) when

$$k' \cos \theta = k \cos \theta'$$

or, since $k' = \sqrt{\epsilon}k$,

$$\sin 2\theta_B = \sin 2\theta'_B. \qquad (3.6.25)$$

Equation (3.6.25), together with Snell's law, has one root: $\theta + \theta' = \pi/2$. Thus, at an angle of incidence θ_B such that the reflected and refracted rays are orthogonal, there will be no reflection of an incident field polarized in the plane of incidence.

One can understand this phenomenon by remembering that reflection consists of radiation by dipole moments of the dielectric. A dipole polarization with direction e' cannot radiate in its direction of polarization (as we shall learn later). Since e' is perpendicular to k' and k_R is also perpendicular to k', k_R is in the direction of e' and, hence, there is no reflection at the Brewster angle. (See Figure 3.3.)

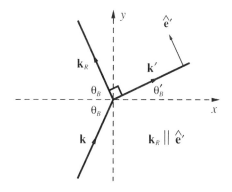

Figure 3.3.

3.7. PROPAGATION IN ANISOTROPIC MEDIA

We now consider a tensor dielectric constant, which we assume to be real and symmetric. (We could as easily take ϵ_{ij} to be Hermitian; however, an anti-Hermitian component—i.e., absorption—is a significant complication that we do not discuss. It would normally be treated perturbatively, and/or numerically.) We assume unit magnetic permeability. Note that a scalar permeability would not be a significant complication, but a tensor permeability would.

The relevant equations are as always (for a fixed wave number)

$$\frac{\omega}{c}\mathbf{b} = \mathbf{k} \times \mathbf{e} \tag{3.7.1}$$

$$\mathbf{k} \times \mathbf{b} = -\frac{\omega}{c}\mathbf{d} \tag{3.7.2}$$

and

$$d_i = \epsilon_{ij}e_j. \tag{3.7.3}$$

We see from (3.7.1–3.7.3) that there are two sets of right-handed coordinate systems associated with these vectors: $\mathbf{k}$, $\mathbf{d}$, and $\mathbf{b}$ on the one hand, and $\mathbf{e}$, $\mathbf{b}$, and $\mathcal{P} = (c/4\pi)\mathbf{e} \times \mathbf{b}$ on the other. Further, since $\mathbf{k}$, $\mathbf{e}$, and $\mathbf{d}$ are all orthogonal to $\mathbf{b}$, they are coplanar. In general, $\mathbf{e} \times \mathbf{b}$ and $\mathbf{k}$ are not in the same direction. Therefore, the direction of energy flow and that of $\mathbf{k}$ will be, in general, different. An exception that we shall see below is the ordinary ray of a monoaxial crystal.

The controlling equation for $\mathbf{e}$ is obtained by substituting (3.7.1) into (3.7.2) to yield

$$[\mathbf{k} \times (\mathbf{k} \times \mathbf{e})]_i = -\frac{\omega^2}{c^2}\epsilon_{ij}e_j \quad\text{or}\quad k^2 e_i - \mathbf{k}\cdot\mathbf{e}k_i = \frac{\omega^2}{c^2}\epsilon_{ij}e_j. \tag{3.7.4}$$

Equation (3.7.4) has an orthogonality property between the solutions with different eigenvalues of ω^2, say, e_i^α with ω_α^2 and e_i^β with ω_β^2, to wit

$$e_i^\alpha \epsilon_{ij} e_j^\beta = d_i^\alpha e_i^\beta = 0, \qquad \omega_\alpha^2 \neq \omega_\beta^2. \tag{3.7.5}$$

This is derived in the usual way by multiplying (3.7.4) for e_i^α by e_i^β, (3.7.4) for e_i^β by e_i^α, and subtracting. The left-hand side vanishes, leaving

$$\frac{(\omega_\alpha^2 - \omega_\beta^2)}{c^2} e_i^\beta \epsilon_{ij} e_j^\alpha = 0.$$

Note, however, that in refraction it is ω that is the same in both media, not $\mathbf{k}$; therefore, this orthogonality does not hold between the two modes of propagation for given ω.

To study (3.7.4), we choose coordinate axes x_α that diagonalize ϵ_{ij}. That is,

$$\sum_\beta \epsilon_{\alpha\beta} e_\beta = \epsilon_\alpha e_\alpha \quad \text{(no sum)} \tag{3.7.6}$$

where ϵ_α are the eigenvalues of ϵ_{ij}, considered here to be known functions of ω. In the x_α coordinate system, (3.7.4) becomes

$$\mathbf{k} \cdot \mathbf{e} k_\alpha - \mathbf{k}^2 e_\alpha = -\frac{\omega^2}{c^2} \epsilon_\alpha e_\alpha \quad \text{(no sum)}. \tag{3.7.7}$$

We expect there to be, for each k, three eigenvalues for ω^2, for which we can solve (3.7.7).

The first eigenvalue is universal and uninteresting: $e_\alpha = k_\alpha$, which gives $\omega = 0$. There remain two eigenvalues. We first "solve" (3.7.7) for e_α:

$$e_\alpha = \frac{\mathbf{k} \cdot \mathbf{e} k_\alpha}{k^2 - \dfrac{\omega^2}{c^2} \epsilon_\alpha} \tag{3.7.8}$$

and, multiplying by k_α and summing, we obtain

$$\mathbf{k} \cdot \mathbf{e} = \sum_\alpha k_\alpha e_\alpha = \mathbf{k} \cdot \mathbf{e} \sum_\alpha \frac{k_\alpha^2}{k^2 - \dfrac{\omega^2}{c^2} \epsilon_\alpha} \tag{3.7.9}$$

so that, for $\mathbf{k} \cdot \mathbf{e} \neq 0$,

$$\sum_\alpha \frac{k_\alpha^2}{k^2 - \dfrac{\omega^2}{c^2} \epsilon_\alpha} = 1 \tag{3.7.10}$$

and, with $1 = \sum_\alpha k_\alpha^2 / k^2$,

$$\sum_\alpha \left(\frac{k_\alpha^2}{k^2 - \dfrac{\omega^2}{c^2} \epsilon_\alpha} - \frac{k_\alpha^2}{k^2} \right) = 0 \quad \text{or} \quad \sum_\alpha \frac{k_\alpha^2 \epsilon_\alpha}{k^2 - \dfrac{\omega^2}{c^2} \epsilon_\alpha} = 0. \tag{3.7.11}$$

Multiplying (3.7.11) by

$$\left(k^2 - \frac{\omega^2}{c^2} \epsilon_1 \right)\left(k^2 - \frac{\omega^2}{c^2} \epsilon_2 \right)\left(k^2 - \frac{\omega^2}{c^2} \epsilon_3 \right)$$

gives a quadratic equation for ω^2 with two roots. Note that if it is necessary to take dispersion in ϵ_α into account, so that ϵ_α itself is a function of ω, the eigenvalues of ω are *not* obtained simply by solving the quadratic equation (3.7.11). In the following we will assume that over the width of the wave packet with which we are dealing the dispersion can be ignored, that is, the ϵ_α may be taken as given positive constants.

Here, we again point out that we have solved the first of two obvious problems. This is to find the propagating mode frequencies and polarizations, given the wave number **k** in the medium. The converse is also straightforward: Given a frequency ω and a direction of propagation $\hat{k}$, find the wave number and polarizations of the two modes. The second, and harder, problem arises in analyzing refraction in a biaxial crystal. There one is given the incident frequency ω and wave vector **k**, and trivially determines the two components of the transmitted wave number that lie in the boundary between the media and, hence, are continuous across the boundary. The problem is then to find the third component of the wave number and polarization of the propagating modes. (See Problem 3.11.)

We turn now to the simple example of a monoaxial crystal, that is, one in which two of the eigenvalues, say, ϵ_1 and ϵ_3, are equal. Then any pair of orthogonal axes in the 1, 3-plane are eigenvectors of the ϵ matrix, and we can directly construct a solution of (3.7.1) and (3.7.2) by choosing the 1-axis so that **k** lies in the 1, 2-plane, **e** along the 3-axis, and **b** perpendicular to **k** in the 1, 2-plane. Since e_3 is an eigenvector of ϵ_{ij} with eigenvalue ϵ_1, **d** is also along the 3-axis, with value $\mathbf{d} = \epsilon_1 \mathbf{e}$. This is illustrated in Figure 3.4.

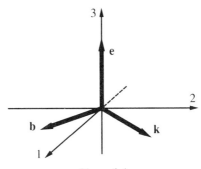

Figure 3.4.

Since $\mathbf{k} \cdot \mathbf{e} = 0$ for this mode, equation (3.7.4) yields

$$\omega_O^2 = \frac{c^2 k^2}{\epsilon_1},\qquad\qquad (3.7.12)$$

where the subscript O stands for ordinary. Further, the direction of energy flow for this ray (the ordinary ray) *is* in the direction of $\mathbf{k}$, so that the ordinary ray behaves as if it were propagating in an isotropic medium.

For given $\mathbf{k}$, the second mode (the extraordinary ray) will have its $\mathbf{d}$ vector $\mathbf{d}_e$ perpendicular to $\mathbf{k}$ and to the $\mathbf{e}$ vector $\mathbf{e}_0$, of the ordinary ray. Its frequency will be given by (3.7.11), that is,

$$\frac{k_1^2 \epsilon_1}{k^2 - \dfrac{\omega^2 \epsilon_1}{c^2}} + \frac{k_2^2 \epsilon_2}{k^2 - \dfrac{\omega^2 \epsilon_2}{c^2}} = 0 \quad \text{or} \quad \omega^2 = c^2 \frac{k_1^2 \epsilon_1 + k_2^2 \epsilon_2}{\epsilon_1 \epsilon_2}. \qquad (3.7.13)$$

The vectorial nature of the extraordinary mode can be easily constructed. With $\mathbf{k} = (k_1, k_2)$, we have $\mathbf{d}_e \cdot \mathbf{k} = 0$, or, in an obvious notation, $\mathbf{e}_e = (e_1, e_2)$ is perpendicular to $\epsilon_{ij} k_j = (\epsilon_1 k_1, \epsilon_2 k_2)$. So,

$$\mathbf{e}_e = (-\epsilon_2 k_2, \epsilon_1 k_1) \times \text{constant}. \qquad (3.7.14)$$

The direction of energy flow is

$$\mathcal{P} \propto \mathbf{e} \times \mathbf{b} \propto \mathbf{e} \times (\mathbf{k} \times \mathbf{e}) = e^2 \mathbf{k} - \mathbf{e} \cdot \mathbf{k} \mathbf{e}$$

or

$$\mathcal{P} \propto (\epsilon_2^2 k_2^2 + \epsilon_1^2 k_1^2)(k_1, k_2) - (\epsilon_2 k_2 k_1 - \epsilon_1 k_2 k_1)(\epsilon_2 k_2, -\epsilon_1 k_1)$$
$$= (\epsilon_1 k_1^2 + \epsilon_2 k_2^2)(\epsilon_1 k_1, \epsilon_2 k_2)$$

so that the direction of energy flow is

$$\hat{\mathcal{P}} = \frac{(\epsilon_1 k_1, \epsilon_2 k_2)}{(\epsilon_1^2 k_1^2 + \epsilon_2^2 k_2^2)^{1/2}}. \qquad (3.7.15)$$

This is also the direction of the group velocity that follows from (3.7.13), still neglecting dispersion:

$$\nabla_k \omega = \frac{c^2}{\epsilon_1 \epsilon_2 \omega} (k_1 \epsilon_1, k_2 \epsilon_2). \qquad (3.7.16)$$

In general, an incident ray on a monoaxial crystal will require, at given ω, both modes to satisfy the boundary conditions. Since the ordinary mode frequency wave number relation is independent of angle, the refracted light will contain both modes, the ordinary ray refracting according to Snell's law, with $\epsilon = \epsilon_1$, the extraordinary ray refracting with a different angular dependence. (See Problem 3.11.)

3.8. HELICITY AND ANGULAR MOMENTUM

We have seen that a polarization vector

$$\mathbf{e}_+ = \frac{\hat{\mathbf{e}}_x + i\hat{\mathbf{e}}_y}{\sqrt{2}} \tag{3.8.1}$$

represents a right-handed circularly polarized wave advancing in the z direction.

$$\mathbf{e}_- = \frac{\hat{\mathbf{e}}_x - i\hat{\mathbf{e}}_y}{\sqrt{2}} \tag{3.8.2}$$

is its left-handed counterpart.

The vectors (3.8.1) and (3.8.2) transform particularly simply under rotations. We consider a primed coordinate system that is rotated clockwise by an angle θ about the z-axis (where x, y, z form a right-handed coordinate system). This is shown in Figure 3.5.

Evidently,

$$\hat{\mathbf{e}}'_x = \hat{\mathbf{e}}_x \cos\theta + \hat{\mathbf{e}}_y \sin\theta \tag{3.8.3}$$

$$\hat{\mathbf{e}}'_y = \hat{\mathbf{e}}_y \cos\theta + \hat{\mathbf{e}}_x \sin\theta \tag{3.8.4}$$

so that

$$\hat{\mathbf{e}}'_x \pm i\hat{\mathbf{e}}'_y = \hat{\mathbf{e}}_x \cos\theta + \hat{\mathbf{e}}_y \sin\theta) \pm i(\hat{\mathbf{e}}_y \cos\theta + \hat{\mathbf{e}}_x \sin\theta)$$

$$= (\hat{\mathbf{e}}_x \pm i\hat{\mathbf{e}}_y) \cos\theta \mp i \sin\theta(\hat{\mathbf{e}}_x \pm i\hat{\mathbf{e}}_y) \tag{3.8.5}$$

or

$$\mathbf{e}'_\pm = e^{\mp i\theta} \mathbf{e}_\pm . \tag{3.8.6}$$

Vectors with the transformation property (3.8.6) are said to have helicity ± 1. There is a corresponding property of the components. Thus, with

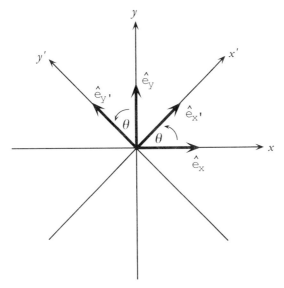

Figure 3.5.

$$\mathbf{e} = \hat{\mathbf{e}}_x e_x + \hat{\mathbf{e}}_y e_y = \hat{\mathbf{e}}'_x e'_x + \hat{\mathbf{e}}'_y e'_y \qquad (3.8.7)$$

we have

$$e'_x = \hat{\mathbf{e}}'_x \cdot \hat{\mathbf{e}}_x e_x + \hat{\mathbf{e}}'_x \cdot \hat{\mathbf{e}}_y e_u \qquad \text{and} \qquad e'_y = \hat{\mathbf{e}}'_y \cdot \hat{\mathbf{e}}_x e_x + \hat{\mathbf{e}}'_y \cdot \hat{\mathbf{e}}_y e_y, \qquad (3.8.8)$$

or, from (3.8.3) and (3.8.4),

$$e'_x = \cos\theta e_x + \sin\theta e_y \qquad (3.8.9)$$

$$e'_y = \cos\theta e_y - \sin\theta e_x \qquad (3.8.10)$$

and from

$$\mathbf{e} = \mathbf{e}_+ e_+ + \mathbf{e}_- e_- = \mathbf{e}'_+ e'_+ + \mathbf{e}'_- e'_- \qquad (3.8.11)$$

and (3.8.6), we find

$$e'_\pm = e_\pm e^{\pm i\theta}. \qquad (3.8.12)$$

Note the opposite definition of $e_\pm$ from that of $\mathbf{e}_\pm$.

There is an intimate connection between the helicity and the angular momentum of a plane wave of wave number $\mathbf{k}$ and frequency ω, propagating in empty space in the z direction. The relation, as we shall now show, is

$$\lambda = \frac{\omega L_z}{W} \tag{3.8.13}$$

where λ is the helicity of the wave, L_z the z-component of the electromagnetic angular momentum, and W the electromagnetic energy. For a single photon, with $W = \hbar\omega$ and $\lambda = \pm 1$, we would find, from (3.8.13),

$$L_z = \pm \hbar, \tag{3.8.14}$$

suggesting that the intrinsic spin angular momentum of a photon is ± 1.

We calculate the angular momentum of an electromagnetic disturbance propagating with approximate wave number $\mathbf{k}_0$, approximate frequency ω_0, and approximate polarization $\hat{e}_x \pm i\hat{e}_y$. These parameters must all be approximate, since for a wave of definite wave number both the energy and angular momentum would be infinite; therefore, the wave must be bounded in all three directions; the polarization must be modified as well to maintain transversality.

We review first the angular momentum integral for charges and currents in free space. We have for the force density on charges

$$\mathbf{f} = \rho\mathbf{E} + \mathbf{j} \times \mathbf{B}, \tag{3.8.15}$$

and from (3.3.6),

$$f_i = -\frac{1}{4\pi c}\frac{\partial}{\partial t}(\mathbf{E} \times \mathbf{B})_i + \partial_k T_{ik} \tag{3.8.16}$$

(where T_{ik} is the Maxwell stress tensor). From (3.3.6), we derived the expression

$$\mathbf{p}_{e.m.} = \int_V d\mathbf{r}\, \frac{\mathbf{E} \times \mathbf{B}}{4\pi c} \tag{3.3.7}$$

for the electromagnetic momentum $\mathbf{p}_{e.m.}$.

Similarly, the torque density is

$$\mathbf{r} = \mathbf{r} \times \mathbf{f} \qquad \text{or} \qquad \tau_i = \epsilon_{ijk}x_j f_k \tag{3.8.17}$$

so that, with $\mathbf{L}_m$ the angular momentum of matter, we have

$$\frac{dL_{mi}}{dt} = -\frac{1}{4\pi c}\frac{d}{dt}\int d\mathbf{r}[\mathbf{r}\times(\mathbf{E}\times\mathbf{B})]_i + \epsilon_{ijk}\int d\mathbf{r}\, x_j\partial_l T_{kl} \quad (3.8.18)$$

or, with

$$\mathbf{L}_{e.m.} = \frac{1}{4\pi c}\int \mathbf{r}\times(\mathbf{E}\times\mathbf{B})\,d\mathbf{r}, \quad (3.8.19)$$

$$\left(\frac{d\mathbf{L}_m}{dt} + \frac{d\mathbf{L}_{e.m.}}{dt}\right)_i = \epsilon_{ijk}\int_V d\mathbf{r}\, x_j\partial_l T_{kl}. \quad (3.8.20)$$

Since $\epsilon_{ilk}T_{kl} = 0$ (ϵ is antisymmetric, T symmetric), we have from (3.8.20) an angular momentum flux density

$$\Phi_{il} = \epsilon_{ijk}x_j T_{kl}, \quad (3.8.21)$$

and for fields that go to zero sufficiently rapidly, a field angular momentum

$$\mathbf{L}_{e.m.} = \int \frac{d\mathbf{r}}{4\pi c}\mathbf{r}\times(\mathbf{E}\times\mathbf{B}). \quad (3.8.22)$$

Our next task is to evaluate the integral for $\mathbf{L}_{e.m.}$ for an approximate monochromatic plane wave. We call ω_0 and $\mathbf{k}_0$ the approximate values of the frequency and wave vector.

We start with the general expression (3.8.22). We express the fields in terms of their Fourier transforms, automatically satisfying the free field equations. Thus, with a convenient normalization,

$$\mathbf{E} = \int \frac{d\mathbf{k}}{(2\pi)}\mathbf{a}(\mathbf{k})\,\omega\, e^{i(\mathbf{k}\cdot\mathbf{x}-\omega t)} + \text{c.c.} \quad (3.8.23)$$

and

$$\mathbf{B} = \int \frac{d\mathbf{k}}{(2\pi)}c\mathbf{k}\times\mathbf{a}(\mathbf{k})\,e^{i(\mathbf{k}\cdot\mathbf{x}-\omega t)} + \text{c.c.} \quad (3.8.24)$$

with $\omega = c|\mathbf{k}|$ and $\mathbf{k}\cdot\mathbf{a} = 0$.

The energy in the electromagnetic field is

$$W = \int d\mathbf{r}\frac{(\mathbf{E}^2 + \mathbf{B}^2)}{8\pi} = \int d\mathbf{k}|\mathbf{a}|^2\omega^2; \quad (3.8.25)$$

the momentum is

$$\mathbf{P} = \int d\mathbf{k} |a|^2 \mathbf{k}\omega; \qquad (3.8.26)$$

they have the approximate ratio

$$\frac{\mathbf{P}}{W} \cong \frac{\mathbf{k}_0 \omega_0}{\omega_0^2} = \frac{\hat{\mathbf{k}}_0}{c}. \qquad (3.8.27)$$

The terms in the integrals for these conserved qualities that come from products of $\mathbf{a}$ with $\mathbf{a}$ or $a*$ with $\mathbf{a}*$ must be zero, since otherwise their time dependence would violate the conservation law.

The angular momentum is a little harder. With $\mathbf{L}$ given by (3.8.19) and $\mathbf{E}$ and $\mathbf{B}$ by (3.8.23) and (3.8.24), we have

$$\mathbf{L} = \frac{1}{4\pi} \frac{1}{(2\pi)^2} \int d\mathbf{r} \, \mathbf{r} \times \left[\int d\mathbf{k}' \mathbf{a}*(\mathbf{k}') \, \omega' \, e^{-i\mathbf{k}'\cdot\mathbf{r}+i\omega't} \right.$$

$$\left. \times \int d\mathbf{k} \, \mathbf{k} \times \mathbf{a}(\mathbf{k}) \, e^{i\mathbf{k}\cdot\mathbf{r}-i\omega t} \right] + \text{c.c.} \qquad (3.8.28)$$

Here again, we have dropped products of $\mathbf{a}$ with $\mathbf{a}$ and $\mathbf{a}*$ with $\mathbf{a}*$.

To carry out the integral in (3.8.28), we replace $\mathbf{r} e^{-i\mathbf{k}'\cdot\mathbf{r}}$ by $i\nabla_{k'} e^{-i\mathbf{k}'\cdot\mathbf{r}}$, do the $\mathbf{r}$ integral, and find

$$\mathbf{L} = \text{Re} \int d\mathbf{k} \, d\mathbf{k}' \, i\nabla_{k'} \delta^3(\mathbf{k} - \mathbf{k}') \times [\mathbf{a}*(\mathbf{k}') \, \omega' \times (\mathbf{k} \times \mathbf{a}(\mathbf{k}))] \, e^{i(\omega'-\omega)t}. \qquad (3.8.29)$$

We integrate by parts, carry out the indicated derivatives, and then set $\mathbf{k} = \mathbf{k}'$, leaving

$$\mathbf{L} = \int d\mathbf{k} \left\{ \omega a_j^*(\mathbf{k}) \, \mathbf{k} \times \frac{\nabla_k}{i} \, a_j(\mathbf{k}) + \frac{1}{i} \mathbf{a}* \times \mathbf{a}\omega \right\}. \qquad (3.8.30)$$

Equation (3.8.30) decomposes the angular momentum into two terms. The first depends in detail on the structure and location of the wave packet; the second does not. We study a wave packet chosen to maintain the monochromatic and planar nature of the field to the maximum extent possible, consistent with the requirement that the total energy in the field be finite. Thus, we let

$$\mathbf{a}\sqrt{\omega} = [(\hat{\mathbf{e}}_x \pm i\hat{\mathbf{e}}_y) k_z - (k_x \pm ik_y)\hat{\mathbf{e}}_z] e^{-i\mathbf{k}\cdot\mathbf{r}_0} b(\mathbf{k}); \qquad (3.8.31)$$

$b(\mathbf{k})$ is a real function of $\mathbf{k} - \mathbf{k}_0$, with widths Δk_x, Δk_y, and Δk_z much smaller than k_0. The vectorial coefficient in (3.8.31) is chosen to ensure the transversality of $\mathbf{a}$. The wave vector $\mathbf{k}_0$ is in the z direction.

Evidently, the explicit phase $-i\mathbf{k}\cdot\mathbf{r}_0$ displaces the wave packet by a vector $\mathbf{r}_0$ and adds an orbital angular momentum $\mathbf{L}_{orb} = \mathbf{r}_0 \times \mathbf{P}$ to the angular momentum. We can now easily calculate the energy, momentum, and angular momentum carried by the wave function (3.8.31), neglecting the spread in wave numbers $\Delta \mathbf{k}$ compared with k_0. We find

$$W = 2ck_0^3 \int d\mathbf{k} |b|^2, \qquad (3.8.32)$$

$$\mathbf{P} = 2k_0^2 \mathbf{k}_0 \int d\mathbf{k} |b|^2, \qquad (3.8.33)$$

and

$$\mathbf{L} = \mathbf{r}_0 \times \mathbf{P} \pm 2k_0 \mathbf{k}_0 \int d\mathbf{k} |b|^2; \qquad (3.8.34)$$

corrections are of order $(\Delta k)^2/k_0^2$. The second term in (3.8.34) corresponds in quantum theory to a spin angular momentum. Our result states that the wave with helicity ± 1 carries an angular momentum

$$L_z = \pm \frac{W}{\omega_0}. \qquad (3.8.35)$$

Note that setting $W = \hbar\omega_0$ leads to $L_z = \pm\hbar$.

CHAPTER 3 PROBLEMS

3.1 At $t = 0$, a charge density $\rho_0(\mathbf{r})$ is found to exist in a medium with constant real dielectric constant ϵ and conductivity σ. Assume that $\int d\mathbf{r}\rho(\mathbf{r}) = Q_0$ is finite.

(a) Show that the charge density decays according to

$$\rho(\mathbf{r}, t) = \rho_0(\mathbf{r}) e^{-(4\pi\sigma/\epsilon)t}$$

(b) Where does the charge go? Show explicitly that the outgoing current correctly accounts for the disappearance of the charge for the special case of a spherically symmetric ρ_0.

3.2 One could try to construct a model of a finite size charged particle by postulating a charge density $\rho(\mathbf{r}, t) = qf(\mathbf{r} - \mathbf{y}(t))$, where $\mathbf{y}(t)$ is the trajectory of a fixed point in the charge distribution and $\mathbf{r}$ the point in space at which one is specifying the charge density. If q is the total charge, we must have

$$\int f(\mathbf{r} - \mathbf{y}(t)) \, d^3r = 1.$$

A point charge would then have $f(\mathbf{x}) = \delta^3(\mathbf{x})$.

We have seen in Problem 2.1 that a suitable current density to go with ρ above would be

$$\mathbf{j}(\mathbf{r}, t) = q \frac{d\mathbf{y}(t)}{dt} f(\mathbf{r} - \mathbf{y}(t)).$$

The limit of a point charge is singular, since the electric field at the position of a point particle is infinite. This singularity does not occur for a finite size charged particle, such as suggested above. The field equations would be, as usual,

$$\nabla \times \mathbf{E} = -\frac{1}{c} \frac{\partial \mathbf{B}}{\partial t}$$

$$\nabla \cdot \mathbf{E} = 4\pi\rho$$

$$\nabla \times \mathbf{B} = \frac{1}{c} \frac{\partial \mathbf{E}}{\partial t} + \frac{4\pi\mathbf{j}}{c}$$

and $\nabla \cdot \mathbf{B} = 0$, with ρ and $\mathbf{j}$ as given above. The particle motion (nonrelativistic) would be given, following Newton and Lorentz, by the equation

$$m\ddot{\mathbf{y}} = \int d\mathbf{r} \, qf(\mathbf{y} - \mathbf{r}) \left[\mathbf{E}(\mathbf{r}, t) + \frac{1}{c} \frac{d\mathbf{y}}{dt} \times \mathbf{B}(\mathbf{r}, t) \right].$$

(a) Suppose there are n charges, each with charge q_i, mass m_i, and coordinate $\mathbf{y}_i$. Give the appropriate generalization of the Maxwell and Newton–Lorentz equations.

(b) Show that there is a conserved energy in a volume V, consisting of the sum of the electromagnetic energy and the kinetic energy of the charged particles contained in V [as in (1.3.11)], provided that the surface integral of the Poynting flux integrated over the bounding surface of the volume vanishes, and that no particle is close enough to the boundary for $\rho(\mathbf{r}, t)$ to be different from zero on it. For this purpose, assume that $f(\mathbf{x})$ vanishes for $|\mathbf{x}| > b$, a particle radius.

We can extend the volume to include all space, provided the energy integral converges and the integrated Poynting flux goes to zero as $r \to \infty$.

(c) Show that there is a conserved momentum in a volume V, again consisting of the sum of the electromagnetic momentum (3.3.7) and the momentum of the charged particles contained in V. Formulate the conditions under which this holds.

*(d) Let $f(\mathbf{x})$ be spherically symmetric, that is, $f(\mathbf{x}) = f(|\mathbf{x}|)$. One might expect in this case that there would be a similar conservation law for angular momentum. Try it. Show that the usual definition works only for a point particle, that is, for $f(\mathbf{x}) = \delta^3(\mathbf{x})$. There is, however, a way of constructing a conserved angular momentum for this case, which we will take up when we come to Lagrangians (Chapter 7).

3.3 Imagine a wave packet incident on a plane dielectric boundary. We study the energy balance in the process for real ϵ. Since $\mathbf{E}_t$ and $\mathbf{H}_t$ are continuous, so is $(\mathbf{E} \times \mathbf{H})_n$. Here, t stands for tangential and n for normal. More interesting is the time and area integral of the Poynting flux, over an appropriate closed surface:

$$I = \int_{t_1}^{t_2} dt \int_S \mathcal{P} \cdot d\mathbf{S}$$

which equals

$$I = W_v(t_2) - W_v(t_1)$$

where $W_v(t)$ is the electromagnetic energy contained in the enclosed volume V at time t. If t_2 and t_1 are, respectively, after the time of departure and before the time of arrival of the wave packet at the surface S, then $I = 0$.

Let W be the incident energy, W_R the reflected energy, and W_T the transmitted energy. Show (using the example worked out in the text) that the condition $I = 0$ is equivalent to $W = W_R + W_T$,

3.4 Calculate the dipole moment of a classical atom in a constant magnetic field $\mathbf{B}_0$. Let the electric field be given by $\mathbf{E} = \mathbf{E}_0 e^{-i\omega t}$, with $\mathbf{E}_0$ constant in space and the atom modeled by an electron bound as an isotropic harmonic oscillator to its center. Let ω_0 be the resonant frequency of the oscillator and m the mass of the electron.

(a) Write the Newton−Lorentz equation of motion for the electron (charge $-e$) in the presence of the fields $\mathbf{E}$ and $\mathbf{B}_0$.

(b) Look for a solution $\mathbf{r} = \mathbf{r}_0 e^{-i\omega t}$ and find the equation satisfied by $\mathbf{r}_0$.

(c) Solve for $\mathbf{r}_0$ by expanding in powers of $\mathbf{B}_0$, keeping terms linear in $\mathbf{B}_0$.

(d) Show that a gas of such atoms, n per unit volume, would have a tensor dielectric constant

$$\epsilon_{ij} = \epsilon_0 \delta_{ij} + i\epsilon_1 \epsilon_{ijk} B_{0k}$$

where the coefficient ϵ_1, the Faraday coefficient, is given by

$$B_0 \epsilon_1 = \omega_L \frac{\partial \epsilon_0}{\partial \omega}$$

with $\omega_L = eB_0/2mc$, the Larmor frequency.

(e) Considering only the zeroth-order term ϵ_0, give an expression for the dielectric constant $\epsilon_0(\omega)$; from it calculate the absorption of energy by the dielectric from an applied field $E(t) = \int d\omega \, f(\omega) \, e^{-i\omega t}$. Remember that causality requires that the integral over ω in the neighborhood of a singularity of ϵ_0 be calculated by circling into the upper half ω plane.

3.5 Rotation of light in a magnetic field: Consider a beam of light moving through a medium whose dielectric constant is

$$\epsilon_{ij} = \epsilon_0 \delta_{ij} + i\epsilon_1 \epsilon_{ij3}.$$

ϵ_0 and ϵ_1 are real numbers.

(a) Show that right and left circularly polarized beams propagate in the 3 direction with no absorption, but different dielectric constants.

(b) Calculate the angle of rotation per unit length of a plane polarized beam propagating in the z direction.

3.6 A monochromatic plane polarized electromagnetic wave in vacuum is normally incident on a flat surface bounding a medium of real permeability μ, real dielectric constant ϵ, and conductivity σ. The circular frequency of the wave is ω. Assume the wave incident from the left. Then the incident wave is

$$\mathbf{E}_i = \mathbf{e}_i \, e^{i(kx - \omega t)}$$

and

$$\mathbf{B}_i = \hat{\mathbf{e}}_x \times \mathbf{e}_i \, e^{i(kx - \omega t)}$$

where $k = \omega/c$, the fixed vector $\mathbf{e}_i$ is in the y, z-plane, and it is understood that the real part is to be taken. The reflected wave is

$$\mathbf{E}_r = R\mathbf{e}_i\, e^{-i(kx+\omega t)}$$

(R is the reflection coefficient) and

$$\mathbf{B}_r = -\hat{\mathbf{e}}_x \times \mathbf{e}_i R\, e^{-i(kx+\omega t)}.$$

The transmitted wave is

$$\mathbf{E}_t = \mathbf{e}_i T\, e^{i(k'x-\omega t)}$$

and

$$\mathbf{B}_t = \hat{\mathbf{e}}_x \times \mathbf{e}_i T\, e^{i(k'x-\omega t)}.$$

T is the transmission coefficient.

(a) Calculate the real and imaginary parts of k' for all positive values of μ, ϵ, and σ.

(b) Calculate the reflection coefficient R and transmission coefficient T for small values of ω/σ.

(c) In the same approximation, calculate the flux of energy into the medium. What happens to it?

(d) Still in the limiting case of small ω/σ, verify that the discontinuity in H_t from outside the conductor to well within the conductor (where the field is zero) is correctly given by

$$\Delta H_t = \frac{4\pi}{c} K = \frac{4\pi}{c} \int_0^\infty dx\, j(x)$$

where K is the surface current and j the current density parallel to the plane boundary.

3.7 Find the reflection and transmission coefficients corresponding to (3.6.23) and (3.6.24) for the case of polarization normal to the plane of incidence.

3.8 (a) Show that the boundary condition $\mathbf{E}_{\text{tangential}}$ continuous between two media guarantees the continuity of $\partial \mathbf{B}_{\text{normal}}/\partial t$, provided $\mathbf{E}$ and $\mathbf{B}$ satisfy Maxwell's equations in both media.

(b) Consider next the boundary condition for the inhomogeneous equations. For media of finite conductivity (i.e., finite $\mathbf{j}$), show that the continuity of $\mathbf{H}_{\text{tangential}}$ guarantees that $\partial/\partial t\, \Delta D_{\text{normal}} = 4\pi\, \partial\sigma/\partial t$ (σ is surface charge density), provided Maxwell's equations hold in both media.

3.9 Consider cylindrical wave guides in the TEM mode.

(a) Show that a cylindrical wave guide consisting of the space inside a single perfectly conducting cylindrical shell cannot support a transverse electromagnetic wave, that is, a wave of the form

$$\mathbf{E}(\mathbf{r}, t) = \mathbf{E}_0 \, e^{i(kz - \omega t)}$$

and

$$\mathbf{B}(\mathbf{r}, t) = \mathbf{B}_0 \, e^{i(kz - \omega t)}$$

where z is the direction of the cylindrical axis, and the vectors $\mathbf{E}_0$ and $\mathbf{B}_0$ are functions of x and y and lie in the x, y-plane.

(b) Show that a wave guide consisting of the space between two perfectly conducting cylinders *can* support such a wave, the functions $\mathbf{E}_0$ and $\mathbf{B}_0$ are unique (up to a scale) for a given shape of the confining cylinders, and the wave propagates with light velocity. The wave is called a TEM (transverse electric and magnetic) mode.

(c) Give these functions $\mathbf{E}$ and $\mathbf{B}$ for the special case of two coaxial circular cylinders.

(d) From your result in (c) above, calculate, in terms of your parameterization, the energy per unit time flowing down the wave guide.

*(e) In the high conductivity ($\sigma / w \gg 1$) and small skin depth limit (Im $k \gg$ curvature of the cylinders, where k is the wave number inside the conducting medium), calculate the energy loss per unit time and length by Poynting flow into the conductors. Treat the conducting surface as planar and calculate the longitudinal electric field by continuity: $H_{\text{tangential}}$ is continuous; so $H_{\text{tangential}}$ outside the conductor $= H_{\text{tangential}}$ inside the conductor; then calculate E_z inside the conductor from Maxwell's equations and E_z just outside the conductor by continuity. From $H_{\text{tangential}}$ and E_z, calculate thePoynting flow into the conductors.

3.10 Consider a cylindrical wave guide in the TE and TM modes.

We consider here a medium of constant permeability μ and dielectric constant ϵ, bounded by a perfect conductor of unspecified cross-sectional shape. It is obviously consistent to separate the fields into longitudinal (z component) and transverse (x, y components). Thus, we write

$$\mathbf{E} = \left[\mathbf{E}_\perp(x, y) + \hat{\mathbf{e}}_z E_z(x, y) \right] f(z, t)$$

and

$$\mathbf{B} = \left[\mathbf{B}_\perp(x, y) + \hat{\mathbf{e}}_z B_z(x, y) \right] g(z, t)$$

Similarly, we write $\nabla = \hat{\mathbf{e}}_z \, (\partial/\partial z) + \nabla_\perp$. We take advantage of the absence of z or t dependence in Maxwell's equations to let $f(z, t) = g(z, t) = e^{i(kz - \omega t)}$, so that $\partial/\partial z$ becomes ik and $\partial/\partial t$ becomes $-i\omega$.

(a) Show that, by eliminating $\mathbf{H}_\perp$ from one transverse equation and substituting in the other (and vice-versa for $\mathbf{E}_\perp$), one can solve for $\mathbf{E}_\perp$ and $\mathbf{B}_\perp$ as functions of E_z, B_z, ω, k, ϵ, and μ, and that E_z and B_z each satisfy the equation

$$(\nabla_\perp^2 - k^2 + \epsilon\mu\omega^2)\binom{E_z}{B_z} = 0.$$

(b) Determine the boundary conditions on E_z and B_z at the perfectly conducting boundary.

(c) Show that these boundary conditions allow solutions with $E_z = 0$ (called TE, or transverse electric) and with $H_z = 0$ (called TM, or transverse magnetic).

*(d) Now derive the correct boundary conditions on E_z and H_z when both the inner and outer media have finite μ and ϵ, and show that the separability into TM and TE modes no longer holds in general.

*(e) Show that a circular geometry still permits TM and TE modes for E_z and H_z independent of azimuthal angle ($m = 0$ mode).

*3.11 A monoaxial crystal with $\mu = 1$ and real, symmetric ϵ_{ij} has a plane boundary perpendicular to the optic axis. The characteristic values of ϵ_{ij} are ϵ_1 and ϵ_2.

(a) Show that the ordinary ray refracts according to Snell's law.

(b) Find the generalization of Snell's law for the extraordinary ray. That is, express $\sin\theta'$ as a function of ϵ_1, ϵ_2, and $\sin\theta$. Remember that the direction of propagation of θ' is given by $\nabla_k\omega$.

3.12 Work out in detail the results (3.8.32), (3.8.33), and (3.8.34) starting from (3.8.25), (3.8.26), (3.8.30), and (3.8.31).

*3.13 Use Cauchy's theorem and more to prove the following statements about $\epsilon(\omega)$. Given that $\epsilon(\omega)$ is analytic in the upper half ω plane, $\epsilon(\omega) \to 1$ as $\omega \to \infty$, $\epsilon^*(\omega) = \epsilon(-\omega)$ on the real axis, and $\text{Im} \, \epsilon(\omega) > 0$ for ω on the real axis and positive, show:

(a) $\text{Im} \, \epsilon(\omega) < 0$ for ω on the real axis and negative.

(b) $\epsilon(\omega)$ has no zeros in the upper half ω plane, so that $\sqrt{\epsilon(\omega)}$ is analytic in the upper half ω plane.

(c) Choose the square root of ϵ so that it has a positive imaginary part for $\omega > 0$ on the real axis. Then show that $\sqrt{\epsilon}$ has a negative imaginary part for $\omega < 0$ on the real axis.

(d) Im $\sqrt{\epsilon}/\mathrm{Re}\,\sqrt{\epsilon} > 0$ for $\omega > 0$ on the real axis, so that $\sqrt{\epsilon}$ also approaches 1 as $\omega \to \infty$.

3.14 Show that the E field defined in (3.5.30) and (3.5.31) propagates causally.

3.15 A plasma is a neutral ionized gas. A simple model for a plasma is a dilute gas of electrons in a uniform positive background (that ensures overall neutrality). Call n the number of electrons per unit volume. It is convenient to define a characteristic plasma frequency ω_p by the equation $\omega_p^2 = 4\pi n e^2/m$, where e is the electron charge and m its mass.

(a) With this model, neglecting interparticle and magnetic interactions and the space dependence of the wave field, find the reaction of the electron gas to a passing electromagnetic wave of circular frequency ω by solving for the steady-state motion of an isolated electron in the electric field of the wave.

(b) Now find the dielectric constant of the dilute gas by calculating the steady-state polarization per unit volume $\mathbf{P}$ in the wave field $\mathbf{E}$ and using the formula

$$\mathbf{D} = \epsilon\mathbf{E} = \mathbf{E} + 4\pi\mathbf{P}.$$

(c) Find the conditions (in terms of e, m, n, ω, and E) for the approximations suggested in (a) above to be valid and compatible with each other.

3.16 The systematic way to deal with a plasma (or for that matter any low-density gas) is to introduce the Boltzmann function $f(\mathbf{v}, \mathbf{x})$, where $\Delta n(\mathbf{v}, \mathbf{x}) = f(\mathbf{v}, \mathbf{x})\,\Delta\mathbf{v}\Delta\mathbf{x}$ is the number of gas particles (electrons in a plasma, with the positive ions approximated as a positive background) in the six-dimensional volume element $\Delta\mathbf{v}\Delta\mathbf{x}$.

(a) By considering the number of particles entering the volume element $\Delta\mathbf{v}\Delta\mathbf{x}$ through the surface $\Delta y\Delta z\Delta\mathbf{v}$ at x and leaving through the surface at $x + \Delta x$, and similarly the number entering through the surface $\Delta v_y\Delta v_z\Delta\mathbf{x}$ at v_x and leaving through the surface at $v_x + \Delta v_x$, and continuing to the other eight surfaces, derive the Boltzmann equation:

$$\frac{\partial f}{\partial t} + \nabla\cdot\mathbf{v}f + \nabla_v\cdot\mathbf{a}f = \left.\frac{\partial f}{\partial t}\right|_{\mathrm{coll}}$$

where $\mathbf{a}(\mathbf{x}, \mathbf{v})$ is the acceleration of a particle at $\mathbf{x}$ with velocity $\mathbf{v}$ and $\partial f/\partial t|_{\mathrm{coll}}\,\Delta\mathbf{x}\Delta\mathbf{v}$ is the number of particles thrown into the volume element $\Delta\mathbf{x}\Delta\mathbf{v}$ through "collisions." The distinction between $-\nabla_\mathbf{v}\cdot\mathbf{a}f$ and $\partial f/\partial t|_{\mathrm{coll}}$ is not absolute. We take $\mathbf{a}$ to respond to the average field acting on the particles; $\partial f/\partial t|_{\text{coll}}$ describes the rest, whatever there may be.

In a rare, approximately collisionless plasma, $\mathbf{a} = e(\mathbf{E} + \mathbf{v}/c \times \mathbf{B})$, and $\partial f/\partial t|_{\text{coll}}$ is set equal to zero; this neglect must be justified *a posteriori* following the calculation of the collisionless motion.

(b) Recover the result of Problem 3.15 by solving for the linear response of the Boltzmann function to a weak applied field $\mathbf{E}(\omega)$. Let f_0 be the Boltzmann function in the absence of the applied field and $f_0 + f_1$ the Boltzmann function in the presence of the field. Then, if we keep only linear terms and neglect collisions, the Boltzmann equation becomes

$$\frac{\partial f_1}{\partial t} + \mathbf{v} \cdot \nabla f_1 + \frac{e\mathbf{E}}{m} \cdot \nabla_v f_0 = 0. \tag{1}$$

Consider now long wavelengths (i.e., $vk \ll \omega$), so that the $\mathbf{v} \cdot \nabla f_1$ term can be dropped. In this approximation, solve Eq. (1) for f_1 and from the solution calculate the induced current $j = e \int d\mathbf{v}\, \mathbf{v} f_1$. Remember that $\int d\mathbf{v}\, f_0(\mathbf{v}, \mathbf{x}) = n_0(\mathbf{x})$, the local density, and that $\mathbf{j} = \partial \mathbf{P}/\partial t$, where $\mathbf{P}$ is the polarization per unit volume.

From this, calculate the dielectric constant and recover the result of the equivalent calculation carried out in Problem 3.15.

(c) In order to include long-range electromagnetic interactions, one must include the plasma as the source of electromagnetic fields (the equations so arrived at are called the Boltzmann–Vlasov equations):

$$\frac{\partial f}{\partial t} + \mathbf{v} \cdot \nabla f + \nabla_v \cdot \mathbf{a} f = 0$$

$$\nabla \times \mathbf{B} = \frac{4\pi \mathbf{j}}{c} + \frac{1}{c}\frac{\partial \mathbf{E}}{\partial t}$$

and

$$\nabla \times \mathbf{E} = -\frac{1}{c}\frac{\partial \mathbf{B}}{\partial t}.$$

Find the normal modes of these equations, for a given wave number $\mathbf{k}$, with the same approximation as in part (b) above, that is, vk/ω, but not ck/ω, negligible and ω of the same order of magnitude as ω_p.

You should find propagating transverse modes with the

dispersion law

$$\omega^2 = \omega_p^2 + c^2 k^2$$

or, with

$$c^2 k^2 = \epsilon \omega^2, \qquad \epsilon = 1 - \frac{\omega_p^2}{\omega^2},$$

just as in part (b) above.

In addition, there is a discrete longitudinal mode, approximately independent of vk: $\omega = \omega_p$.

A careful treatment taking the small vk dependence into account shows that for a normal velocity distribution function f_0, this mode actually damps, transferring its energy to particle motion.

3.17 (a) Formulate a set of Maxwell's equations that would take into account the existence of a magnetic current and charge density $\mathbf{j}_m$ and ρ_m, corresponding to the familiar electric quantities $\mathbf{j}_e$ and ρ_e. Clearly, $\mathbf{j}_m$ and ρ_m would go on the right-hand side of the homogeneous equations:

$$\nabla \cdot \mathbf{B} = 4\pi \alpha \rho_m, \qquad \nabla \times \mathbf{E} = -\frac{1}{c}\frac{\partial \mathbf{B}}{\partial t} + \beta \mathbf{j}_m,$$

where α and β are to be determined. The volume force law is

$$\mathbf{F} = \int \mathbf{f}_v \, d\mathbf{r},$$

and

$$\mathbf{f}_v = \rho_e \mathbf{E} + \frac{\mathbf{j}_e}{c} \times \mathbf{B} + \rho_m \mathbf{B} + \gamma \frac{\mathbf{j}_m}{c} \times \mathbf{E}$$

where γ is to be determined. The positive sign of $\rho_m \mathbf{B}$ in the force equation is a definition. The coefficients are to be determined by the requirement of internal consistency and the existence of conservation laws for energy, momentum, and angular momentum. Thus, you must find for the change of field energy:

(i) $$\frac{\partial}{\partial t}\frac{(\mathbf{B}^2 + \mathbf{E}^2)}{8\pi} = -\nabla \cdot \frac{c\mathbf{E} \times \mathbf{B}}{4\pi} - (\mathbf{j}_e \cdot \mathbf{E} + \mathbf{j}_m \cdot \mathbf{B}),$$

and for the total force on matter inside the volume V:

(ii)
$$\int_V d\mathbf{r}\, f_{V_j} = -\frac{d}{dt}\int_V d\mathbf{r}\, \frac{(\mathbf{E} \times \mathbf{B})}{4\pi c}\, j + \int_V d\mathbf{r}\, \partial_i T_{ij}$$

where

$$T_{ij} = \frac{E_i E_j + B_i B_j}{4\pi} - \frac{(\mathbf{B}^2 + \mathbf{E}^2)\,\delta_{ij}}{8\pi};$$

and
(iii) a field angular momentum

$$\mathbf{L} = \int d\mathbf{r}\, \frac{\mathbf{r} \times (\mathbf{E} \times \mathbf{B})}{4\pi c}$$

satisfying an equation analogous to (ii) above.

(b) Finally, calculate the angular momentum of a stationary magnetic pole p at $\mathbf{r}_1$ and electric charge q at $\mathbf{r}_2$:

$$\mathbf{L} = \int \frac{d\mathbf{r}}{4\pi c}\, \mathbf{r} \times \left[\frac{q(\mathbf{r} - \mathbf{r}_2)}{|\mathbf{r} - \mathbf{r}_2|^3} \times \frac{p(\mathbf{r} - \mathbf{r}_1)}{|\mathbf{r} - \mathbf{r}_1|^3}\right]$$

(i) Show first that $\mathbf{L}$ is independent of the origin of $\mathbf{r}$.
(ii) Now choose $\mathbf{r}_2$ as the origin and recognize both fields as gradients. Thus,

$$\mathbf{L} = \int d\mathbf{r}\, \frac{\mathbf{r} \times (\nabla \phi_1 \times \nabla \phi_2)}{4\pi c}.$$

Integrate appropriately by parts (i.e., $\nabla \phi_1$ or $\nabla \phi_2$) and then use the Legendre polynomial expansion to find the answer. Note on dimensional grounds that $\mathbf{L}$ must be proportional to

$$\frac{qp}{c}\, \frac{\mathbf{r}_1 - \mathbf{r}_2}{|\mathbf{r}_1 - \mathbf{r}_2|}.$$

(c) From the formula for $\mathbf{L}$ and the quantization of angular momentum follows the quantization of electric charge, which must hold if a magnetic pole exists. This semiclassical result also holds in quantum theory.

3.18 It is amusing that Maxwell's equations take a particularly simple form in terms of the vector $\mathbf{Q} = \mathbf{E} + i\mathbf{B}$. For the free field equation,

$$\frac{1}{c}\mathbf{Q} = -i\nabla \times \mathbf{Q}.$$

Verify this and, following the work of 3.17, find the current that goes on the right-hand side in the presence of magnetic poles.

CHAPTER 4

Radiation by Prescribed Sources

$\mathbf{I}$n this chapter we study the emission of radiation by a given source (charge and current densities). The complete solution of a physics problem will, in principle, require the simultaneous calculation of the effect of the radiation on the source. In many cases, however, the radiation reaction is relatively weak so that it can be taken into account in a series of successive approximations. Those are the cases we will be dealing with in this chapter. Chapter 5, on scattering, deals with the more general problem of interacting fields and sources.

4.1. VECTOR AND SCALAR POTENTIALS

The equations for $\mathbf{E}$ and $\mathbf{B}$ can be solved most easily by introducing a vector and scalar potential. Since $\nabla \cdot \mathbf{B} = 0$, there exists a vector potential $\mathbf{A}$, such that $\mathbf{B} = \nabla \times \mathbf{A}$. The homogeneous equation

$$\nabla \times \mathbf{E} = -\frac{1}{c} \frac{\partial \mathbf{B}}{\partial t} \quad \text{or} \quad \nabla \times \left(\mathbf{E} + \frac{1}{c} \frac{\partial \mathbf{A}}{\partial t} \right) = 0 \qquad (4.1.1)$$

implies that

$$\mathbf{E} + \frac{1}{c} \frac{\partial \mathbf{A}}{\partial t} = -\nabla \phi, \qquad (4.1.2)$$

that is, that there exists a scalar potential ϕ. Thus, $\mathbf{E}$ and $\mathbf{B}$ are given by the equations

$$\mathbf{B} = \nabla \times \mathbf{A} \qquad (4.1.3)$$

and

$$\mathbf{E} = -\frac{1}{c}\frac{\partial \mathbf{A}}{\partial t} - \nabla \phi. \tag{4.1.4}$$

We have seen in the quasistatic case that $\mathbf{E}$ and $\mathbf{B}$ do not uniquely determine $\mathbf{A}$ and ϕ. That is still true here. The substitutions

$$\mathbf{A}' = \mathbf{A} + \nabla \chi \tag{4.1.5}$$

$$\phi' = \phi - \frac{1}{c}\frac{\partial \chi}{\partial t} \tag{4.1.6}$$

evidently leave $\mathbf{B}$ and $\mathbf{E}$ unchanged and so correspond to the same physical fields as $\mathbf{A}$ and ϕ. This property is called *gauge invariance*. In classical field theory the potentials are introduced as a calculational convenience, and their gauge similarly chosen for convenience.[1]

The inhomogeneous equations are now

$$\nabla \cdot \mathbf{E} = -\frac{1}{c}\nabla \cdot \frac{\partial \mathbf{A}}{\partial t} - \nabla^2 \phi = 4\pi\rho \tag{4.1.7}$$

and

$$\nabla \times \mathbf{B} - \frac{1}{c}\frac{\partial \mathbf{E}}{\partial t} = \nabla \times (\nabla \times \mathbf{A}) + \frac{1}{c}\left(\frac{1}{c}\frac{\partial^2 \mathbf{A}}{\partial t^2} + \nabla\frac{\partial \phi}{\partial t}\right) = \frac{4\pi\mathbf{j}}{c}. \tag{4.1.8}$$

There are two especially useful gauges to work in. The first is called the transverse, or Coulomb gauge. It is defined by

$$\nabla \cdot \mathbf{A} = 0 \tag{4.1.9}$$

and can clearly be reached by the proper choice of χ in (4.1.5), starting from any $\mathbf{A}_0$, by solving the equation $\nabla \cdot \mathbf{A}_0 + \nabla^2\chi = 0$ for χ. Equation (4.1.7) then becomes

$$\nabla^2 \phi = -4\pi\rho, \tag{4.1.10}$$

the familiar Poisson equation, which produces a scalar potential

$$\phi(\mathbf{r}, t) = \int d\mathbf{r}' \frac{\rho(\mathbf{r}', t)}{|\mathbf{r} - \mathbf{r}'|} \tag{4.1.11}$$

---

[1]In quantum theory the potentials play an essential role.

which is instantaneously connected to the charge density. This solution appears to violate causality—the finite propagation velocity of electromagnetic signals—but does not, since the potential ϕ is not a physical field. The test of causality must come in the field strengths $\mathbf{E}$ and $\mathbf{B}$, where the vector potential must make a contribution that will cancel the acausal contribution of $\nabla\phi$. This comes about from (4.1.8) for $\mathbf{A}$:

$$\nabla \times (\nabla \times \mathbf{A}) + \frac{1}{c^2}\frac{\partial^2 \mathbf{A}}{\partial t^2} = \frac{4\pi\mathbf{j}}{c} - \frac{1}{c}\nabla\frac{\partial\phi}{\partial t} \tag{4.1.12}$$

with ϕ given by (4.1.11).

Note that (4.1.12) is purely transverse; that is, the divergence of both sides is zero, since

$$\nabla \cdot \left(\mathbf{j} - \frac{\nabla}{4\pi}\frac{\partial\phi}{\partial t} \right) = \nabla \cdot \mathbf{j} - \frac{\partial}{\partial t}\frac{\nabla^2\phi}{4\pi}$$

$$= -\frac{\partial\rho}{\partial t} - \frac{\partial}{\partial t}(-\rho) = 0.$$

We see that $(1/4\pi)\nabla(\partial\phi/\partial t)$ is the longitudinal part of $\mathbf{j}$, so that even with $\mathbf{j}$ and ρ spatially confined, the longitudinal part of $\mathbf{j}$ falls off only like an inverse power of r as $r \to \infty$. (See Problem 4.1.) The contribution of this term to $\mathbf{A}$ and $\partial\mathbf{A}/\partial t$ at large $\mathbf{r}$ will have to cancel the unwanted contribution to $\mathbf{E}$ from $-\nabla\phi$. (See Problem 4.2.)

To show that this must happen, it is convenient to use the second gauge, called the Lorentz gauge. To introduce the Lorentz gauge, we rewrite (4.1.7) and (4.1.8) as

$$-\nabla^2\phi + \frac{1}{c^2}\frac{\partial^2\phi}{\partial t^2} + \frac{1}{c}\frac{\partial}{\partial t}\left(-\frac{1}{c}\frac{\partial\phi}{\partial t} - \nabla \cdot \mathbf{A} \right) = 4\pi\rho \tag{4.1.13}$$

and

$$-\nabla^2\mathbf{A} + \frac{1}{c^2}\frac{\partial^2\mathbf{A}}{\partial t^2} + \nabla\left(\nabla \cdot \mathbf{A} + \frac{1}{c}\frac{\partial\phi}{\partial t} \right) = \frac{4\pi\mathbf{j}}{c}. \tag{4.1.14}$$

The obvious choice of gauge now is

$$\nabla \cdot \mathbf{A} + \frac{1}{c}\frac{\partial\phi}{\partial t} = 0 \tag{4.1.15}$$

which decouples the different vectors from each other and allows the three components of $\mathbf{A}$ and the one component of ϕ to be calculated as if they were independent scalars. To see that this gauge can always be chosen, start with $\mathbf{A}_0$, ϕ_0 and solve the equation

$$\nabla \cdot (\mathbf{A}_0 + \nabla\chi) + \frac{1}{c}\frac{\partial}{\partial t}\left(\phi_0 - \frac{1}{c}\frac{\partial\chi}{\partial t}\right) = 0$$

or

$$\left(\nabla^2 - \frac{1}{c^2}\frac{\partial^2}{\partial t^2}\right)\chi = -\left(\nabla \cdot \mathbf{A}_0 + \frac{1}{c}\frac{\partial\phi_0}{\partial t}\right). \tag{4.1.16}$$

Equation (4.1.16) can be solved, in general, by iteration, given χ and $\partial\chi/\partial t$ at $t = 0$. Of course, the solution is not unique, since one can still add any solution of

$$\left(\nabla^2 - \frac{1}{c^2}\frac{\partial^2}{\partial t^2}\right)\chi = 0$$

to a particular solution of (4.1.16).

We have now to solve the decoupled equations

$$\nabla^2\mathbf{A} - \frac{1}{c^2}\frac{\partial^2\mathbf{A}}{\partial t^2} \equiv \Box\mathbf{A} = -\frac{4\pi\mathbf{j}}{c} \tag{4.1.17}$$

and

$$\nabla^2\phi - \frac{1}{c^2}\frac{\partial^2\phi}{\partial t^2} \equiv \Box\phi = -4\pi\rho. \tag{4.1.18}$$

4.2. GREEN'S FUNCTIONS FOR THE RADIATION EQUATION

We will want a solution of the equation

$$\Box\psi = -4\pi f(\mathbf{r}, t) \tag{4.2.1}$$

with f a given function of space and time. The Green's function $G(\mathbf{r}, t; \mathbf{r}', t')$ that gives

$$\psi = \int d\mathbf{r}'\, dt'\, G(\mathbf{r}, t; \mathbf{r}', t') f(\mathbf{r}', t') \tag{4.2.2}$$

must satisfy the equation

$$\Box G = \left(\nabla^2 - \frac{1}{c^2} \frac{\partial^2}{\partial t^2} \right) G(\mathbf{r}, t; \mathbf{r}', t') = -4\pi \delta^3(\mathbf{r} - \mathbf{r}') \delta(t - t').$$

$$(4.2.3)$$

We look for a solution that is a function only of $\mathbf{r} - \mathbf{r}'$ and $t - t'$, which must satisfy the equation

$$\Box G(\mathbf{r}, t) = -4\pi \delta^3(\mathbf{r}) \delta(t). \qquad (4.2.4)$$

Equation (4.2.4) invites a Fourier transform:

$$G(\mathbf{r}, t) = \int d\mathbf{k} \, d\omega \, e^{i(\mathbf{k} \cdot \mathbf{r} - \omega t)} g(\mathbf{k}, \omega), \qquad (4.2.5)$$

which, with

$$\delta^3(\mathbf{r}) \delta(t) = \int \frac{d\mathbf{k} \, d\omega}{(2\pi)^4} e^{i(\mathbf{k} \cdot \mathbf{r} - \omega t)},$$

gives

$$\left(\frac{\omega^2}{c^2} - \mathbf{k}^2 \right) g = -\frac{4\pi}{(2\pi)^4}, \qquad (4.2.6)$$

or

$$g = -\frac{1}{4\pi^3} \frac{1}{\frac{\omega^2}{c^2} - \mathbf{k}^2}, \qquad (4.2.7)$$

so that

$$G(\mathbf{r}, t) = -\frac{c^2}{4\pi^3} \int d\mathbf{k} \, e^{i\mathbf{k} \cdot \mathbf{r}} \int \frac{d\omega \, e^{-i\omega t}}{\omega^2 - c^2 k^2}. \qquad (4.2.8)$$

We have seen denominators of this type before in connection with the causality of atomic polarizability; see (3.4.24)–(3.4.30). Recall that $G(\mathbf{r}, t)$ will vanish for negative t if we take ω in the denominator of (4.2.8) to approach the real axis from above. That definition corresponds to a retarded potential. That is, the source of (4.2.8) is a δ function pulse at $t = 0$ and $\mathbf{r} = 0$. Requiring G to vanish for $t < 0$ is equivalent to requiring that the radiation from the source pulse come after the pulse. Thus, we have the retarded Green's function

$$G_R(\mathbf{r}, t) = - \lim_{\epsilon \to 0} \frac{c^2}{4\pi^3} \int d\mathbf{k}\, e^{i\mathbf{k}\cdot\mathbf{r}} \int_{-\infty}^{\infty} \frac{d\omega\, e^{-i\omega t}}{(\omega + i\epsilon)^2 - c^2 k^2}; \qquad (4.2.9)$$

the advanced Green's function (corresponding to a time-reversed world) is

$$G_A = - \lim_{\epsilon \to 0} \frac{c^2}{4\pi^3} \int d\mathbf{k}\, e^{i\mathbf{k}\cdot\mathbf{r}} \int \frac{d\omega\, e^{-i\omega t}}{(\omega - i\epsilon)^2 - c^2 k^2}. \qquad (4.2.10)$$

To calculate G_R for $t > 0$, we must close the ω integral (4.2.9) in the lower half ω plane, where there are poles at $\omega = -i\epsilon \pm ck$. The result is

$$\lim_{\epsilon \to 0} \int_{-\infty}^{\infty} \frac{d\omega\, e^{-i\omega t}}{(\omega + i\epsilon)^2 - c^2 k^2} = \begin{cases} -\dfrac{2\pi}{ck} \sin ckt & t > 0 \\[2mm] 0 & t < 0 \end{cases}$$

and so

$$G_R = \frac{c}{2\pi^2} \int \frac{d\mathbf{k}}{k}\, e^{i\mathbf{k}\cdot\mathbf{r}} \sin ckt, \qquad (4.2.11)$$

always for $t > 0$.

The integral over angles of $\mathbf{k}$ gives

$$G_R = \frac{2c}{\pi r} \int_0^{\infty} dk \sin kr \sin ckt \qquad (4.2.12)$$

$$= \frac{c}{\pi r} \int_0^{\infty} dk[\cos k(r - ct) - \cos k(r + ct)]$$

$$= \frac{c}{2\pi r} \int_{-\infty}^{\infty} dk[e^{ik(r - ct)} - e^{ik(r + ct)}]$$

$$= \frac{c}{r} (\delta(r - ct) - \delta(r + ct)). \qquad (4.2.13)$$

The second term is zero (both r and t are positive) so the final result is

$$G_R(\mathbf{r}, t) = \frac{c}{r} \delta(r - ct). \qquad (4.2.14)$$

Having found the solution, we can verify directly that (4.2.14) satisfies (4.2.4). (See Problem 4.9.)

Returning to the original equation (4.2.1), we find the retarded solution

$$\psi(\mathbf{r}, t) = \int \frac{d\mathbf{r}'\, dt'}{|\mathbf{r} - \mathbf{r}'|} \delta\left(t - t' - \frac{|\mathbf{r} - \mathbf{r}'|}{c}\right) f(\mathbf{r}', t') \qquad (4.2.15)$$

or

$$\psi(\mathbf{r}, t) = \int \frac{d\mathbf{r}'}{|\mathbf{r} - \mathbf{r}'|} f(\mathbf{r}', t_R) \qquad (4.2.16)$$

where the retarded time $t_R = t - |(\mathbf{r} - \mathbf{r}')|/c|$ is the time at which a signal must be emitted at $\mathbf{r}'$ to reach $\mathbf{r}$ at the time t.

A useful way to picture the content of (4.2.16) is to imagine a spherical wave converging at time t on the point $\mathbf{r}$. The retarded time t_R is the time the wave crosses the point $\mathbf{r}'$, picking up its contribution from the source $f(\mathbf{r}', t_R)$ as it does so.

The fields $\mathbf{A}$ and ϕ radiated by the current and charge density $\mathbf{j}$ and ρ as described in (4.1.17) and (4.1.18) are now given by (4.2.16):

$$\mathbf{A} = \frac{1}{c} \int \frac{d\mathbf{r}'}{|\mathbf{r} - \mathbf{r}'|} \mathbf{j}(\mathbf{r}', t_R) \qquad (4.2.17)$$

and

$$\phi = \int \frac{d\mathbf{r}'}{|\mathbf{r} - \mathbf{r}'|} \rho(\mathbf{r}', t_R). \qquad (4.2.18)$$

In this gauge, the absence of the instantaneous electric field encountered in (4.1.11) is obvious, since all the potentials and fields are retarded.

4.3. RADIATION FROM A FIXED FREQUENCY SOURCE

We suppose $\mathbf{j}$ and ρ are given by

$$\mathbf{j}(\mathbf{r}, t) = \mathbf{j}_0(\mathbf{r})e^{-i\omega t} \qquad (4.3.1)$$

and

$$\rho(\mathbf{r}, t) = \rho_0(\mathbf{r})e^{-i\omega t}, \qquad (4.3.2)$$

with

$$\nabla \cdot \mathbf{j}_0 - i\omega\rho_0 = 0.$$ (4.3.3)

We calculate the potentials as

$$\mathbf{A}(\mathbf{r}, t) = \frac{1}{c}\int \frac{d\mathbf{r}'}{|\mathbf{r} - \mathbf{r}'|}\mathbf{j}_0(\mathbf{r})e^{-i\omega t_R}$$ (4.3.4)

and

$$\phi(\mathbf{r}, t) = \int \frac{d\mathbf{r}'}{|\mathbf{r} - \mathbf{r}'|}\rho_0(\mathbf{r}')e^{-i\omega t_R}.$$ (4.3.5)

The signature of radiation is that, for sufficiently large r, $\mathbf{E}$ and $\mathbf{B}$ go like $1/r$ so that the energy radiated through a solid angle $d\Omega$:

$$d\mathbf{S} \cdot \mathcal{P} = \frac{c}{4\pi}\mathbf{E} \times \mathbf{B} \cdot d\mathbf{S}$$

$$= \frac{c}{4\pi}\mathbf{E} \times \mathbf{B} \cdot \hat{\mathbf{r}}\, r^2\, d\Omega,$$

goes like a constant (independent of r). The region where the $1/r$ behavior dominates the fields is called the wave zone. The radiated power per unit solid angle is given by the coefficient of $1/r^2$ in the Poynting vector, so that the calculation of radiation requires only the leading term in $1/r$ in (4.3.4) and (4.3.5). Thus,

$$\frac{1}{|\mathbf{r} - \mathbf{r}'|} \sim \frac{1}{r},$$

and

$$t_R = t - \frac{|\mathbf{r} - \mathbf{r}'|}{c}$$

$$= t - \frac{r}{c}\left(1 - \frac{2\mathbf{r} \cdot \mathbf{r}'}{r^2} + \mathcal{O}\left(\frac{r'^2}{r^2}\right)\right)^{1/2}$$

$$= t - \frac{r}{c}\left(1 - \frac{\mathbf{r} \cdot \mathbf{r}'}{r^2} + \mathcal{O}\left(\frac{r'^2}{r^2}\right)\right)$$

$$= t - \frac{r}{c} + \frac{\hat{\mathbf{r}} \cdot \mathbf{r}'}{c} + \mathcal{O}\left(\frac{r'^2}{rc}\right).$$ (4.3.6)

The last term is negligible in $e^{-i\omega t_R}$, provided the wavelength λ is such that

$$\frac{\omega r'^2}{rc} \cong \frac{r'^2}{r\lambda} \ll 1. \tag{4.3.7}$$

Note that a macroscopic source could require kilometers to reach the wave zone as defined above. Clearly, the method we are discussing here does not apply to such a case.

There remain, as $r \to \infty$,

$$\mathbf{A} = \frac{e^{-i\omega(t-r/c)}}{cr} \int d\mathbf{r}' \exp\left(-i\frac{\omega}{c}\hat{\mathbf{r}} \cdot \mathbf{r}'\right)\mathbf{j}_0(\mathbf{r}') \tag{4.3.8}$$

and

$$\phi = \frac{e^{-i\omega(t-r/c)}}{r} \int d\mathbf{r}' \exp\left(-i\frac{\omega}{c}\hat{\mathbf{r}} \cdot \mathbf{r}'\right)\rho_0(\mathbf{r}'). \tag{4.3.9}$$

We recognize that $\mathbf{k} = \omega\hat{\mathbf{r}}/c$ is the radiated wave number. In differentiation, $(\partial/\partial x_i)\hat{\mathbf{r}}$ goes like $1/r$, so that $\mathbf{k}$ can be treated as a constant. Thus,

$$\mathbf{A} = \frac{e^{i(kr-\omega t)}}{cr}\mathbf{j}_\mathbf{k} \tag{4.3.10}$$

and

$$\phi = \frac{e^{i(kr-\omega t)}}{r}\rho_\mathbf{k} \tag{4.3.11}$$

where

$$\mathbf{j}_\mathbf{k} = \int d\mathbf{r}' \, e^{-i\mathbf{k}\cdot\mathbf{r}'}\mathbf{j}_0(\mathbf{r}') \tag{4.3.12}$$

and

$$\rho_\mathbf{k} = \int d\mathbf{r}' \, e^{-i\mathbf{k}\cdot\mathbf{r}'} \rho_0(\mathbf{r}'). \tag{4.3.13}$$

Equation (4.3.3) leads to

$$\mathbf{k} \cdot \mathbf{j}_\mathbf{k} = \omega\rho_\mathbf{k}. \tag{4.3.14}$$

We find the fields from $\mathbf{A}$ and ϕ. To leading order in $1/r$,

$$\frac{\partial \mathbf{A}}{\partial x_i} = \frac{e^{i(kr-\omega t)}}{cr} \mathbf{j_k} \frac{ikx_i}{r} \tag{4.3.15}$$

and

$$\frac{\partial \phi}{\partial x_i} = \frac{e^{i(kr-\omega t)}}{r} \rho_\mathbf{k} \frac{ikx_i}{r}, \tag{4.3.16}$$

so that

$$\mathbf{E} = -\frac{1}{c}\frac{\partial \mathbf{A}}{\partial t} - \nabla \phi$$

$$= \frac{i\omega}{c^2 r} \mathbf{j_k} e^{i(kr-\omega t)} - \frac{i\mathbf{k}}{r} \rho_\mathbf{k} e^{i(kr-\omega t)} \tag{4.3.17}$$

and

$$\mathbf{B} = \nabla \times \mathbf{A} = i\mathbf{k} \times \frac{\mathbf{j_k} e^{i(kr-\omega t)}}{cr}. \tag{4.3.18}$$

Observe that both **B** and **E** as given in (4.3.17) and (4.3.18) are transverse, **B** obviously and **E** from (4.3.14). The second term in (4.3.17) serves merely to cancel the longitudinal part of $\mathbf{j_k}$, leaving only the transverse part,

$$\mathbf{j}_{kT} = \mathbf{j_k} - \frac{\mathbf{k}\mathbf{k} \cdot \mathbf{j_k}}{k^2}, \tag{4.3.19}$$

which of course would have appeared more naturally in the transverse gauge. Thus, the **E** and **B** fields have the properties of propagating fields with polarization vector $\mathbf{j}_{kT}$, wave number **k**, and satisfying the correct right-hand rule relating **k**, **E**, and **B**.

We calculate the radiated energy [averaged over a cycle as described in (3.5.19)] as

$$\frac{dW}{dt\, d\Omega} = \overline{\mathscr{P} \cdot \hat{\mathbf{r}} r^2} \tag{4.3.20}$$

$$= \frac{c}{4\pi} \overline{\mathbf{E} \times \mathbf{B} \cdot \hat{\mathbf{r}} r^2}$$

$$= \operatorname{Re} \frac{c}{8\pi} \frac{\omega}{c^3} \mathbf{j}_{kT}^* \times (\mathbf{k} \times \mathbf{j}_{kT}) \tag{4.3.21}$$

$$= \frac{\omega k}{8\pi c^2} \mathbf{j}_{kT}^* \cdot \mathbf{j}_{kT} \tag{4.3.22}$$

$$= \frac{\omega^2}{8\pi c^3} \hat{\mathbf{k}} \mathbf{j}_{kT}^* \cdot \mathbf{j}_{kT}$$

and

$$\mathbf{j}_{kT}^* \cdot \mathbf{j}_{kT} = \mathbf{j}_k^* \cdot \mathbf{j}_k - \frac{1}{k^2} \mathbf{j}_k^* \cdot \mathbf{k} \mathbf{j}_k \cdot \mathbf{k} . \tag{4.3.23}$$

These formulas are very general and are correct to order $1/r$. In the next two sections, we will apply this general theory to specific cases.

4.4. RADIATION BY A SLOWLY MOVING POINT PARTICLE

We consider the radiation from a slowly moving point particle, with current density

$$\mathbf{j}(\mathbf{r}, t) = q\mathbf{v}(t) \, \delta^3(\mathbf{r} - \mathbf{r}(t)) \tag{4.4.1}$$

where $\mathbf{r}(t)$ and $\mathbf{v}(t)$ are the position and velocity of the particle at time t, and q is the particle's electric charge. Slow means $v/c \ll 1$.

To find the vector potential produced by $\mathbf{j}$, we Fourier transform

$$\mathbf{j}(\mathbf{r}, t) = \int d\omega \, e^{-i\omega t} \mathbf{j}_\omega(\mathbf{r}) \tag{4.4.2}$$

[$\mathbf{j}_\omega$ here denotes the object corresponding to $\mathbf{j}_0(\mathbf{r})$ in Section 4.3]. Equation (4.3.8) then gives $\mathbf{A}$ in the radiation zone:

$$\mathbf{A}(\mathbf{r}, t) = \int d\omega \, \frac{e^{-i\omega(t - r/c)}}{rc} \int d\mathbf{r}' \, \mathbf{j}_\omega(\mathbf{r}') \, e^{-i\mathbf{k} \cdot \mathbf{r}'} \tag{4.4.3}$$

where $\mathbf{k} = (\omega/c)\hat{\mathbf{r}}$.

The exponential $e^{-i\mathbf{k} \cdot \mathbf{r}'} = e^{-i\hat{\mathbf{r}} \cdot \mathbf{r}'(\omega/c)}$ is approximately constant for a slowly moving particle, since $\Delta(\mathbf{r}'\omega/c) \sim v/c$. If we make the obvious choice of origin, (4.4.3) becomes

$$\mathbf{A}(\mathbf{r}, t) = \frac{1}{rc} \int d\omega \, e^{-i\omega(t - r/c)} \int d\mathbf{r}' \, \mathbf{j}_\omega(\mathbf{r}') = \frac{1}{rc} \int d\mathbf{r}' \, \mathbf{j}\left(\mathbf{r}', t - \frac{r}{c}\right) \tag{4.4.4}$$

which with (4.4.1) gives

$$\mathbf{A}(\mathbf{r}, t) = \frac{q}{rc} \mathbf{v}\left(t - \frac{r}{c}\right) + \mathcal{O}\left(\frac{1}{r^2}\right). \tag{4.4.5}$$

Thus, **A** is given by the radiating particle's velocity at its retarded time. We immediately find

$$\mathbf{E} = -\frac{1}{c}\frac{\partial \mathbf{A}}{\partial t}\bigg|_T = -\frac{q}{rc^2}\mathbf{a}\left(t - \frac{r}{c}\right)_T + \mathcal{O}\left(\frac{1}{r^2}\right) \qquad (4.4.6)$$

where

$$\mathbf{a}_T = \mathbf{a} - \hat{\mathbf{r}}\hat{\mathbf{r}} \cdot \mathbf{a} \qquad (4.4.7)$$

is the transverse component of **a**, the acceleration at the retarded time. Remember that $\hat{\mathbf{r}}$ is the direction of observation.

The magnetic field is

$$\mathbf{B} = \nabla \times \mathbf{A} = \frac{q}{rc^2}(-\nabla r) \times \mathbf{a} = \hat{\mathbf{r}} \times \mathbf{E} + \mathcal{O}\left(\frac{1}{r^2}\right). \qquad (4.4.8)$$

The instantaneous Poynting vector $\mathcal{P}$ is

$$\mathcal{P} = \frac{c}{4\pi}\mathbf{E} \times \mathbf{B}$$

$$= \frac{q^2}{4\pi r^2 c^3}\left|\mathbf{a}_T\left(t - \frac{r}{c}\right)\right|^2 \hat{\mathbf{r}}$$

$$= \frac{q^2}{4\pi r^2 c^3}\left[\mathbf{a}^2 - (\mathbf{a} \cdot \hat{\mathbf{r}})^2\right]\hat{\mathbf{r}}, \qquad (4.4.9)$$

where **a** is evaluated at the retarded time $t - r/c$.

If we average over a time T (e.g., a cycle of simple harmonic motion), we have for the average rate of radiation per unit area and time

$$\mathcal{P} = \frac{q^2}{4\pi r^2 c^3}\left[\overline{\mathbf{a}}^2 - \overline{(\mathbf{a} \cdot \hat{\mathbf{r}})}^2\right]\hat{\mathbf{r}}, \qquad (4.4.10)$$

in agreement with (4.3.22), recognizing that $q^2\overline{\mathbf{a}}^2 = \omega^2|\mathbf{j}_\mathbf{k}|^2/2$ for simple harmonic motion and $\mathbf{k} \cong 0$. Equation (4.4.10) tells us that the time-average power radiated per unit solid angle, $dP/d\Omega$, is independent of r:

$$d\Omega\frac{d\overline{P}}{d\Omega} = \int\limits_{\text{over } d\Omega} d\mathbf{S} \cdot \mathcal{P} = d\Omega\frac{q^2}{4\pi c^3}\left(\overline{\mathbf{a}}^2 - \overline{(\mathbf{a} \cdot \hat{\mathbf{r}})}^2\right). \qquad (4.4.11)$$

Note that in this approximation (called electric dipole), plane polarized light is produced by a linear particle trajectory, the plane of polarization being the plane containing the particle motion $\mathbf{r}(t)$ and the direction of observation, $\hat{r} = \hat{k}$. With any other than a linear trajectory, the nature of the polarization depends on the direction $\hat{r}$ of observation. Except for special directions, the polarization is, in general, elliptic.

The dependence on the direction of observation of the intensity of radiation is given by (4.4.10). For a linear trajectory, it is $\sin^2 \theta$, where θ is the angle between the trajectory direction and the direction of observation. The total radiated energy per unit time (power) is given by the integral of $\overline{\mathcal{P}}$ over a distant surface

$$\frac{dW}{dt} = \int_S d\mathbf{S} \cdot \mathcal{P} = \int \frac{\overline{\mathbf{a}}^2}{4\pi} \frac{q^2}{c^3} d\Omega \sin^2 \theta = \frac{2}{3} \frac{q^2}{c^3} \overline{\mathbf{a}}^2. \qquad (4.4.12)$$

4.5. ELECTRIC AND MAGNETIC DIPOLE AND ELECTRIC QUADRUPOLE RADIATION

The approximation made in (4.4.4), $\mathbf{k} \cdot \mathbf{r}' \ll 1$, can be applied to a more general current distribution than the one given in (4.4.1). It forms the basis of a multipole expansion analogous to the electrostatic and magnetostatic multipole expansions discussed earlier in Chapters 1 and 2. For small kr', that is, for dimensions of the radiating system much smaller than the radiated wavelength, the first few nonvanishing terms in the expansion provide a good approximation to the radiation amplitude $\mathbf{j_k}$ of (4.3.12). For kr' not small, the multipole expansion can still be carried out, as we shall see in Chapter 5, but is not equivalent to an expansion in kr', and may converge slowly. For the moment, we confine ourselves to the first few terms, which we evaluate by expanding in kr'.

We consider, then, a given frequency ω and wave number $\mathbf{k} = \hat{r}(\omega/c)$ and evaluate the radiated amplitude of (4.3.12)

$$\mathbf{j_k} = \int d\mathbf{r}' \, e^{-i\mathbf{k} \cdot \mathbf{r}'} \mathbf{j}_0(\mathbf{r}') \qquad (4.3.12)$$

where from (4.3.3)

$$\nabla \cdot \mathbf{j}_0(\mathbf{r}') = i\omega \rho_0(\mathbf{r}'). \qquad (4.3.3)$$

We proceed by expanding the exponential

$$e^{-i\mathbf{k}\cdot\mathbf{r}'} = 1 - i\mathbf{k}\cdot\mathbf{r}' + \cdots. \tag{4.5.1}$$

The first term gives

$$\mathbf{j}_k^{(1)} = \int d\mathbf{r}'\mathbf{j}_0(\mathbf{r}'). \tag{4.5.2}$$

For a confined current,

$$\mathbf{j}_k^{(1)} = -\int d\mathbf{r}'\,\mathbf{r}'\nabla'\cdot\mathbf{j}_0(\mathbf{r}') \tag{4.5.3}$$

or

$$\mathbf{j}_k^{(1)} = -i\omega\int d\mathbf{r}'\,\rho_0(\mathbf{r}')\mathbf{r}'. \tag{4.5.4}$$

The question of origin of coordinates does not enter into (4.5.4), since $\int \rho(\mathbf{r}')\,d\mathbf{r}'$ is conserved and, hence, has no component with ω different from zero. The term "electric dipole" is now clear—the electric dipole moment of the charge distribution $\mathbf{p}_E$ is given by

$$\mathbf{p}_E = \int d\mathbf{r}'\,\rho_0(\mathbf{r}')\,\mathbf{r}' \tag{4.5.5}$$

and so

$$\mathbf{j}_k^{(1)} = -i\omega\mathbf{p}_E \tag{4.5.6}$$

independent of the direction of radiation.

The complex electric field vector is given by (4.3.17) and (4.3.19) as

$$\mathbf{E}^{(1)} = \frac{i\omega}{c^2 r}e^{i(kr-\omega t)}(-i\omega)(\mathbf{p}_E - \hat{\mathbf{r}}\hat{\mathbf{r}}\cdot\mathbf{p}_E), \tag{4.5.7}$$

and the magnetic field as $\hat{\mathbf{r}}\times\mathbf{E}$:

$$\mathbf{B}^{(1)} = \hat{\mathbf{r}}\times\mathbf{E}^{(1)}. \tag{4.5.8}$$

The angular distribution of the radiation is given by (4.3.22) and (4.3.23) as

$$\frac{dW}{dt\,d\Omega} = \frac{\omega^4}{8\pi c^3}\left(|\mathbf{p}_E|^2 - \mathbf{p}_E^*\cdot\hat{\mathbf{r}}\mathbf{p}_E\cdot\hat{\mathbf{r}}\right). \tag{4.5.9}$$

We make contact with the formula (4.4.11) for the radiating point particle by recognizing that for motion described by

$$\mathbf{r}(t) = \text{Re}\left(\mathbf{r}_0\, e^{-i\omega t}\right) \tag{4.5.10}$$

the acceleration $\mathbf{a}$ will be given by

$$\mathbf{a}(t) = -\omega^2\, \text{Re}\left(\mathbf{r}_0\, e^{-i\omega t}\right)$$

and the complex electric dipole moment corresponding to (4.5.5) by

$$\mathbf{p}_E = q\mathbf{r}_0. \tag{4.5.11}$$

These connections lead back to (4.4.11). The apparent factor of two difference comes about because $|\bar{\mathbf{a}}^2| = \tfrac{1}{2}\omega^4|\mathbf{r}_0|^2$.

Concluding, we see that for a system with small kr', radiation emission will be largely determined by the system's electric dipole moment $\mathbf{p}_E$. $\mathbf{p}_E$ is a complex vector, independent of the details of the charge distribution; for example, as we have seen, it does not distinguish between a moving point charge and an oscillating continuum charge distribution ρ.

We turn next to the second term in the expansion (4.5.1), giving for the next approximation to $\mathbf{j_k}$,

$$\mathbf{j_k}^{(2)} = -i\int d\mathbf{r}'\, \mathbf{k}\cdot\mathbf{r}'\, \mathbf{j}_0(\mathbf{r}'). \tag{4.5.12}$$

Normally, $\mathbf{j_k}^{(2)}$ will be smaller than $\mathbf{j_k}^{(1)}$ by $kr' \sim v/c$; the exception is usually when the electric dipole moment vanishes for reasons of symmetry. In atomic and nuclear physics, for example, this happens for transitions with $\Delta J > 1$, or $\Delta J = 1$ and no change of parity.

We manipulate (4.5.12) in a familiar way. We write

$$-i\mathbf{k}\cdot\mathbf{r}'\,j_{0\ell}(\mathbf{r}') = -ik_i x_i' j_\ell(\mathbf{r}') = -ik_i\left[\frac{(x_i' j_{0\ell} - x_\ell' j_{0i})}{2} + \frac{x_i' j_{0\ell} + x_\ell' j_{0i}}{2}\right]$$

so

$$\left(\mathbf{j_k}^{(2)}\right)_\ell = \int -i\mathbf{k}\cdot\mathbf{r}'\,j_{0\ell}(\mathbf{r}') = i\left[\mathbf{k}\times\int d\mathbf{r}'\,\frac{\mathbf{r}'\times\mathbf{j}_0(\mathbf{r}')}{2}\right]_\ell$$

$$+ i\frac{k_i}{2}\int d\mathbf{r}'\, x_i' x_\ell' \nabla'\cdot\mathbf{j}_0$$

or

$$\left(\mathbf{j_k^{(2)}}\right)_\ell = i\left(c\mathbf{k} \times \mathbf{p}_M\right)_\ell - \frac{\omega k_i}{2} \int d\mathbf{r}' \, x_i' x_\ell' \rho_0(\mathbf{r}'). \tag{4.5.13}$$

In (4.5.13), $\mathbf{p}_M$ is the recognizable magnetic dipole moment

$$\mathbf{p}_M = \int d\mathbf{r}' \frac{\mathbf{r}' \times \mathbf{j}_0(\mathbf{r}')}{2c}; \tag{4.5.14}$$

the integral $\frac{1}{2} \int d\mathbf{r}' \, x_i' x_\ell' \rho_0(\mathbf{r}')$ can be replaced by the traceless quadrupole tensor

$$Q_{E_{i\ell}} = \frac{1}{2} \int d\mathbf{r}' \rho_0 \left(x_i' x_\ell' - \frac{1}{3} \delta_{i\ell} r'^2 \right) \tag{4.5.15}$$

since $\delta_{i\ell}$ inserted into (4.5.13) will produce a longitudinal contribution to $\mathbf{j_k^{(2)}}$ and, hence, will not contribute to radiation.

We compare the form of magnetic dipole radiation with that of electric dipole radiation.

The electric vectors $(\mathbf{j}_T)$ are given for the two cases by

$$\mathbf{E}_E^{(1)} \propto \mathbf{j}_{ET}^{(1)} = -i\omega(\mathbf{p}_E - \mathbf{p}_E \cdot \hat{\mathbf{r}}\hat{\mathbf{r}}) \tag{4.5.16}$$

and

$$\mathbf{E}_M^{(2)} \propto \mathbf{j}_M^{(2)} = i\omega\hat{\mathbf{r}} \times \mathbf{p}_M. \tag{4.5.17}$$

The corresponding magnetic vectors are given by $\mathbf{B} = \hat{\mathbf{r}} \times \mathbf{E}$ or

$$\mathbf{B}_E^{(1)} \propto -i\omega\hat{\mathbf{r}} \times \mathbf{p}_E \tag{4.5.18}$$

and

$$\mathbf{B}_M^{(2)} \propto -i\omega(\mathbf{p}_M - \hat{\mathbf{r}}\hat{\mathbf{r}} \cdot \mathbf{p}_M). \tag{4.5.19}$$

Thus, the transformation from electric radiation to magnetic radiation is

$$\mathbf{E}_E \to \mathbf{B}_M, \qquad \mathbf{B}_E \to -\mathbf{E}_M. \tag{4.5.20}$$

Equation (4.5.20) reflects a general symmetry: Maxwell's equations for propagating electric and magnetic fields in the absence of sources are invariant under the transformation $\mathbf{E} \to \mathbf{B}$, $\mathbf{B} \to -\mathbf{E}$. Note the importance of the minus sign. Without it, the Poynting vector would go in the wrong direction!

Finally, we consider the electric quadrupole radiation (magnetic quadrupole will come in the next order of kr'):

$$j_{E\ell}^{(2)} = -\omega k_i Q_{i\ell} \tag{4.5.21}$$

and

$$E_{E\ell}^{(2)} = \frac{i\omega^2}{2c^2 r} e^{i(kr - \omega t)} \left\{ k_i Q_{i\ell} - k_i Q_{im} \frac{x_m}{r} \frac{x_\ell}{r} \right\}. \tag{4.5.22}$$

The angular distribution of power radiated will be

$$\frac{dW_{EQ}}{dt\, d\Omega} \propto \frac{x_i Q_{i\ell}^* x_j Q_{j\ell}}{r^2} - \frac{(x_i Q_{ij}^* x_j)}{r^2} \frac{(x_k Q_{km} x_m)}{r^2}. \tag{4.5.23}$$

Note that there is no general rule against interference of different radiation multipoles, although in some quantum transitions it may be forbidden. For example, in a $J = 1$ to $J = 0$ transition, ED and MD will not interfere, since ED requires a parity change and MD requires no parity change in the system. However, in $J = 2 \rightarrow J = 1$ with no parity change, MD and EQ can and generally will interfere. In contrast, in scattering problems at large kr' there is usually interference between many multipoles. We shall return to this question when we discuss general multipole radiation in Chapter 5.

4.6. FIELDS OF A POINT CHARGE MOVING AT CONSTANT HIGH VELOCITY v: EQUIVALENT PHOTONS

We write equations in the Lorentz gauge

$$\left(\nabla^2 - \frac{1}{c^2} \frac{\partial^2}{\partial t^2} \right) \mathbf{A} = -\frac{4\pi \mathbf{j}}{c} = -\frac{4\pi}{c} q \mathbf{v} \delta^3(\mathbf{x} - \mathbf{v}t) \tag{4.6.1}$$

and

$$\left(\nabla^2 - \frac{1}{c^2} \frac{\partial^2}{\partial t^2} \right) \phi = -4\pi \rho = -4\pi q \delta^3(\mathbf{r} - \mathbf{v}t). \tag{4.6.2}$$

We look for a solution $\mathbf{A}(\mathbf{r} - \mathbf{v}t)$, $\phi(\mathbf{r} - \mathbf{v}t)$. With $\mathbf{v}$ and $\mathbf{A}$ in the z-direction, these satisfy the equations

$$\left[\frac{\partial^2}{\partial x^2} + \frac{\partial^2}{\partial y^2} + \frac{\partial^2}{\partial z^2}\left(1 - \frac{v^2}{c^2}\right)\right]\binom{A_z}{\phi} = -4\pi q\,\delta^3(\mathbf{r} - \mathbf{v}t)\binom{\frac{v}{c}}{1}. \quad (4.6.3)$$

Change variables to

$$z' = \frac{z - vt}{\sqrt{1 - \dfrac{v^2}{c^2}}}.$$

Then

$$\left(\frac{\partial^2}{\partial x^2} + \frac{\partial^2}{\partial y^2} + \frac{\partial^2}{\partial z'^2}\right)\binom{A_z}{\phi} = -4\pi q v \delta^2(\boldsymbol{\rho})\,\delta\!\left(z'\sqrt{1 - \frac{v^2}{c^2}}\right)\binom{\frac{v}{c}}{1}$$

$$= -4\pi q\,\frac{\delta^3(\mathbf{r}')}{\sqrt{1 - \dfrac{v^2}{c^2}}}\binom{\frac{v}{c}}{1} \quad (4.6.4)$$

where $\boldsymbol{\rho} = \hat{\mathbf{e}}_x x + \hat{\mathbf{e}}_y y$ and $\mathbf{r}' = \boldsymbol{\rho} + \hat{\mathbf{e}}_z z'$, so that

$$r' = \left(\rho^2 + \frac{(z - vt)^2}{1 - \dfrac{v^2}{c^2}}\right)^{1/2}. \quad (4.6.5)$$

The solutions of (4.6.4) are immediately given by the Coulomb potential:

$$A_z = q\,\frac{v}{c}\,\frac{1}{\sqrt{1 - \dfrac{v^2}{c^2}}}\,\frac{1}{r'} \quad (4.6.6)$$

and

$$\phi = \frac{q}{\sqrt{1 - \dfrac{v^2}{c^2}}}\,\frac{1}{r'} \quad (4.6.7)$$

The electric field is

$$\mathbf{E} = -\frac{1}{c}\frac{\partial \mathbf{A}}{\partial t} - \nabla\phi$$

$$= \hat{\mathbf{e}}_z\left(-\frac{1}{c}\cdot\frac{qv}{c}\frac{1}{r'^3}\frac{(z-vt)v}{1-\dfrac{v^2}{c^2}} + q\frac{z-vt}{1-\dfrac{v^2}{c^2}}\frac{1}{r'^3}\right)\frac{1}{\sqrt{1-\dfrac{v^2}{c^2}}} + \frac{q\boldsymbol{\rho}}{r'^3}\frac{1}{\sqrt{1-\dfrac{v^2}{c^2}}}$$

or

$$\mathbf{E} = q\,\frac{(\mathbf{r}-\mathbf{v}t)}{r'^3\left(1-\dfrac{v^2}{c^2}\right)^{1/2}}. \tag{4.6.8}$$

The magnetic field is

$$\mathbf{B} = \nabla\times\mathbf{A} = \left(\hat{\mathbf{e}}_x\frac{\partial}{\partial x} + \hat{\mathbf{e}}_y\frac{\partial}{\partial y}\right)\times\frac{\hat{\mathbf{e}}_z vq}{c}\frac{1}{\sqrt{1-\dfrac{v^2}{c^2}}}\frac{1}{r'} = \frac{\mathbf{v}}{c}\times\mathbf{E}. \tag{4.6.9}$$

Remarkably, the electric field at $\mathbf{r}$ points from the present position of the charge (vt) to the field point, as shown by (4.6.8). Note also the first hint we have seen that $v > c$ would cause major problems.[2]

The fields $\mathbf{E}$ and $\mathbf{B}$ look remarkably like a light wave if the particle velocity v is very close to c. First, the field packet moves with a velocity very close to c. Second, it is concentrated near $z = vt$, so that E_z is small compared with $\mathbf{E}_\perp$; thus, $\mathbf{E}$ is almost transverse, $\mathbf{B}$ is exactly transverse, orthogonal to $\mathbf{E}$, and almost equal to $\mathbf{E}$ in magnitude. This circumstance can be exploited to relate a process induced by fast charged particles to the same process induced by low-frequency photons.

To do this, we calculate the radiant energy incident per unit area, time, and frequency by Fourier-transforming the electric and magnetic fields (4.6.8) and (4.6.9) as

---

[2]The fields $\mathbf{E}$ and $\mathbf{B}$ can also be calculated by Lorentz-transforming the Coulomb field of a charged particle at rest. (See Problem 6.4.)

$$E(\mathbf{r}, \omega) = \int_{-\infty}^{\infty} \frac{dt}{\sqrt{2\pi}} e^{i\omega t} E(\mathbf{r}, t) \tag{4.6.10}$$

and

$$B(\mathbf{r}, \omega) = \frac{\mathbf{v}}{c} \times E(\mathbf{r}, \omega). \tag{4.6.11}$$

Since E is effectively transverse and $v/c \cong 1$, the integrated Poynting vector will be, with $E(\mathbf{r}, \omega) = e^{i\omega(z/c)}E(\boldsymbol{\rho}, \omega)$ and $B(\mathbf{r}, \omega) = e^{i\omega(z/c)}B(\boldsymbol{\rho}, \omega)$,

$$\int dt\, \mathcal{P}(\boldsymbol{\rho}, t) \cong \frac{c}{4\pi} \int E^*(\boldsymbol{\rho}, \omega) \times B(\boldsymbol{\rho}, \omega)\, d\omega \tag{4.6.12}$$

$$\cong \frac{c}{4\pi}\, \hat{\mathbf{v}} \int |E(\boldsymbol{\rho}, \omega)|^2\, d\omega \tag{4.6.13}$$

and the number of photons per unit frequency per charged particle incident will be

$$\frac{dN}{d\omega} = \frac{2c}{4\pi} \int d^2\boldsymbol{\rho}\, \frac{|E(\boldsymbol{\rho}, \omega)|^2}{\hbar\omega}. \tag{4.6.14}$$

The factor of two comes from adding negative to positive frequencies.

How are we to interpret (4.6.14) in quantum theory? Since quantum theory predicts probabilities, the number of photons in the range $\omega_1 < \omega < \omega_2$, or $\int_{\omega_1}^{\omega_2} (dN/d\omega)\, d\omega$, must be interpreted, if small, as the probability $p(\omega_2, \omega_1)$, of finding a photon in that range with a single incident charge. That is, if N_e is the (large) number of incident charged particles, the number of photons emitted in the calculated frequency interval will be

$$N_\gamma = N_e p(\omega_2, \omega_1),$$

thus giving the same effective answer as (4.6.14).

If the calculated probability $p(\omega_2, \omega_1)$ turns out to be large, doubt is cast on the calculation; the reaction of the target system on the charged particle must be taken into account. This will be the case, for example, if we consider very low frequencies, where the factor $1/\omega$ in (4.6.14) becomes large.

We should comment here on the validity of using classical field theory to calculate effects associated with the radiation of low-frequency quanta.

We learn in quantum electrodynamics that classical field theory is valid when the quantum state contains many photons per volume λ^3 (with λ the wavelength). This is certainly not the case for the problem we are dealing with here. Quite the contrary, we consider the radiation of one photon at a time. There is, however, another regime in which the classical equations are applicable. That is a régime where we can limit ourselves to a linear approximation in the field strengths. In that case, since the quantum equations of motion are the same as the classical ones, classical solutions hold as well for the quantum field operators. Thus, here and in Section 4.8, where we discuss low-frequency bremsstrahlung, the specifically quantum properties of electromagnetic fields may be ignored.

Of course, we cannot expect this classical calculation to hold for all frequencies ω and radii ρ. The frequency must be small enough so that the energy quantum $\hbar\omega$ is negligible compared to the energy of the incident particle; the radius ρ must be larger than the wavelength of the particle, $\lambda = \hbar/p$, since otherwise one cannot give classical meaning to the location ρ. With those caveats, we go ahead and calculate

$$\mathbf{E}(\boldsymbol{\rho}, \omega) \cong \frac{1}{\sqrt{2\pi}} \int dt\, e^{i\omega t}\, \frac{q\boldsymbol{\rho}}{\left(1 - \dfrac{v^2}{c^2}\right)^{1/2}} \frac{1}{r'^3}. \tag{4.6.15}$$

The appropriate change of variables in (4.6.15) is

$$z' = \frac{vt - z}{\sqrt{1 - \dfrac{v^2}{c^2}}}, \tag{4.6.16}$$

leading, with $v/c \approx 1$, to

$$\mathbf{E}(\mathbf{r}, \omega) = \frac{q\boldsymbol{\rho}}{\sqrt{2\pi}} \frac{e^{i\omega(z/c)}}{c} \int \frac{dz'\, e^{i\omega(z'/c)} \sqrt{1 - \dfrac{v^2}{c^2}}}{(\rho^2 + z'^2)^{3/2}} \tag{4.6.17}$$

and with $z' = \rho u$,

$$\mathbf{E}(\mathbf{r}, \omega) = \frac{q}{\sqrt{2\pi}} \frac{\boldsymbol{\rho}}{c\rho^2} e^{i\omega(z/v)} \int_{-\infty}^{\infty} \frac{du}{(1 + u^2)^{3/2}} e^{i\omega(\rho/c)\sqrt{1 - \frac{v^2}{c^2}}u}. \tag{4.6.18}$$

The factor $\exp\{i\omega(\rho/c)\sqrt{1 - v^2/c^2}\,u\}$ produces a classical cut-off for ρ in (4.6.14): For $\rho(\omega/c)\sqrt{1 - v^2/c^2} > 1$, the oscillating exponential will decrease the u integral from its value at $\omega = 0$. This comes about because at a transverse location ρ, the electromagnetic pulse has a characteristic time of passage

$$\Delta t \sim \rho \frac{\sqrt{1 - \frac{v^2}{c^2}}}{c},$$

so that the characteristic classical frequency ω_{cl} in the Fourier transform will be $\omega_{cl} \sim c/\rho\sqrt{1 - v^2/c^2}$.

For $\omega < \omega_{cl}$, the approximation $\omega \sim 0$ can be made. For $\omega > \omega_{cl}$, or $\rho > c/\omega\sqrt{1 - v^2/c^2}$, the Fourier transform will fall off, as shown explicitly in (4.6.18). We thus have, for

$$\rho < \rho_{max} = \frac{c}{\omega\sqrt{1 - \frac{v^2}{c^2}}},$$

$$\mathbf{E}(z, \boldsymbol{\rho}, \omega) = \frac{q}{\sqrt{2\pi}} \frac{\boldsymbol{\rho}}{\rho^2} \frac{e^{i\omega(z/c)}}{c} \int_{-\infty}^{\infty} \frac{du}{(1 + u^2)^{3/2}}$$

$$= \frac{2q}{\sqrt{2\pi}} \frac{\boldsymbol{\rho}}{\rho^2} \frac{\epsilon^{i\omega(z/c)}}{c}. \qquad (4.6.19)$$

The equivalent photon spectrum is given by substituting (4.6.19) in (4.6.14):

$$\frac{dN}{d\omega} = \frac{2q^2}{\pi\hbar\omega c} \cdot \int_{\rho_{min}}^{\rho_{max}} \frac{d\rho}{\rho}. \qquad (4.6.20)$$

As discussed earlier,

$$\rho_{max} = \frac{c}{\omega\sqrt{1 - \frac{v^2}{c^2}}} \qquad \text{and} \qquad \rho_{min} = \frac{\hbar}{p}$$

with p the incident particle momentum. The final result is

$$\frac{dN}{d\omega} = \frac{2}{\pi}\frac{q^2}{\hbar c} \cdot \frac{1}{\omega} \log \frac{cp}{\hbar\omega \sqrt{1 - \frac{v^2}{c^2}}}.$$ (4.6.21)

Note that since v is close to c and $\hbar\omega \ll cp$ (the particle energy), the argument of the logarithm is very large, and therefore the log will be insensitive to the precise value of these cut-offs. Thus, (4.6.21) makes quantitative, not just qualitative, sense, since in addition the factor $q^2/\hbar c$ (equal to 1/137 for electrons) allows the probability $\int_{\omega_1}^{\omega_2} d\omega(dN/d\omega)$ to be small.

Equation (4.6.21) gives directly the relation of a fast charged particle induced cross section $d\sigma_p$, with energy loss $\hbar\omega$, to the photon induced cross section $d\sigma_\gamma$ at frequency ω. If the particle is an electron, it is for a range $\Delta\omega$ of frequency

$$d\sigma_e = \frac{2}{\pi}\frac{e^2}{\hbar c} \int_{\Delta\omega} \frac{d\omega}{\omega} \log \frac{cp}{\hbar\omega \sqrt{1 - \frac{v^2}{c^2}}} d\sigma_\gamma(\omega).$$ (4.6.22)

This relation was discovered in the early days of quantum theory by C. F. Weizsäcker and E. J. Williams.

4.7. A POINT CHARGE MOVING WITH ARBITRARY VELOCITY LESS THAN c: THE LIÉNARD–WIECHERT POTENTIALS

We return to the general form (4.2.17) and (4.2.18). This gives

$$\mathbf{A}(\mathbf{r}, t) = \frac{1}{c}\int \frac{d\mathbf{r}'dt'}{|\mathbf{r} - \mathbf{r}'|} \delta\left(t - t' - \frac{|\mathbf{r} - \mathbf{r}'|}{c}\right) \mathbf{j}(\mathbf{r}', t')$$ (4.7.1)

and

$$\phi(\mathbf{r}, t) = \int \frac{d\mathbf{r}'dt'}{|\mathbf{r} - \mathbf{r}'|} \delta\left(t - t' - \frac{|\mathbf{r} - \mathbf{r}'|}{c}\right) \rho(\mathbf{r}', t').$$ (4.7.2)

The current and charge density are those of a point particle:

$$\mathbf{j}(\mathbf{r}', t') = q\mathbf{v}(t')\,\delta^3(\mathbf{r}' - \mathbf{r}(t')) \tag{4.7.3}$$

and

$$\rho(\mathbf{r}', t') = q\,\delta^3(\mathbf{r}' - \mathbf{r}(t')) \tag{4.7.4}$$

where $\mathbf{r}(t')$ and $\mathbf{v}(t')$ are the coordinate and velocity vectors of the particle at time t'. We carry out the $d\mathbf{r}'$ integral first. There results

$$\mathbf{A}(\mathbf{r}, t) = \frac{q}{c}\int \frac{dt'}{|\mathbf{r} - \mathbf{r}(t')|}\,\delta\!\left(t - t' - \frac{|\mathbf{r} - \mathbf{r}(t')|}{c}\right)\mathbf{v}(t') \tag{4.7.5}$$

and

$$\phi(\mathbf{r}, t) = q\int \frac{dt'}{|\mathbf{r} - \mathbf{r}(t')|}\,\delta\!\left(t - t' - \frac{|\mathbf{r} - \mathbf{r}(t')|}{c}\right). \tag{4.7.6}$$

To carry out the dt' integration, we are first required, given a field point $\mathbf{r}$ and a time t, to find the retarded time $t' = t_R$, such that

$$t_R = t - \frac{|\mathbf{r} - \mathbf{r}(t_R)|}{c}. \tag{4.7.7}$$

Equation (4.7.7) has only one solution, provided the particle velocity is less than c. To see this, imagine again a spherical light wave aimed to converge on the point $\mathbf{r}$ at time t. It will cross every charged particle at some time t_R and only cross each particle once, since it is moving with velocity $c > v$. Clearly, the time the spherical wave crosses the particle trajectory is the retarded time for that particle. In general, one cannot solve for t_R analytically, but the argument just given shows that a numerical calculation [given $\mathbf{r}(t)$, of course] can succeed. In the special case of uniform motion, (4.7.7) leads to a quadratic equation for t_R, which can be solved algebraically. In fact, the procedure we are about to follow here could be used as an alternative way of finding the fields of a particle moving with constant velocity. Both the retarded and advanced fields of a uniformly moving charge are equal to each other and to the convective fields described in the last section.

Assuming we have found the solution of (4.7.7) for t_R, we must do the integrals in (4.7.5) and (4.7.6). To do that, we change variables to the argument of the δ function, that is,

$$\tau = t' + \frac{|\mathbf{r} - \mathbf{r}(t')|}{c}. \tag{4.7.8}$$

Then

$$d\tau = dt'\left(1 - \frac{\mathbf{r} - \mathbf{r}(t')}{c|\mathbf{r} - \mathbf{r}(t')|} \cdot \frac{d\mathbf{r}(t')}{dt'}\right)$$

or

$$\frac{dt'}{d\tau} = \frac{1}{1 - \frac{\hat{\mathbf{r}} \cdot \mathbf{v}}{c}} \tag{4.7.9}$$

where $\hat{\mathbf{r}}$ is the unit vector pointing from the retarded position[3] of the particle $\mathbf{r}(t_R)$ to the field point $\mathbf{r}$, and $\mathbf{v}$ is the velocity $d\mathbf{r}(t')/dt'$ at $t' = t_R$.

The integrals (4.7.5) and (4.7.6) can now be done using the τ variable:

$$\mathbf{A}(\mathbf{r}, t) = \frac{q}{c} \int \frac{d\tau}{|\mathbf{r} - \mathbf{r}(t_R)|} \delta(\tau) \frac{\mathbf{v}(t_R)}{1 - \frac{\mathbf{v}(t_R)}{c} \cdot \frac{\mathbf{r} - \mathbf{r}(t_R)}{|\mathbf{r} - \mathbf{r}(t_R)|}}$$

or

$$\mathbf{A} = \frac{q\mathbf{v}}{cs} \tag{4.7.10}$$

and

---

[3] A point to keep in mind for possible future reference is that the equivalent denominator for the advanced solution is

$$\frac{dt'}{d\tau_A} = \frac{1}{1 + \hat{\mathbf{r}} \cdot \frac{\mathbf{v}}{c}}$$

where $\hat{\mathbf{r}}$ and $\mathbf{v}$ are now calculated at the advanced time:

$$t_A = t + \frac{|\mathbf{r} - \mathbf{r}(t_A)|}{c}.$$

$$\phi = \frac{q}{s} \tag{4.7.11}$$

where

$$s = |\mathbf{r} - \mathbf{r}(t_R)| - \frac{(\mathbf{r} - \mathbf{r}(t_R)) \cdot \mathbf{v}(t_R)}{c}. \tag{4.7.12}$$

4.8. LOW-FREQUENCY BREMSSTRAHLUNG[4]

Before taking up the **E** and **B** fields, we consider the radiation of low-frequency photons in the course of a scattering event. As in our discussion of equivalent photons in Section 4.6, we must confine our calculations to low enough frequencies so that the quantum corrections will not be significant. That means that $\hbar\omega$ must as a matter of principle be small compared to characteristic energies of the radiationless scattering process; for example, we must have

$$\hbar\omega \ll W \tag{4.8.1}$$

where W is the incident energy of the charged projectile.

As a matter of practice, we will consider ω also smaller than the characteristic classical frequencies of the motion, for example, the classical frequency $\omega_c \sim v/b$, where v is the incident particle velocity and b the impact parameter (assuming that b is within the range of the force). This is because a calculation of the frequency dependence of the process must be specific to the system being considered; we are interested here in general results, including the case of nonclassical particle motion.

What we will do therefore is to calculate the zero frequency limit of radiation by a system that we imagine to be correctly described—either by classical or by quantum equations, whichever is called for.

We imagine a scattering event (see Figure 4.1) in which the observation of electromagnetic radiation is made at $\mathbf{r}$, between times t_1 and t_2, where t_1 and t_2 are such that t_{1R} is before the particle has entered the force field of the scatterer and t_{2R} is after the particle has left the force field of the scatterer. (This can always be done: Choose t_{1R}, $\mathbf{r}_1$ and t_{2R}, $\mathbf{r}_2$ first; then find t_1 and t_2 by clocking rays from $\mathbf{r}_1$ to $\mathbf{r}$ and from $\mathbf{r}_2$ to $\mathbf{r}$.)

We calculate the electric field as the transverse part of $-1/c \; \partial\mathbf{A}/\partial t$.

---

[4]*Bremsstrahlung* is German for braking radiation.

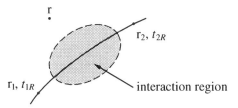

Figure 4.1.

Then

$$\mathbf{E} = -\frac{1}{c}\frac{\partial \mathbf{A}_T}{\partial t}, \tag{4.8.2}$$

and the Fourier transform of $\mathbf{E}$ is

$$\mathbf{E}(\mathbf{r}, \omega) = \frac{1}{\sqrt{2\pi}}\int_{t_1}^{t_2} dt\, e^{i\omega t}\left(-\frac{1}{c}\frac{\partial}{\partial t}\mathbf{A}_T\right). \tag{4.8.3}$$

We do not have to integrate before t_1 or after t_2 since there will be no radiation field (i.e., no $\mathbf{E} \sim 1/r$) at those times.

Given $\mathbf{E}(\mathbf{r}, \omega)$, we know how to calculate the energy radiated per unit area. It is

$$\frac{dW}{dS} = \int_{-\infty}^{\infty} \mathscr{P}\cdot \hat{r}\, d\omega \tag{4.8.4}$$

where $\mathscr{P}$ is the Poynting vector,

$$\mathscr{P}(\omega) = \frac{c}{4\pi}\mathbf{E}^*(\mathbf{r}, \omega) \times \mathbf{B}(\mathbf{r}, \omega) \tag{4.8.5}$$

$$= \frac{c}{4\pi}|\mathbf{E}^*(\mathbf{r}, \omega)|^2\,\hat{\mathbf{r}}, \tag{4.8.6}$$

and where $\hat{\mathbf{r}}$ is the unit vector pointing from the target. Remember that r is asymptotic—that is, the unit vector

$$\hat{r} = \frac{r - r(t_R)}{|r - r(t_R)|}$$

is independent of t_R during the scattering.

We consider only ω near zero, as discussed earlier. More precisely, we calculate the energy spectrum at $\omega = 0$. It is

$$\frac{dW}{dS\, d\omega} = 2\mathcal{P}(0) \cdot \hat{k} \tag{4.8.7}$$

where the factor 2 takes into account both signs of the frequency. We observe that $E(r, \omega = 0)$ can be calculated from (4.8.3). It is

$$E(r, \omega = 0) = -\frac{1}{\sqrt{2\pi}\,c}[A_T(r, t_2) - A_T(r, t_1)] \tag{4.8.8}$$

and from (4.7.10) and (4.7.12),

$$A(r, t_2) = \frac{q v_2}{rc\left(1 - \dfrac{v_2 \cdot \hat{r}}{c}\right)} \tag{4.8.9}$$

and

$$A(r, t_1) = \frac{q v_1}{rc\left(1 - \dfrac{v_1 \cdot \hat{r}}{c}\right)} \tag{4.8.10}$$

where v_1 and v_2 are the velocities of the charged particle before and after the scattering. Thus,

$$E(r, \omega = 0) = -\frac{1}{rc^2}\frac{1}{\sqrt{2\pi}}q\left(\frac{v_2}{1 - \dfrac{v_2 \cdot \hat{r}}{c}} - \frac{v_1}{1 - \dfrac{v_1 \cdot \hat{r}}{c}}\right)_T. \tag{4.8.11}$$

The total energy radiated per unit solid angle is

$$\frac{dW}{d\Omega} = \frac{c}{2\pi}\frac{1}{\Delta\Omega}\int_{\Delta\Omega} dS \cdot \hat{r}\int_0^\infty d\omega\, E^*(r, \omega) \cdot E(r, \omega) \tag{4.8.12}$$

and per unit solid angle and frequency is, for ω near zero,

$$\frac{dW}{d\Omega\,d\omega} = \frac{q^2}{(2\pi)^2 c^3}\left|\left(\frac{\mathbf{v}_2}{1 - \dfrac{\mathbf{v}_2\cdot\hat{\mathbf{r}}}{c}} - \frac{\mathbf{v}_1}{1 - \dfrac{\mathbf{v}_1\cdot\hat{\mathbf{r}}}{c}}\right)_T\right|^2.$$

The number of photons in a frequency range[5] $d\omega$ is thus

$$\frac{dN}{d\Omega\,d\omega} = \frac{dW}{d\Omega\,d\omega\,\hbar\omega} = \frac{q^2}{(2\pi)^2\hbar c\omega}\left|\left(\frac{\dfrac{\mathbf{v}_2}{c}}{1 - \dfrac{\mathbf{v}_2\cdot\hat{\mathbf{r}}}{c}} - \frac{\dfrac{\mathbf{v}_1}{c}}{1 - \dfrac{\mathbf{v}_1\cdot\hat{\mathbf{r}}}{c}}\right)_T\right|^2.$$

$$(4.8.13)$$

The meaning of (4.8.13) is the following: *If* the charged particle comes in with velocity $\mathbf{v}_1$ and is scattered with velocity $\mathbf{v}_2$, the number of photons it will radiate in the frequency range $\Delta\omega$ and solid angle $\Delta\Omega$ is

$$N_{\Delta\Omega\Delta\omega} = \int_{\Delta\Omega} d\Omega \int_{\Delta\omega} d\omega\, \frac{dN}{d\Omega\,d\omega}. \qquad (4.8.14)$$

The probability of that charged particle event is the cross section for the event times the incident number of particles per unit area. Therefore, the cross section for producing a photon in the frequency range $\Delta\omega$ and angular interval $\Delta\Omega$ is given by (4.8.13) in terms of the radiationless cross section $d\sigma/d\tau$, where $d\tau$ describes the final particle state; for example, in a scattering, $d\tau$ might be $d\Omega$ of the final particle. The relation is

$$\frac{d\sigma_\gamma}{d\Omega\,d\omega\,d\tau} = \frac{1}{(2\pi)^2}\cdot\frac{q^2}{\hbar c\omega}\frac{1}{}\left|\left(\frac{\dfrac{\mathbf{v}_2}{c}}{1 - \dfrac{\mathbf{v}_2\cdot\hat{\mathbf{r}}}{c}} - \frac{\dfrac{\mathbf{v}_1}{c}}{1 - \dfrac{\mathbf{v}_1\cdot\hat{\mathbf{r}}}{c}}\right)_T\right|^2 \frac{d\sigma}{d\tau}. \qquad (4.8.15)$$

Equation (4.8.15) can be generalized to several incoming and outgoing charged particles. The squared bracket times q^2 is simply replaced by

---

[5]Remember, as discussed in Section 4.6, that the quantum mechanical translation of "number of photons in a range" is the "probability of radiating a photon in a range."

$$\frac{d\sigma_\gamma}{d\Omega \, d\omega \, d\tau} = \frac{1}{(2\pi)^2 \hbar c \omega} \left| \left(\sum_f \frac{q_f \dfrac{\mathbf{v}_f}{c}}{1 - \dfrac{\mathbf{v}_f \cdot \hat{\mathbf{r}}}{c}} - \sum_i \frac{q_i \dfrac{\mathbf{v}_i}{c}}{1 - \dfrac{\mathbf{v}_i \cdot \hat{\mathbf{r}}}{c}} \right)_T \right|^2 \frac{d\sigma}{d\tau}$$

(4.8.16)

where f denotes final, i initial. The transverse component squared of a vector $\mathbf{j}$ is, of course,

$$(\mathbf{j}_T)^2 = (\mathbf{j} - \mathbf{j} \cdot \hat{\mathbf{r}} \hat{\mathbf{r}})^2 = \mathbf{j}^2 - (\mathbf{j} \cdot \hat{\mathbf{r}})^2. \qquad (4.8.17)$$

We see from (4.8.15) that the radiation of fast particles peaks strongly near the direction of $\mathbf{v}_1$ or $\mathbf{v}_2$. Thus, although

$$\frac{1}{v_1^2} \left| \frac{\mathbf{v}_1 - \mathbf{v}_1 \cdot \hat{\mathbf{r}} \hat{\mathbf{r}}}{1 - \dfrac{\mathbf{v}_1 \cdot \hat{r}}{c}} \right|^2 = \frac{\sin^2 \theta}{\left(1 - \dfrac{v_1}{c} \cos \theta \right)^2} \qquad (4.8.18)$$

vanishes at $\theta = 0$, it peaks strongly at $\cos \theta = v_1/c$, where it has the value

$$\frac{1}{1 - \left(\dfrac{v_1}{c} \right)^2}. \qquad (4.8.19)$$

The low-frequency radiation thus comes out mainly in two sprays near $\mathbf{v}_1$ and $\mathbf{v}_2$. Integrated over solid angle, each of these sprays gives

$$\int d\varphi \int \frac{\sin^2 \theta \, d\theta \, \sin \theta}{\left(1 - \dfrac{v_1}{c} \cos \theta \right)^2} = 2\pi \int_{-1}^{1} \frac{(1 - x^2) \, dx}{\left(1 - \dfrac{v_1 x}{c} \right)^2} \approx 4\pi \log \frac{1}{1 - \dfrac{v_1}{c}} \qquad (4.8.20)$$

for v/c close to 1.

An important consequence of the finiteness of $\mathbf{E}(\mathbf{r}, \omega = 0)$ is the divergence of the cross section for photon emission at low frequencies, since $\int d\omega/\omega$ diverges at $\omega = 0$. The high ω divergence in the integral is a consequence of our low-frequency approximation. It turns out in quantum theory that the meaning of the low-frequency divergence is that no charged particles can interact without radiating—perhaps no big surprise. That means that one cannot define an "elastic" amplitude which includes charged particles in the initial or final state, since there is always accompanying radiation. One can however define and measure a cross section for a charged particle to scatter with a finite energy resolution for the scattered particle. The cross section will be a function of the incident energy, the scattered angle, and the resolution ΔE. Because of the zero rest mass of the photon, no matter how small ΔE, any number of low-energy photons could be produced in the process. The low-frequency divergence of the bremsstrahlung cross section is a signal that as $\Delta E \to 0$, the 'elastic' cross section goes to zero. The meaning of 'elastic' is charged particle energy loss less than ΔE.

The mathematical working out of the problem[6] makes use of the fact that the total cross section for a finite resolution, including all radiation, is finite. Thus,

$$
\sigma_{\text{elastic}} + \int_0^{\Delta E} d\omega \, \frac{d\sigma_\gamma}{d\omega}
$$

should be finite. But $d\sigma_\gamma/d\omega$, as we have seen, goes like $1/\omega$, so $\int_0 (d\sigma_\gamma/d\omega)d\omega$ diverges at $\omega = 0$. Therefore, σ_{elastic} must have a canceling divergence:

$$
\sigma_{\text{elastic}} = \sigma_0 \left(1 - \int_0^\infty f(\omega) \, d\omega \right),
$$

where σ_0 is the lowest-order calculation (in $q^2/\hbar c$) and $\sigma_0 f(\omega) \to d\sigma_\gamma/d\omega$ as $\omega \to 0$. This has the embarrassing problem of producing a negative elastic cross section. The remedy is found in quantum electrodynamics, where it is shown that an exact calculation would replace

$$
1 - \int f(\omega) \, d\omega \qquad \text{by} \qquad \exp\left(- \int f(\omega) \, d\omega \right) = 0,
$$

since the integral is positive and divergent. This is the way the elastic cross section is made to vanish. The measured cross section with a resolution ΔE will be, in lowest order,

---

[6]This paragraph is impressionistic and must be read with that in mind. The formulas given are not mathematics. However, the final result is correct and important for experiments with charged particles.

$$\sigma_{\text{measured}} \cong \sigma_{\text{elastic}} + \int_0^{\Delta E} d\omega \, \sigma_{d\sigma_\gamma/d\omega}.$$

or

$$\sigma_{\text{measured}} \cong \sigma_0 \left(1 - \int_0^\infty f(\omega) \, d\omega \right) + \sigma_0 \int_0^{\Delta E} f(\omega) \, d\omega$$

$$\cong \sigma_0 \left(1 - \int_{\Delta E}^\infty f(\omega) \, d\omega \right)$$

which depends logarithmically on ΔE, but is always finite. The exact formula for small ΔE will be

$$\sigma_0 \exp\left(- \int_{\Delta E}^\infty f(\omega) \, d\omega \right).$$

The appearance in (4.8.15) of $q^2/\hbar c \sim 1/137$ keeps the correction from being large except for *very* small ΔE's.

4.9. LIÉNARD–WIECHERT FIELDS

We now use (4.7.10–4.7.12) for $\mathbf{A}$ and ϕ to calculate the $\mathbf{E}$ and $\mathbf{B}$ fields of a charged particle moving with arbitrary velocity $\mathbf{v}$ ($|\mathbf{v}| < 1$).[7] To do so, we must be able to calculate space and time derivatives of t_R. First, $\partial t_R/\partial t$. Since $t_R = \mathrm{t} - |\mathbf{x} - \mathbf{y}(t_R)|$,

$$\frac{\partial t_R}{\partial t} = 1 + \hat{\mathbf{r}} \cdot \mathbf{v} \frac{\partial t_R}{\partial t}$$

---

[7]From now on we use $\mathbf{x}$ to designate the field point, $\mathbf{y}$ the particle coordinate, and $\mathbf{r}$ to designate $\mathbf{x} - \mathbf{y}(t_R)$; thus, $|\mathbf{r}| = |\mathbf{x} - y(t_R)|$, $\hat{\mathbf{r}} = (\mathbf{x} - \mathbf{y}(t_R))/r$, etc. The velocity and acceleration of the radiating particle are always taken at the retarded time. In addition, we choose units in which $c = 1$. This saves a lot of writing and prevents a lot of trivial errors. The final answer to any problem can always be expressed in conventional units by dimensional analysis.

so that

$$\frac{\partial t_R}{\partial t} = \frac{1}{1 - \hat{\mathbf{r}} \cdot \mathbf{v}} = \frac{r}{s} \tag{4.9.1}$$

where all symbols stand for the retarded values. Next, $\nabla t_R = -\hat{\mathbf{r}} + \hat{\mathbf{r}} \cdot \mathbf{v} \nabla t_R$, so

$$\nabla t_R = -\frac{\hat{\mathbf{r}}}{1 - \hat{\mathbf{r}} \cdot \mathbf{v}} = -\frac{\mathbf{r}}{s}. \tag{4.9.2}$$

We can now proceed to **E** and **B**. From (4.7.10) and (4.7.11),

$$-\nabla \phi = \frac{q}{s^2} \nabla s = \frac{q}{s^2}\left[\hat{\mathbf{r}} - \mathbf{v} + \frac{\partial s}{\partial t_R} \nabla t_R\right], \tag{4.9.3}$$

and

$$-\frac{\partial \mathbf{A}}{\partial t} = -q\left(\frac{\mathbf{a}}{s} - \frac{\mathbf{v}}{s^2}\frac{\partial s}{\partial t_R}\right)\frac{\partial t_R}{\partial t}, \tag{4.9.4}$$

where **a** is $d\mathbf{v}(t_R)/dt_R$. Thus,

$$\mathbf{E} = -\nabla \phi - \frac{\partial \mathbf{A}}{\partial t}$$

$$= \frac{q}{s^2}\left(\hat{\mathbf{r}} - \mathbf{v} + \frac{\partial s}{\partial t_R}\left(-\frac{\mathbf{r}}{s}\right)\right) - q\left(\frac{\mathbf{a}}{s} - \frac{\mathbf{v}}{s^2}\frac{\partial s}{\partial t_R}\right)\frac{r}{s} \tag{4.9.5}$$

which, with

$$\frac{\partial s}{\partial t_R} = -\hat{\mathbf{r}} \cdot \mathbf{v} + v^2 - \mathbf{r} \cdot \mathbf{a}, \tag{4.9.6}$$

gives

$$s^3 \frac{\mathbf{E}}{q} = (\mathbf{r} - r\mathbf{v})\left(1 - v^2\right) + \mathbf{r} \times [(\mathbf{r} - r\mathbf{v}) \times \mathbf{a}]. \tag{4.9.7}$$

Notice that for large r the first contribution goes like $1/r^2$; the second goes like $1/r$ and is transverse.

Turning to the magnetic field, we obtain

$$\frac{\mathbf{B}}{q} = \nabla \times \frac{\mathbf{A}}{q} = \nabla \times \left(\frac{\mathbf{v}(t_R)}{s} \right), \tag{4.9.8}$$

$$= \frac{\nabla t_R \times \mathbf{a}}{s} - \frac{\nabla s}{s^2} \times \mathbf{v}; \tag{4.9.9}$$

with

$$\nabla t_R = -\frac{\mathbf{r}}{s}, \qquad \frac{\partial t_R}{\partial t} = \frac{r}{s}$$

and

$$\nabla s = \hat{\mathbf{r}} - \mathbf{v} + \frac{\partial s}{\partial t_R} \nabla t_R$$

we find, after some algebra,

$$\mathbf{B} = \hat{\mathbf{r}} \times \mathbf{E}. \tag{4.9.10}$$

An aid to memory in (4.9.7) is to define a "virtual present radius," $\mathbf{r}_v = \mathbf{r} - \mathbf{v}r$, that appears twice in (4.9.7). Thus,

$$\frac{s^3 \mathbf{E}}{q} = \mathbf{r}_v(1 - v^2) + \mathbf{r} \times (\mathbf{r}_v \times \mathbf{a}) \tag{4.9.11}$$

and $\mathbf{B}$ is still $\hat{\mathbf{r}} \times \mathbf{E}$.

We call $\mathbf{r}_v$ the virtual present radius because it is the value $\mathbf{r}$ would have at time t if the radiating particle kept on the course it was following at time t_R for the time $t - t_R = r$. This should be clear from Figure 4.2.

The radiation fields are thus given by

$$\mathbf{E}_r = \frac{q}{s^3} \mathbf{r} \times [\mathbf{r}_v \times \mathbf{a}] \tag{4.9.12}$$

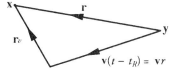

Figure 4.2.

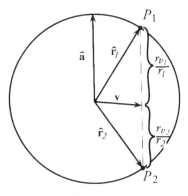

Figure 4.3.

and

$$\mathbf{B}_r = \hat{\mathbf{r}} \times \mathbf{E}_r. \tag{4.9.13}$$

Of course, $\mathbf{E}_r$ and $\mathbf{B}_r$ are transverse to $\hat{\mathbf{r}}$ and orthogonal to each other.

For small velocities, (4.9.12) and (4.9.13) reduce to our previous results for electric dipole radiation, (4.4.6–4.4.8).

For $v \sim 1$, one sees that the factor $1/(1 - \mathbf{v} \cdot \hat{\mathbf{r}})^3$ peaks the radiation sharply in the direction of $\mathbf{v}$, even though the amplitude vanishes quite close to $\theta = 0$. It follows from (4.9.12) that $\mathbf{E}_r$ and $\mathbf{B}_r$ vanish whenever $\mathbf{a}$ and $\mathbf{r}_v$ are parallel. That this always happens for two values of $\hat{\mathbf{r}}$ can be seen geometrically, as shown in Figure 4.3.

The two segments of the dotted line give two positions of $\hat{\mathbf{r}}$ such that $\mathbf{r}_v$ is parallel to $\mathbf{a}$ and, hence, for which $\mathbf{E}$ and $\mathbf{B}$ vanish. There is no other direction in which $\mathbf{E}$ vanishes. Note that as $v \to 1$, the intersections P_1, P_2 and the vectors $\hat{\mathbf{r}}_1$ and $\hat{\mathbf{r}}_2$ move to the direction of $\mathbf{v}$.

The fact that $\mathbf{E}$ vanishes near $\theta = 0$ does not prevent the radiation from peaking forward (as we have already shown in our discussion of bremsstrahlung in Section 4.8.) We will discuss this for the simple case of $\mathbf{a}$ parallel to $\mathbf{v}$, so that

$$\mathbf{E} = \frac{q}{s^3} \mathbf{r} \times (\mathbf{r} \times \mathbf{a}), \tag{4.9.14}$$

very much like the low v electric dipole radiation, but with the factor $(1 - v \cos \theta)^3$ in the denominator.

The intensity of radiation crossing the distant sphere is

$$\frac{dW}{dt\,d\Omega} = r^2 \mathcal{P} \cdot \hat{\mathbf{r}} = \frac{1}{4\pi}\,\mathbf{a}^2 \frac{q^2}{s^6}\,r^6 \sin^2\theta \tag{4.9.15}$$

which vanishes exactly for $\theta = 0$ and π, but whose angular dependence for small θ and $v \sim 1$ is

$$\frac{dW}{dt\,d\Omega} \propto \frac{\theta^2}{\left[1 - v + \dfrac{\theta^2}{2}\right]^6}, \tag{4.9.16}$$

producing a sharp maximum at $\theta^2 = 2(1 - v)/5$.

The calculation of the total radiated energy is elementary, but complicated. The case of $\mathbf{a}$ parallel to $\mathbf{v}$ (4.9.15), however, is quite simple.

We choose to calculate the rate of radiation by the particle, that is, $dW/dt_R\,d\Omega$, rather than the rate of radiation through the distant sphere, $dW/dt\,d\Omega$. These two rates are different, since $\partial t_R/\partial t = r/s$. Of course, integrated over time, they are equal, since

$$\int \frac{dW}{dt}\,dt = \int \frac{dW}{dt_R}\,dt_R; \tag{4.9.17}$$

However, the rate of energy loss by the particle in its trajectory is generally the more interesting question. We calculate, for $\mathbf{a}$ parallel to $\mathbf{v}$,

$$\int d\Omega\, \frac{dW_\parallel}{dt_R\,d\Omega} = \int d\Omega\, \frac{dW_\parallel}{dt\,d\Omega}\, \frac{\partial t}{\partial t_R}$$

$$= \frac{q^2 \mathbf{a}^2}{4\pi} \int \frac{d\Omega \sin^2\theta}{(1 - v\cos\theta)^5} \tag{4.9.18}$$

$$= \frac{2}{3}\frac{\mathbf{a}^2 q^2}{(1 - v^2)^3}. \tag{4.9.19}$$

It is easy to see from (4.9.12) that the parallel and perpendicular components of $\mathbf{a}$ do not interfere in the total energy radiation rate after integration over the azimuthal angle φ. The total radiation rate from the perpendicular component of $\mathbf{a}$ is not so simple an integral as (4.9.18). We give the result:

$$\int d\Omega\, \frac{dW_\perp}{dt_R\,d\Omega} = \frac{2}{3}\frac{(\mathbf{a}_\perp)^2 q^2}{(1 - v^2)^2}. \tag{4.9.20}$$

We shall see later (see Problem 6.3) that (4.9.19) and (4.9.20) are simple consequences of the relativistic transformation properties of acceleration.

We consider qualitatively one more topic in this section: radiation by a fast particle in a circular orbit, as in a cyclotron. Strictly, the spectrum is a line spectrum at the fundamental cyclotron frequency ω_0 plus overtones $n\omega_0$. Obviously, very high overtones will dominate, since the pulse of forward radiation sweeps rapidly by the observer. We reason as follows: As the radiation sweeps by the observer, it has an angular width [as we have seen in (4.9.16)] of order $\Delta\theta \sim \sqrt{1 - v}$. It sweeps by in time

$$\Delta t \sim \frac{\Delta\theta}{\dfrac{d\theta}{dt}}. \tag{4.9.21}$$

However, it is not $d\theta/dt$ but $d\theta/dt_R$ that is controlled at the accelerator. Since $d\theta/dt_R = \omega_0$,

$$\Delta t \cong \frac{\Delta\theta}{\omega_0 \dfrac{\partial t_R}{\partial t}} \cong \frac{\sqrt{1 - v}}{\omega_0}(1 - v\cos\theta),$$

and since $\theta \sim 1$, the observed frequency will be predominantly in the range

$$\omega \sim \frac{\omega_0}{(1 - v)^{3/2}} \sim \frac{\omega_0}{(1 - v^2)^{3/2}}. \tag{4.9.22}$$

4.10. CERENKOV RADIATION

A charged particle moving at constant velocity v in a medium in which c_M, the phase velocity of light, is smaller than v radiates energy.

That something peculiar happens under these circumstances can be seen from the Liénard–Wiechart potentials, where the denominator $1 - \hat{\mathbf{r}} \cdot \mathbf{v}/c_M$ is zero at an angle $\cos\theta_c = c_M/v < 1$, and the corresponding potentials become infinite. Of course, the singularity is not really there; it appears as a consequence of assuming a dielectric constant that is independent of frequency, so that there is no high-frequency cut-off. In practice, as $\omega \to \infty$, the dielectric constant $\epsilon \to 1$ and $c_M \to c > v$, leaving the total radiation finite. This makes it clear that we must consider frequency-dependent dielectric constants $\epsilon(\omega)$. We take μ, the magnetic permeability, equal to 1.

We consider a given frequency ω. Maxwell's equations for the ωth components are[8]

$$\nabla \times \mathbf{B} = -i\omega \frac{\mathbf{D}}{c} + 4\pi \frac{\mathbf{j}}{c}, \qquad \mathbf{D} = \epsilon \mathbf{E} \qquad (4.10.1)$$

and

$$\nabla \times \mathbf{E} = \frac{i\omega}{c} \mathbf{B} \qquad (4.10.2)$$

which for $\omega \neq 0$ impose the constraint equations

$$\nabla \cdot \mathbf{D} = 4\pi\rho \qquad (4.10.3)$$

and

$$\nabla \cdot \mathbf{B} = 0. \qquad (4.10.4)$$

We introduce the potentials as usual:

$$\mathbf{B} = \nabla \times \mathbf{A} \qquad (4.10.5)$$

and

$$\mathbf{E} = \frac{i\omega \mathbf{A}}{c} - \nabla\phi \qquad (4.10.6)$$

that, with (4.10.1) and (4.10.3), yield, for spatially constant ϵ,

$$\nabla^2 \mathbf{A} + \frac{\omega^2}{c^2} \epsilon \mathbf{A} = -\frac{4\pi \mathbf{j}}{c} - \nabla\left(\frac{\omega \epsilon i}{c} \phi - \nabla \cdot \mathbf{A}\right) \qquad (4.10.7)$$

and

$$\nabla^2 \phi + \frac{\omega^2}{c^2} \epsilon \phi = -\frac{4\pi\rho}{\epsilon} - \frac{i\omega}{c}\left(\frac{\omega \epsilon i}{c} \phi - \nabla \cdot \mathbf{A}\right). \qquad (4.10.8)$$

The Lorentz gauge here is evidently achieved by setting the terms in parentheses in (4.10.7) and (4.10.8) equal to zero.

---

[8]For obvious reasons, we reinstate c in our equations.

The charge and current densities are given by

$$\rho_\omega(\mathbf{r}) = q \int_{-\infty}^{\infty} dt\, e^{i\omega t} \delta(z - \upsilon t)\, \delta^2(\boldsymbol{\rho}) \qquad (4.10.9)$$

and

$$\mathbf{j}_\omega(\mathbf{r}) = q\hat{\mathbf{e}}_z \upsilon \int_{-\infty}^{\infty} dt\, e^{i\omega t} \delta(z - \upsilon t)\, \delta^2(\boldsymbol{\rho}) \qquad (4.10.10)$$

where $\boldsymbol{\rho} = \hat{\mathbf{e}}_y y + \hat{\mathbf{e}}_x x$ and q is the charge of the (point) particle. We have deliberately omitted the conventional factors $1/\sqrt{2\pi}$ from (4.10.9) and (4.10.10) to save writing. They are reinserted in (4.10.34). Please note that ρ (the charge density) and $\boldsymbol{\rho}$ (the radius in the x, y-plane) are totally disconnected entities. (We drop the ω subscript from now on.)

Thus,

$$\rho = \frac{q}{\upsilon} \exp\!\left(\frac{i\omega z}{\upsilon}\right) \delta^2(\boldsymbol{\rho}) \qquad (4.10.11)$$

and

$$\mathbf{j} = \hat{\mathbf{e}}_z q \exp\!\left(\frac{i\omega z}{\upsilon}\right) \delta^2(\boldsymbol{\rho}). \qquad (4.10.12)$$

The vector and scalar potentials satisfy the equations

$$(\nabla^2 + k^2)\mathbf{A} = -\frac{4\pi q}{c} \hat{\mathbf{e}}_z \delta^2(\boldsymbol{\rho}) \exp\!\left(\frac{i\omega z}{\upsilon}\right) \qquad (4.10.13)$$

and

$$(\nabla^2 + k^2)\phi = -\frac{4\pi q}{\epsilon\upsilon} \delta^2(\boldsymbol{\rho}) \exp\!\left(\frac{i\omega z}{\upsilon}\right) \qquad (4.10.14)$$

with

$$k^2 = \frac{\epsilon\omega^2}{c^2} = \frac{\omega^2}{c_M^2}. \qquad (4.10.15)$$

The solution of the equation

$$(\nabla^2 + k^2)\phi = -4\pi\rho \qquad (4.10.16)$$

corresponding to outgoing waves we have already seen by Fourier transformation of the retarded Green's function. It is

$$\phi = \int d\mathbf{r}' \frac{e^{ik|\mathbf{r}-\mathbf{r}'|}}{|\mathbf{r}-\mathbf{r}'|} \rho(\mathbf{r}'). \tag{4.10.17}$$

Solving (4.10.13) and (4.10.14) via (4.10.17), we find

$$\mathbf{A} = \frac{q}{c}\hat{\mathbf{e}}_z \int \frac{e^{ik|\mathbf{r}-\mathbf{r}'|}}{|\mathbf{r}-\mathbf{r}'|} \delta^2(\boldsymbol{\rho}') \exp\left(\frac{i\omega z'}{v}\right) d\mathbf{r}' \tag{4.10.18}$$

and

$$\phi = \frac{q}{v\epsilon} \int \frac{e^{ik|\mathbf{r}-\mathbf{r}'|}}{|\mathbf{r}-\mathbf{r}'|} \delta^2(\boldsymbol{\rho}') \exp\left(\frac{i\omega z'}{v}\right) d\mathbf{r}'. \tag{4.10.19}$$

We carry out the trivial $\boldsymbol{\rho}'$ integral and change variables to $z' - z = \rho u$. There results

$$\mathbf{A} = \frac{q\hat{\mathbf{e}}_z}{c} \exp\left(\frac{i\omega z}{v}\right) I \tag{4.10.20}$$

and

$$\phi = \frac{q}{v\epsilon} \exp\left(\frac{i\omega z}{v}\right) I \tag{4.10.21}$$

where

$$I = \int_{-\infty}^{\infty} du \, \frac{\exp\left[ik\rho\left[(1+u^2)^{1/2} + \frac{c_M}{v}u\right]\right]}{(1+u^2)^{1/2}}. \tag{4.10.22}$$

Since we are looking for radiation, we go to large ρ and approximate I by the method of stationary phase. That is, we look for the value of u, u_0, for which the phase of the exponential is stationary. If there is no such point, then the integral goes like $1/\rho$ for large ρ, whereas with the cylindrical geometry, fields must go like $1/\rho^{1/2}$ to radiate. The stationary point is given by

$$\frac{\partial}{\partial u}\left[(1+u^2)^{1/2} + \frac{c_M}{v}u\right]\Big|_{u_0} = 0 \tag{4.10.23}$$

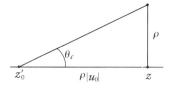

Figure 4.4.

or

$$\frac{c_M}{v} = -\frac{u_0}{(1 + u_0^2)^{1/2}}.$$ (4.10.24)

Thus, to have a stationary point, we must have $v > c_M$ and

$$u_0 = -\frac{c_M}{(v^2 - c_M^2)^{1/2}}.$$ (4.10.25)

Evidently, u_0 corresponds to the Cerenkov cone. Recall that $z_0' - z = \rho u_0$, where z_0' is the point from which radiation emerges to arrive at ρ, z as shown in Figure 4.4.

Since

$$\tan \theta_c = -\frac{1}{u_0},$$

$$\cos \theta_c = \frac{1}{\left(1 + \dfrac{1}{u_0^2}\right)^{1/2}} = \frac{|u_0|}{\sqrt{1 + u_0^2}} = \frac{c_M}{v}$$

as expected.

We expand the phase about u_0. Setting $u = u_0 + s$, we find for the integral

$$I = \int \frac{\exp\left(ik\rho\left[\dfrac{(v^2 - c_M^2)^{1/2}}{v} + \dfrac{s^2}{2}\dfrac{(v^2 - c_M^2)^{3/2}}{v^3} + O(s^3) + \cdots\right]\right)}{[1 + u_0^2 + 2u_0 s + \cdots]^{1/2}} \, ds.$$

(4.10.26)

In the form (4.10.26), it is clear that only values of s of order of or less than $1/\sqrt{\rho}$ make significant contributions to I. Hence, we can drop the extra terms in the denominator and the ρs^3 terms in the exponent; these give corrections of order $1/\sqrt{\rho}$. The final answer is then

$$I = \exp\left(ik\rho\frac{(v^2 - c_M^2)^{1/2}}{v}\right)\frac{(v^2 - c_M^2)^{1/2}}{v}\int ds \exp\left(i\frac{k\rho s^2}{2}\frac{(v^2 - c_M^2)^{3/2}}{v^3}\right),$$

$$(4.10.27)$$

or

$$I = \frac{(v^2 - c_M^2)^{1/2}}{v}(1 + i)\left[\frac{\pi}{k\rho}\frac{v^3}{(v^2 - c_M^2)^{3/2}}\right]^{1/2}\exp\left(ik\rho\frac{(v^2 - c_M^2)^{1/2}}{v}\right),$$

$$(4.10.28)$$

which is accurate to order $1/\sqrt{\rho}$.

The final expressions for the potentials are

$$\mathbf{A} = \frac{q\hat{\mathbf{e}}_z}{c}\exp\left(i\frac{\omega z}{v}\right)\exp\left(ik\rho\frac{(v^2 - c_M^2)^{1/2}}{v}\right)I_0 \qquad (4.10.29)$$

and

$$\phi = \frac{q}{v\epsilon}\exp\left(i\frac{\omega z}{v}\right)\exp\left(ik\rho\frac{(v^2 - c_M^2)^{1/2}}{v}\right)I_0 \qquad (4.10.30)$$

with

$$I_0 = (1 + i)\left[\frac{\pi}{k\rho}\frac{v}{(v^2 - c_M^2)^{1/2}}\right]^{1/2}. \qquad (4.10.31)$$

The fields are given in the $\rho \to \infty$ limit by

$$\mathbf{B} = \nabla \times \mathbf{A} = i\frac{q}{c}k\frac{(v^2 - c_M^2)^{1/2}}{v}I_0 e^{i\psi}\hat{\rho} \times \hat{\mathbf{e}}_z$$

and

$$\mathbf{E} = i\frac{\omega}{c}\mathbf{A} - \nabla\phi$$

or

$$\mathbf{E} = \left(\frac{i\omega q}{c^2}\hat{\mathbf{e}}_z - \frac{i\omega}{v}\frac{q}{v\epsilon}\hat{\mathbf{e}}_z\right)I_0 e^{i\psi} \qquad (4.10.32)$$

$$= \frac{i\omega q}{c^2 v^2}\hat{\mathbf{e}}_z(v^2 - c_M^2)I_0 e^{i\psi}, \qquad (4.10.33)$$

where

$$\psi = \frac{\omega z}{v} + k\rho \frac{(v^2 - c_M^2)^{1/2}}{v}.$$

The Poynting vector, integrated over time, is with our normalization of Fourier components[9]

$$\mathscr{P} = \frac{c}{4\pi} \int \mathbf{E}^* \times \mathbf{B} \frac{dw}{2\pi}. \tag{4.10.34}$$

The energy flux per unit length through a cylinder at radius ρ is then independent of ρ and equal to

$$\frac{dW}{dz} = \frac{q^2}{2} \int_{-\infty}^{\infty} d\omega |\omega| \left(1 - \frac{c_M^2}{v^2}\right) \frac{1}{c^2}, \tag{4.10.35}$$

where the integration over ω is limited to values of ω for which $c_M^2/v^2 < 1$. The absolute value $|\omega|$ comes about because the expression for $|I_0|^2$ has a term $1/k$ that must be interpreted as $|k|$.

Integrating over positive frequencies only, we may drop the factor of 1/2. The number of photons radiated per unit frequency and length is obtained by dividing by $\hbar|w|$:

$$\frac{dN}{dz \, d\omega} = \frac{q^2}{\hbar c} \left(1 - \frac{c_M^2}{v^2}\right) \cdot \frac{1}{c}, \qquad v > c_m$$

$$= 0, \qquad v < c_m. \tag{4.10.36}$$

CHAPTER 4 PROBLEMS

4.1. (a) Show that a function ψ that satisfies $\nabla^2 \psi = 0$ in a region can have no maximum or minimum in the region.

(b) From this, show that a finite function that satisfies $\nabla^2 \psi = 0$ everywhere and approaches zero as $r \to \infty$ is zero everywhere.

---

[9]We carry out the calculation inside a dielectric cylinder. Since the tangential components of $\mathbf{E}$ and $\mathbf{B}$ are continuous at the dielectric boundary, the $\mathbf{E} \times \mathbf{B}$ flux through any cylinder correctly calculates the radiated energy. Remember that $\mu = 1$.

(c) From this, show that a vector field whose divergence and curl both vanish, and which approaches zero at ∞, is zero.

(d) From this, show that a vector field that vanishes sufficiently rapidly and smoothly as $r \to \infty$ can be written as the sum of a longitudinal field (with zero curl) and a transverse field with zero divergence: $\mathbf{V} = \mathbf{V}_{\ell} + \mathbf{V}_t$.

(e) For such a function, give a general integral formula for $\mathbf{V}_{\ell}$ and $\mathbf{V}_t$ as functions of $\nabla \cdot \mathbf{V}$ and $\nabla \times \mathbf{V}$, and show the limiting behavior of $\mathbf{V}_{\ell}$ and $\mathbf{V}_t$ as $r \to \infty$. Give sufficient conditions on the large r behavior for your results to hold.

4.2. Verify explicitly that the $\mathbf{E}$ and $\mathbf{B}$ fields calculated from $\mathbf{A}$ and ϕ in the transverse gauge are equal (for all $\mathbf{r}$) to those obtained from $\mathbf{A}$ and ϕ in the Lorentz gauge.

4.3. Write an integral formula, analogous to (4.3.10–4.3.13) for the $1/r^2$ correction to $\mathbf{A}$ and ϕ at large r. If the characteristic radius of the charge and current distribution is b, estimate the order of magnitude of the correction compared to the $1/r$ term.

4.4. A unit point charge oscillates in one dimension with amplitude b and frequency ω_0:

$$x = b \cos \omega_0 t.$$

The charge density is a periodic function of time:

$$\rho(x, t) = \delta(x - b \cos \omega_0 t).$$

Expand $\rho(x, t)$ in a Fourier series:

$$\rho(x, t) = \sum_{n=\infty}^{\infty} a_n e^{in\omega_0 t}$$

and find a general formula for a_n. Check your algebra by calculating the monopole, dipole, and quadrapole amplitudes:

$$M = \int_{-b}^{b} dx \, \delta(x - b \cos \omega_0 t) = 1,$$

$$D = \int_{-b}^{b} x \, \delta(x - b \cos \omega_0 t) = b \cos \omega_0 t,$$

$$Q = \int_{-b}^{b} x^2 \, \delta(x - b \cos \omega_0 t) = b^2 \cos^2 \omega_0 t.$$

4.5. A point charge q oscillates along the z-axis: $z = b \cos \omega_0 t$, $y = 0$,

$x = 0$. Consider radiation in the direction θ, φ (the usual spherical coordinates). Assuming $\omega_0 b/c \ll 1$, give the angular distribution of the radiated power:

(a) At frequency ω_0.

(b) At frequency $2\omega_0$.

4.6. A small magnetic dipole rotates in the x–y plane following the formula

$$\mathbf{M} = \hat{\mathbf{e}}_x \cos \omega_0 t + \hat{\mathbf{e}}_y \sin \omega_0 t.$$

Give the electric field radiated in the direction $\mathbf{k}$, or at angle θ, φ.

(a) Give the polarization state of the electric field for $k_x = k_y = 0$, $k_z \neq 0$.

(b) Do the same for $k_z = 0$, k_x, $k_y \neq 0$.

(c) Determine the angular distribution of power radiated.

4.7. Calculate the retarded potentials of a point charge moving with uniform velocity $\mathbf{v}$ and show that the result is the same as obtained in (4.6.6) and (4.6.7).

4.8. The rate of energy radiation by a slowly moving charged particle is given by

$$\frac{dW}{dt} = \frac{2}{3} \frac{q^2}{c^3} \left(\frac{d\mathbf{v}}{dt} \right)^2.$$

This energy must show up as a loss of energy by the radiating particle. Show that a radiation reaction force

$$\mathbf{f}_r = \frac{2}{3} \frac{q^2}{c^3} \frac{d^2\mathbf{v}}{dt^2}$$

inserted in the equation of motion of a confined particle will account *on the average* for energy loss by the particle, as long as the velocity and acceleration of the particle are bounded. Show however that f_r inserted into the free particle equation of motion has unacceptable solutions. These are discussed in Section 5.9.

4.9. Show directly that

$$\nabla^2 \frac{1}{r} f(r) = -4\pi \delta^3(\mathbf{r}) f(r) + \frac{1}{r} \frac{d^2 f}{dr^2}$$

and hence that

$$\left(\nabla^2 - \frac{1}{c^2}\frac{d^2}{dt^2}\right)\frac{\delta\left(t - \frac{r}{c}\right)}{r} = -4\pi\delta^3(\mathbf{r})f(t).$$

4.10. Consider the electromagnetic field in a vacuum inside of a perfectly conducting cavity or wave guide. With the vector potential in the transverse gauge, find the boundary conditions on the vector potential at the conducting wall. From this, find the normal modes of a rectangular perfectly conducting cavity with sides a, b, and c.

4.11. A charge distribution oscillates according to the formula

$$\rho(\mathbf{r}, t) = \left(\frac{3x^2}{2} - \frac{r^2}{2}\right)F(r)\cos\omega t,$$

where $F(r) \to 0$ rapidly as $r \to \infty$. Give the angular distribution of the emitted radiation to lowest nonvanishing order in $\omega b/c$, where b is the length scale of the charge distribution.

4.12. A point electron of charge e moves in a given path $\mathbf{r}_p(t) = \hat{\mathbf{e}}_x a \cos\omega t + \hat{\mathbf{e}}_y b \sin\omega t(\hat{\mathbf{e}}_x, \hat{\mathbf{e}}_y$ are orthogonal unit vectors).

(a) Write formulas for the charge and current densities $\rho(\mathbf{r}', t)$ and $\mathbf{j}(\mathbf{r}', t)$.

(b) Write an exact integral formula for $\mathbf{j}_n(\mathbf{r}')$, where

$$\mathbf{j}(\mathbf{r}', t) = \sum_{n=-\infty}^{\infty} \mathbf{j}_n(\mathbf{r}') e^{-in\omega t}.$$

(c) Each current $\mathbf{j}_n$ now radiates a frequency $\omega_n = n\omega$, with a corresponding wave number $\mathbf{k}_n = n(w/c)\hat{\mathbf{r}}$. The relevant amplitude $\mathbf{j}_{\mathbf{k}_n}$ will be

$$\mathbf{j}_{\mathbf{k}_n} = \int e^{-i\mathbf{k}_n\cdot\mathbf{r}'}\mathbf{j}_n(\mathbf{r}')\,d\mathbf{r}'.$$

Evaluate the $\mathbf{r}'$ integral to obtain $\mathbf{j}_{\mathbf{k}_n}$ expressed as a time integral over one period of the motion.

(d) Do the final t integral for $n = 0$, 1, and 2, in each case to lowest nonvanishing order in ka and kb, where $k = \omega/c$.
From the $n = 1$ electric dipole vector potential,

$$\mathbf{A}_1 = \frac{e^{i(kr - \omega t)}}{cr}\mathbf{j}_{\mathbf{k}_1},$$

calculate:

(e) The electric field (in terms of ω, a, b, $\mathbf{k}$, etc.).

(f) The same for the magnetic field.

(g) The Poynting vector averaged over a cycle.

(h) The polarization of the radiated light is normally elliptic. Are there one or more directions of observation $\hat{k}$ for which it is plane-polarized? If so, what are they? For which is it circularly polarized? If so, what are they?

4.13. Two electrons, each with charge e, move oppositely along the x-axis with simple harmonic motion $x_1 = a \cos \omega t$, $x_2 = -a \cos \omega t$. Suppose $\omega a/c \ll 1$. Calculate to lowest order in $\omega a/c$ the radiated electric and magnetic fields and the angular distribution of radiated power.

CHAPTER 5

Scattering

Almost all physics experiments can be described as scattering processes: We start with initial objects (fields or particles) approaching each other; we end with final objects separating. Among the processes we took up in Chapter 4 several can be thus described: For example, in bremsstrahlung, an initial charged particle approaches a target; a final charged particle emerges, accompanied by a radiated electromagnetic field. In Cerenkov radiation the target is the dielectric. In most of the other topics, the connection to scattering is less obvious, but it is still present. Therefore, scattering is important in physics, and it makes sense to treat it as a separate topic. This is true although no new principles are involved. Indeed, the reader may omit this entire chapter without experiencing any consequent difficulty in understanding the rest of the text.

The author's recommendation to the interested, but not devoted, reader is to compromise by omitting Sections 10–12. In the first six sections, we study the general theory of scattering, illustrated by the case of a scalar field (or in quantum theory a spin zero particle). Included is a discussion of partial wave amplitudes that decouple when the system being discussed has spherical symmetry. Sections 7–9 concern the general formulation of scattering of the electromagnetic field, with two simple applications to weak field scattering, by a harmonic oscillator and by a dielectric with $\epsilon - 1 \ll 1$. The remaining three sections, Sections 10–12, address the vector partial wave expansion and apply it to scattering by a dielectric sphere. This method is very important for numerical work in many cases where approximate methods are invalid. However, the discussion given here involves much more detailed algebra than the rest of the text and can be omitted easily in a first reading.

5.1. SCALAR FIELD

The electromagnetic field is most conveniently described by a vector potential **A** and the accompanying scalar potential ϕ. It is called a vector field. We consider here first a theory that depends on a single scalar potential $\psi(\mathbf{x}, t)$, which we call a scalar field. Although there is no such field known in nature, the theory provides a simple model in which the mathematics and physics are more transparent than for the more realistic vector and tensor fields. Nevertheless, many of the essential physics elements that characterize the vector and tensor fields are present. It is only a minor complication to deal with a massive scalar field (quantum language; classically, we would say a field with a finite Compton wavelength), so we will do so.

The wave equation for the field away from sources and scatterers (which we will always assume to be spatially confined) is taken to be

$$\left(\nabla^2 - \mu^2 - \frac{\partial^2}{\partial t^2}\right)\psi(\mathbf{x}, t) = 0, \qquad (5.1.1)$$

where $1/\mu$ is the Compton wavelength of the field. As before, $c = 1$. The form (5.1.1) is, of course, suggested by the corresponding equation for the components of the electromagnetic potentials in the Lorentz gauge. We include the term μ^2 since that permits the particles associated with the quantum field to be massive, with mass $\mu_0 = \hbar\mu/c$. We will also see in Chapter 7 that (5.1.1) is the simplest nontrivial Lorentz invariant equation that we can write.

In the presence of sources and scatterers, the right-hand side of (5.1.1) will be different from zero. However, in a scattering event, both the initial and final field configurations are far away from the sources, so that (5.1.1) is sufficient for our general discussion.

The elementary, fixed wave number and fixed frequency solutions of (5.1.1) are

$$\phi_{\mathbf{k}}(\mathbf{x}, t) = e^{i(\mathbf{k}\cdot\mathbf{x} - \omega(\mathbf{k})t)} \qquad (5.1.2)$$

where $\omega = \sqrt{k^2 + \mu^2}$; conventionally, we call the minus sign in $e^{-i\omega t}$ positive frequency. Of course, if the field ψ is real, ψ must consist in a superposition of at least two of the elementary solutions, $\phi_{\mathbf{k}}$ and $\phi_{\mathbf{k}}^*$.

A conserved energy functional of the scalar field ψ, and a corresponding locally conserved energy density and energy flux, are permitted by (5.1.1).[1] The energy density is given in arbitrary units by

$$u = \frac{1}{8\pi}\left[\left(\frac{\partial\psi}{\partial t}\right)^2 + (\nabla\psi)^2 + \mu^2\psi^2\right] \qquad (5.1.3)$$

---

[1]We shall learn general rules for constructing such conserved quantities in Chapter 7.

and the energy flux (the equivalent of the Poynting vector) by

$$\mathcal{P} = -\frac{1}{4\pi} \frac{\partial \psi}{\partial t} \nabla \psi. \tag{5.1.4}$$

u and $\mathcal{P}$ satisfy the conservation equation

$$\frac{\partial u}{\partial t} + \nabla \cdot \mathcal{P} = 0 \tag{5.1.5}$$

leading to a conserved energy in a volume V:

$$W_V = \int_V d\mathbf{r}\, u \tag{5.1.6}$$

provided there are no sources of ψ inside the volume and the flux through the boundary surfaces is zero:

$$\int_S \mathcal{P} \cdot d\mathbf{S} = 0. \tag{5.1.7}$$

A scattering problem must specify an incident wave packet heading toward the target T, as shown in Figure 5.1. The vertical lines are meant to represent maxima of the amplitude within the envelope; thus, the distance between the lines is roughly λ_0, where $\lambda_0 = 2\pi/k_0$ is the mean wavelength of the incident field. The incident field $\psi_0(\mathbf{x}, t)$ is taken to be

$$\psi_0(\mathbf{x}, t) = \int a(\mathbf{k} - \mathbf{k}_0)\, e^{i[\mathbf{k}\cdot(\mathbf{x}-\mathbf{x}_0) - \omega(t - t_0)]}\, d\mathbf{k} + \text{c.c.}. \tag{5.1.8}$$

Here, $\mathbf{k}_0$ is the central wave number of the packet; we would refer to the scattering of this packet as the scattering at wave number $\mathbf{k}_0$, even though the packet involves a superposition of a continuum of wave numbers. For this terminology to make sense, the packet spread in wave numbers Δk

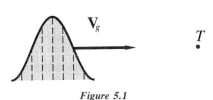

Figure 5.1

must be small compared to the characteristic wave number of the source; that is, $\Delta x \sim 1/\Delta k$ must be much larger than the source size, or force range. Note that this creates special problems for Coulomb scattering.

We choose $a(\mathbf{k} - \mathbf{k}_0)$ for simplicity to be a real, smooth function, symmetric in $\mathbf{k} - \mathbf{k}_0 \to -(\mathbf{k} - \mathbf{k}_0)$, with width Δk as stated above. Then at time $t = t_0$ and with $\mathbf{q} = \mathbf{k} - \mathbf{k}_0$,

$$\psi_0(\mathbf{x}, 0) = e^{i\mathbf{k}_0 \cdot (\mathbf{x} - \mathbf{x}_0)} h(\mathbf{x} - \mathbf{x}_0) + \text{c.c.} \tag{5.1.9}$$

where

$$h(\mathbf{x} - \mathbf{x}_0) = \int d\mathbf{q}\, a(\mathbf{q})\, e^{i\mathbf{q} \cdot (\mathbf{x} - \mathbf{x}_0)} \tag{5.1.10}$$

where h is real and symmetric under the reflection of $\mathbf{x} - \mathbf{x}_0$. The mean value of x_i, defined by

$$\overline{\mathbf{x}}_i = \frac{\displaystyle\int d\mathbf{x}\, x_i\, |\psi_0(\mathbf{x}, 0)|^2}{\displaystyle\int d\mathbf{x}\, |\psi_0(\mathbf{x}, 0)|^2},$$

will be x_{0i}. The mean value of k_i, defined by

$$\overline{k}_i = \frac{\displaystyle\int d\mathbf{k}\, k_i\, |a(\mathbf{k} - \mathbf{k}_0)|^2}{\displaystyle\int d\mathbf{k}\, |a(\mathbf{k} - \mathbf{k}_0)|^2},$$

will be k_{0i}. The root mean square spread in x_i will be

$$\Delta x_i = \left[\frac{\displaystyle\int d\mathbf{x}(x_i - x_{0i})^2 |\psi_0(\mathbf{x}, 0)|^2}{\displaystyle\int d\mathbf{x}\, |\psi_0(\mathbf{x}, 0)|^2} \right]^{1/2}. \quad (\text{No sum over } i)$$

The root mean square spread in k_i will be

$$\Delta k_i = \left[\frac{\int d\mathbf{k}(k_i - k_{0i})^2 |a(\mathbf{k} - \mathbf{k}_0)|^2}{\int d\mathbf{k} |a(\mathbf{k} - \mathbf{k}_0)|^2} \right]^{1/2} . \quad \text{(No sum over } i\text{)}.$$

Of course Δx_i and Δk_i have the uncertainty property

$$\Delta x_i \cdot \Delta k_i \geq \frac{1}{2}.$$

To study the time dependence of (5.1.8), we expand ω in powers of $\mathbf{q} = \mathbf{k} - \mathbf{k}_0$; thus,

$$\omega(\mathbf{k}) = \omega(\mathbf{k}_0) + \mathbf{q} \cdot \nabla_{\mathbf{k}_0}\omega(\mathbf{k}_0) + \frac{1}{2}\mathbf{q} \cdot \nabla_{\mathbf{k}_0}\mathbf{q} \cdot \nabla_{\mathbf{k}_0}\omega(\mathbf{k}_0) + \cdots \quad (5.1.11)$$

and

$$\psi_0(\mathbf{x}, t) = e^{i[\mathbf{k}_0 \cdot (\mathbf{x} - \mathbf{x}_0) - \omega(\mathbf{k}_0)(t - t_0)]} \int d\mathbf{q}\, a(\mathbf{q})$$

$$\times e^{i\{\mathbf{q} \cdot [\mathbf{x} - \mathbf{x}_0 - \mathbf{v}_g(t - t_0)] - (1/2)\mathbf{q} \cdot \nabla_{\mathbf{k}_0}\mathbf{q} \cdot \nabla_{\mathbf{k}_0}\omega(\mathbf{k}_0)(t - t_0) + \cdots\}} + \text{c.c.} \quad (5.1.12)$$

where the group velocity

$$\mathbf{v}_g = \nabla_{\mathbf{k}_0}\omega(k_0) = \frac{\mathbf{k}_0}{\omega(\mathbf{k}_0)}. \quad (5.1.13)$$

The last term in the exponent can be neglected if

$$\frac{q^2}{\omega}(t - t_0) \ll 1 \quad \text{or} \quad \frac{L \cdot (\Delta k)^2}{k_0} \ll 1,$$

where L is the distance we may allow the packet to travel between observations. We recall that $\Delta x \cong 1/\Delta k \gg$ size of the target; hence, since $\Delta k/k_0 \ll 1$, we can always choose L so that the initial (and final) distances to the target are much larger than the target size. Neglecting the last term in the exponent, we find for (5.1.12)

$$\psi_0(\mathbf{x}, t) = e^{i[\mathbf{k}_0 \cdot (\mathbf{x} - \mathbf{x}_0) - \omega_0(t - t_0)]} h(\mathbf{x} - \mathbf{x}_0 - \mathbf{v}_g(t - t_0)) + \text{c.c.} \quad (5.1.14)$$

$$= 2\cos(\mathbf{k}_0 \cdot (\mathbf{x} - \mathbf{x}_0) - \omega_0(t - t_0))h(\mathbf{x} - \mathbf{x}_0 - \mathbf{v}_g(t - t_0)). \quad (5.1.15)$$

Thus, the packet envelope moves rigidly, without changing shape.

To find the energy incident per unit area, dW/dA, we must integrate the flux $\mathcal{P}$ over time:

$$\frac{d\mathbf{W}}{dA} = -\frac{1}{4\pi} \int_{-\infty}^{\infty} dt \, \frac{\partial \psi_0}{\partial t} \, \nabla \psi_0. \tag{5.1.16}$$

We note that since $k_0 \gg 1/\Delta x$, the gradient in (5.1.16) acting on h is negligible; similarly, since $\omega_0 = k_0/v_g \gg \Delta k/v_g \approx (1/\Delta x)v_g \sim (1/h)\partial h/\partial t$, the time derivative in (5.1.16) acting on h is negligible. Finally, since the envelope function h varies negligibly in a period $1/\omega_0$, the integral in (5.1.16) averages the trigonometric function over time. There remains

$$\frac{d\mathbf{W}}{dA} = 2\omega_0 \mathbf{k}_0 \frac{1}{4\pi} \int_{-\infty}^{\infty} dt [h(\mathbf{x} - \mathbf{x}_0 - \mathbf{v}_g(t - t_0))]^2 \tag{5.1.17}$$

If we take the target to be located at $\mathbf{x} = 0$, $\mathbf{x}_0$ and $\mathbf{v}_g$ must be parallel; otherwise, the packet will miss the target. Call that direction z. Then

$$h(\mathbf{x} - \mathbf{x}_0 - \mathbf{v}_g(t - t_0)) = h(\boldsymbol{\rho}, z - z_0 - v_g(t - t_0))$$

and

$$\frac{d\mathbf{W}}{dA} = \frac{\omega_0 \mathbf{k}_0}{2\pi} \int dt [h(\boldsymbol{\rho}, z - z_0 - v_g(t - t_0))]^2$$

$$= \frac{\omega_0^2 \hat{\mathbf{k}}_0}{2\pi} \int_{-\infty}^{\infty} dz \, h^2(\boldsymbol{\rho}, z) \tag{5.1.18}$$

where

$$\boldsymbol{\rho} = \hat{\mathbf{e}}_x x + \hat{\mathbf{e}}_y y.$$

With $\boldsymbol{\rho}$ located at the target transverse coordinate, that is, $\boldsymbol{\rho} = 0$, we have

$$\frac{d\mathbf{W}}{dA} = \frac{\omega_0^2 \hat{\mathbf{k}}_0}{2\pi} \int_{-\infty}^{\infty} dz \, h^2(\mathbf{0}, z) \tag{5.1.19}$$

energy incident on the target.

A shorthand for obtaining the result (5.1.19) is to consider only the positive frequency part of ψ_0, ψ_{0+}, and then calculate the integrated flux as

$$\frac{d\mathbf{W}}{dA} = -2\,\text{Re}\,\frac{1}{4\pi}\int\limits_{-\infty}^{\infty} dt\,\frac{\partial \psi_0^*}{\partial t}\,\nabla\psi_{0+}. \tag{5.1.20}$$

We shall use that procedure from now on. Thus, we consider in the following only the positive frequency part of ψ:

$$\psi_+ = \int d\mathbf{k}\,a(\mathbf{k}-\mathbf{k}_0)\,e^{i[\mathbf{k}\cdot(\mathbf{x}-\mathbf{x}_0)-\omega(t-t_0)]} \simeq e^{i[\mathbf{k}_0\cdot(\mathbf{x}-\mathbf{x}_0)-\omega_0(t-t_0)]}$$

$$\times\,h(\mathbf{x}-\mathbf{x}_0-\mathbf{v}_g(t-t_0)).$$

After the scattering is over, there will be an outgoing spherical wave ψ_{Sc} and the forward-going residue of the incoming field ψ_{0+}. Figure 5.2 illustrates the configuration.

The wave field far from the source after the collision is given by the retarded Green's function, $\Delta_R(\mathbf{x}-\mathbf{x}',t-t')$, acting on the source,

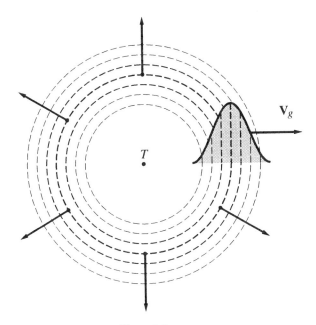

Figure 5.2

whatever that source may be. Δ_R is the retarded solution of the inhomogeneous equation

$$\left(\nabla^2 - \mu^2 - \frac{\partial^2}{\partial t^2}\right)\Delta_R = -4\pi\delta^3(\mathbf{x} - \mathbf{x}')\delta(t - t') \qquad (5.1.21)$$

which is shown in the next section to be

$$\Delta_R(\mathbf{r}, t) = \int\limits_{-\infty}^{\infty} \frac{d\omega}{2\pi} e^{-i\omega t}\Delta_\omega(r) \qquad (5.1.22)$$

where

$$\Delta_\omega(r) = \frac{e^{ikr}}{r} \qquad (5.1.23)$$

with $k^2 = \omega^2 - \mu^2$, and $k/\omega > 0$ for $|\omega| > \mu$, $k = i\sqrt{\mu^2 - \omega^2}$ for $|\omega| < \mu$.

5.2. GREEN'S FUNCTION FOR MASSIVE SCALAR FIELD

The equation to be satisfied is (5.1.21):

$$\left(\nabla^2 - \mu^2 - \frac{\partial^2}{\partial t^2}\right)\Delta_R(\mathbf{r}, t) = -4\pi\delta^3(\mathbf{r})\delta(t). \qquad (5.1.21)$$

As in Section 4.2, we proceed by carrying out a four-dimensional Fourier transform:

$$\Delta_R(\mathbf{r}, t) = \int \frac{d\mathbf{k}\, d\omega}{(2\pi)^4} e^{i(\mathbf{k}\cdot\mathbf{r} - \omega t)} \tilde{\Delta}_R(\mathbf{k}, \omega) \qquad (5.2.1)$$

and

$$\delta^3(\mathbf{r})\delta(t) = \int \frac{d\mathbf{k}\, d\omega}{(2\pi)^4} e^{i(\mathbf{k}\cdot\mathbf{r} - \omega t)}. \qquad (5.2.2)$$

There results

$$\tilde{\Delta}_R(\mathbf{k}, \omega) = -\frac{4\pi}{(\omega^2 - \mu^2 - \mathbf{k}^2)} \tag{2.2.3}$$

and

$$\Delta_R(\mathbf{r}, t) = \lim_{\epsilon \to 0+} = -4\pi \int \frac{d\mathbf{k}\, d\omega}{(2\pi)^4} \frac{e^{i(\mathbf{k}\cdot\mathbf{r} - \omega t)}}{(\omega + i\epsilon)^2 - \mu^2 - \mathbf{k}^2} \tag{5.2.4}$$

where the $\epsilon \to 0+$ limit ensures a retarded solution.

Unfortunately, the integral in (5.2.4) leads to a more complicated function than the zero mass case. However, the ω Fourier transform is very simple. Thus, instead of integrating over ω, we integrate over $\mathbf{k}$:

$$\Delta_R(\mathbf{r}, t) = -\frac{1}{\pi^2 r} \lim_{\epsilon \to 0} \int_{-\infty}^{\infty} d\omega\, e^{-i\omega t} \int_{0}^{\infty} \frac{k\, dk \sin kr}{(\omega + i\epsilon)^2 - \mu^2 - k^2}$$

$$= \lim_{\epsilon \to 0} \frac{1}{2\pi^2 ri} \int_{-\infty}^{\infty} d\omega\, e^{-i\omega t} \int_{-\infty}^{\infty} \frac{k\, dk\, e^{ikr}}{k^2 - [(\omega + i\epsilon)^2 - \mu^2]} \tag{5.2.5}$$

We proceed by carrying out the k integration. The integrand has poles at

$$k = \sqrt{(\omega + i\epsilon)^2 - \mu^2} \cong \sqrt{\omega^2 - \mu^2} + i\epsilon'\omega$$

and at $k = -(\sqrt{\omega^2 - \mu^2} + i\epsilon'\omega)$, where the infinitesimal ϵ' has the same sign as ϵ. Since we intend to close the contour above, only poles in the upper half-plane will contribute. These are, for $\omega > \mu$, $k = \sqrt{\omega^2 - \mu^2}$; for $\omega < -\mu$, $k = -\sqrt{\omega^2 - \mu^2}$; and for $-\mu < \omega < \mu$, $k = i\sqrt{\mu^2 - \omega^2}$. The end result is

$$\Delta_R(\mathbf{r}, t) = \frac{1}{r} \int_{-\infty}^{\infty} \frac{d\omega}{2\pi} e^{-i\omega t + ikr} \tag{5.2.6}$$

where $k = (\omega^2 - \mu^2)^{1/2}$ for $\omega > \mu$, $k = -(\omega^2 - \mu^2)^{1/2}$ for $\omega < -\mu$, and $k = i(\mu^2 - \omega^2)^{1/2}$ for $-\mu < \omega < \mu$. The function k so defined is analytic in the entire plane except for branch points at $\omega = \pm\mu$. The cut is taken between the branch points; the definition of k informs us that the ω integral goes *above* the cut.

We list a few properties of (5.2.6). First note that for $\mu = 0$, $\Delta_R = (1/r)\, \delta(t - r)$, as it must.

Second, for $r > t$, including all $t < 0$, the integral vanishes. This is shown by closing the ω contour in the upper half-plane.

Finally, we note that for $t > r$, Δ_R is nonzero and not particularly simple.

We are now in a position to discuss the scattering as a radiation of a scalar field by a source $S(\mathbf{x}, t)$. Equation (5.1.1) becomes

$$\left(\nabla^2\psi - \frac{\partial^2\psi}{\partial t^2} - \mu^2\right)\psi = -4\pi S \tag{5.2.7}$$

where S may be a given source, in which case we could study the radiation from the source; or if we are considering scattering of the field by a potential, the source S would be a linear function of the field itself, and (5.2.7) would become a Schrödinger-like equation for the wave function (field amplitude).

The retarded propagation problem posed by (5.2.7) is solved by integration:

$$\psi(\mathbf{x}, t) = \psi_0(\mathbf{x}, t) + \int dt' \, d\mathbf{x}' \, \Delta_R(\mathbf{x} - \mathbf{x}, t - t') \, S(\mathbf{x}', t'), \tag{5.2.8}$$

where ψ_0 satisfies the free equation (5.1.1). In a scattering problem, ψ_0 would describe the incident wave.

Suppose now that $S(\mathbf{x}', t')$ contains only one frequency, so that

$$S(\mathbf{x}', t') = e^{-i\omega t'} S(\mathbf{x}').$$

Assigning the same frequency ω to ψ and ψ_0, we have the result

$$\psi(\mathbf{x}, t) = \phi(\mathbf{x}) \, e^{-i\omega t}, \qquad \psi_0 = \phi_0(\mathbf{x}) \, e^{-i\omega t}, \tag{5.2.9}$$

and

$$\phi(\mathbf{x}) = \phi_0(\mathbf{x}) + e^{i\omega t} \int dt' \, d\mathbf{x}' \, \frac{d\omega'}{2\pi} \, \frac{e^{-i\omega'(t - t') + ik'|\mathbf{x} - \mathbf{x}'|}}{|\mathbf{x} - \mathbf{x}'|} \, S(\mathbf{x}') \, e^{-i\omega t'}$$

or

$$\phi(\mathbf{x}) = \phi_0(\mathbf{x}) + \int d\mathbf{x}' \, \frac{e^{ik|\mathbf{x} - \mathbf{x}'|}}{|\mathbf{x} - \mathbf{x}'|} \, S(\mathbf{x}'). \tag{5.2.10}$$

At this point, we have made contact with the work on electromagnetic radiation in Chapter 4; in that case, there was no incoming field ϕ_0, and the source $S(\mathbf{x}, t)$ consisted of given charge and current densities. The field at distant **x** was given by replacing

$$\frac{e^{ik|\mathbf{x} - \mathbf{x}'|}}{|\mathbf{x} - \mathbf{x}'|} \quad \text{by} \quad \frac{e^{ikr}}{r} e^{-ik\hat{\mathbf{x}}\cdot\mathbf{x}'},$$

accurate to order $1/r$. Evidently, the same expansion works for radiation of the ψ field by a given source. In a scattering process, the source is affected by the incoming field.

5.3. FORMULATION OF THE SCATTERING PROBLEM

In a scattering problem, ϕ_0 would specify the incoming field—position, velocity, wave number and shape of the packet, as described in Section 5.1. In a linear system, the source S would have its frequency determined by the frequency of ψ_0. A Schrödinger-like model, for example, would have the wave equation

$$\left(\nabla^2 - (\mu + U)^2 - \frac{\partial^2}{\partial t^2}\right)\psi = 0 \tag{5.3.1}$$

where μ is the particle mass. For $U \ll \mu$ and the frequency $E = \mu + W$, $W \ll \mu$, (5.3.1) becomes

$$\left(-\frac{\nabla^2}{2\mu} + U\right)\psi = W\psi \tag{5.3.2}$$

which is the Schrödinger equation. The source $S(\mathbf{x})$ from (5.3.1) is

$$S = -\frac{2\mu U + U^2}{4\pi}\psi \tag{5.3.3}$$

and (5.2.10) becomes an integral equation for ϕ.

The exponential dependence e^{ikr} is called an outgoing wave (remember that the time-dependent factor $e^{-i\omega t}$ is appended to the wave function and that k has the same sign as ω; hence, an outgoing wave).

The standard procedure is to solve the integral equation (5.2.10) for an incident plane wave, $\phi_0 = e^{i\mathbf{k}\cdot\mathbf{x}}$. In principle, one would then construct a wave packet superposition of the plane waves, as described in Section 5.1. In practice, as we shall see, the calculated cross section is substantially independent of the structure of the wave packet.

We proceed as follows: The asymptotic $(r \to \infty)$ solution of (5.2.10) is

$$\phi(\mathbf{x}) = e^{i\mathbf{k}\cdot\mathbf{x}} + \frac{e^{ikr}}{r} f(\hat{r}, k) + \mathcal{O}\left(\frac{1}{r^2}\right). \tag{5.3.4}$$

The function $f(\hat{r}, k)$ is called the elastic scattering amplitude at wave number k and angle $\hat{r}$.

The wave packet superposition is given, as in (5.1.8), by

$$\psi_0(\mathbf{x}, t) = \int d\mathbf{k}\, e^{i[\mathbf{k}\cdot(\mathbf{x}-\mathbf{x}_0) - \omega(t-t_0)]} a(\mathbf{k} - \mathbf{k}_0) + \text{c.c.}.$$

We have seen in (5.1.15) that, for times of interest, the wave packet moves rigidly without changing shape:

$$\psi_0(\mathbf{x}, t) = e^{i[\mathbf{k}_0(\mathbf{x}-\mathbf{x}_0) - \omega_0(t-t_0)]} h(\mathbf{x} - \mathbf{x}_0 - \mathbf{v}_g(t - t_0)) + \text{c.c.} \tag{5.3.5}$$

and that the total energy incident per unit area at the target ($\rho = 0$) is

$$-\frac{1}{4\pi} \int_{-\infty}^{\infty} dt\, \frac{\partial \psi_0}{\partial t} \nabla \psi_0 = \frac{1}{2\pi} \omega_0^2 \hat{k}_0 \int_{-\infty}^{\infty} |h(\mathbf{0}, z)|^2\, dz, \tag{5.1.19}$$

provided $S \to 0$ rapidly enough as $r \to \infty$. We assume that to be the case, here and in the following.

The scattered wave packet will be given by the superposition of outgoing waves:

$$\psi_{sc} = \int d\mathbf{k}\, e^{-i\mathbf{k}\cdot\mathbf{x}_0 + i\omega t_0} a(\mathbf{k} - \mathbf{k}_0) \frac{e^{i(kr-\omega t)}}{r} f(\hat{r}, k) + \text{c.c.}. \tag{5.3.6}$$

We carry out the same expansion about $\mathbf{k} = \mathbf{k}_0$ for ψ_{sc} as we did for ψ_0. The difference appears in the expansion of k vs. that of $\mathbf{k}$:

$$k = \sqrt{\mathbf{k}^2} = \sqrt{(\mathbf{k}_0 + \mathbf{k} - \mathbf{k}_0)^2} = \sqrt{k_0^2 + 2\mathbf{k}_0 \cdot (\mathbf{k} - \mathbf{k}_0) + (\mathbf{k} - \mathbf{k}_0)^2} \tag{5.3.7}$$

$$= k_0\left[1 + \frac{\mathbf{k}_0 \cdot (\mathbf{k} - \mathbf{k}_0)}{k_0^2} + \mathcal{O}\left(\frac{(\mathbf{k} - \mathbf{k}_0)^2}{k_0^2}\right)\right] \tag{5.3.8}$$

$$= k_0 + \hat{\mathbf{k}}_0 \cdot (\mathbf{k} - \mathbf{k}_0) + \mathcal{O}\left(\frac{(\mathbf{k} - \mathbf{k}_0)^2}{k_0}\right). \tag{5.3.9}$$

We neglect the quadratic term in (5.3.9), as we did earlier in (5.1.12). Note that the neglected term here is $[(\Delta \mathbf{k})^2/k_0] r$; in (5.1.13) it was the equivalent expression $[(\Delta k)^2/\omega_0](t - t_0)$. We also assume that $f(\hat{r}, k)$ varies negligibly over the width of the wave packet. This is consistent with our earlier assumption that the coordinate spread in the wave packet be much larger than the size of the target. The result for ψ_{sc} is

$$\psi_{sc} = \frac{h(\hat{k}_0 r - \mathbf{x}_0 - \mathbf{v}_g(t - t_0)) f(\hat{r}, k_0)}{r} e^{i[k_0 r - \mathbf{k}_0 \cdot \mathbf{x}_0 - \omega_0(t - t_0)]} + \text{c.c.}$$

(5.3.10)

Note the role of the wave packet function h. Since $\mathbf{k}_0$, $\mathbf{x}_0$, and $\mathbf{v}_g = \{k_0/\omega_0$ are all in the same z direction, the function h is evaluated at $\rho = 0$, just as in the incident packet. The outgoing integrated energy flux per unit area is

$$\int \mathcal{P}_{sc} \, dt = |f(\hat{r}, k_0)|^2 \frac{\omega_0 k_0 \hat{\mathbf{r}}}{2\pi r^2} \int |h(\mathbf{0}, r - z_0 - v_g(t - t_0))|^2 \, dt$$

$$= |f(\hat{r}, k_0)|^2 \frac{\omega_0^2 \hat{\mathbf{r}}}{2\pi r^2} \int_{-\infty}^{\infty} |h(\mathbf{0}, z)|^2 \, dz.$$

(5.3.11)

The cross section $d\sigma$ for scattering into a solid angle $d\Omega$ is defined as the ratio of the energy scattered into $d\Omega$ divided by the energy incident per unit area on the target. The differential cross-section $d\sigma/d\Omega$ is defined by the equation $d\sigma = (d\sigma/d\Omega) \, d\Omega$ (note that $d\sigma/d\Omega$ is not a derivative!), so that

$$d\Omega \frac{d\sigma}{d\Omega} = d\Omega \frac{\int_{-\infty}^{\infty} \mathcal{P}_{sc} \, dt \cdot \hat{r} r^2}{\int_{-\infty}^{\infty} \mathcal{P}_0 \, dt \cdot \hat{\mathbf{k}}_0} = d\Omega |f(\hat{\mathbf{r}}, k_0)|^2.$$

(5.3.12)

Thus, the scattering problem can be stated in two ways:

1. Solve the integral equation

$$\phi(\mathbf{x}) = e^{i\mathbf{k} \cdot \mathbf{x}} + \int \frac{e^{ik|\mathbf{x} - \mathbf{x}'|}}{|\mathbf{x} - \mathbf{x}'|} \overline{S}(\mathbf{x}') \, d\mathbf{x}'$$

(5.3.13)

with $\overline{S}$ a linear function of ϕ as in (5.3.3), with $\overline{S}$ replacing S and ϕ replacing ψ; from $\phi(\mathbf{x})$ determine $\overline{S}$ and from $\overline{S}$ calculate the scattering amplitude

$$f(\hat{r}, k) = \int e^{-ik\hat{r}\cdot\mathbf{x}}\,\overline{S}(\mathbf{x}')\,d\mathbf{x}'. \tag{5.3.14}$$

Or, equivalently,

2. Solve the differential equation

$$(\nabla^2 - \mu^2 + \omega^2)\,\phi(\mathbf{x}) = -4\pi\overline{S}(\mathbf{x}) \tag{5.3.15}$$

subject to the boundary condition

$$\psi_{r\to\infty} \to e^{i\mathbf{k}\cdot\mathbf{x}} + \frac{e^{ikr}}{r}f(\hat{r}, k);$$

from the solution determine the function f, which is the scattering amplitude.

5.4. THE OPTICAL THEOREM

An important theorem, applied to elastic scattering, relates the imaginary part of the forward ($\theta = 0$) scattering amplitude to the total elastic scattering cross section

$$\operatorname{Im} f(\theta = 0) = \frac{k}{4\pi}\int \frac{d\sigma}{d\Omega}\,d\Omega. \tag{5.4.1}$$

In fact, the theorem is more general. On the right of (5.4.1) in the case where there is absorption of energy, there should be the total instead of the elastic cross section. That is, the general theorem says

$$\operatorname{Im} f(\theta = 0) = \frac{k}{4\pi}\sigma_T \tag{5.4.2}$$

where $f(\theta = 0)$ is the forward elastic scattering amplitude and σ_T the total cross section.

In the case of vector (or higher) fields, forward signifies not only zero deflection angle θ, but identical polarization to that of the incoming field. We will see this explicitly when we discuss scattering of electric and magnetic fields.

The theorem follows directly from the energy conservation equation (5.1.5)

$$\nabla \cdot \mathcal{P} + \frac{\partial u}{\partial t} = 0 \qquad (5.1.5)$$

or

$$\int_S \mathcal{P} \cdot d\mathbf{S} = -\frac{d}{dt} \int_V u \, d\mathbf{r}. \qquad (5.4.3)$$

If we consider a given frequency

$$\psi = \psi_+ + \psi_-$$

with

$$\psi_+ = e^{-i\omega t}\phi \qquad \text{and} \qquad \psi_- = e^{i\omega t}\phi*$$

then the time average over a cycle of (5.4.3) tells us that

$$\int \overline{\mathcal{P}} \cdot d\mathbf{S} = 0 \qquad (5.4.4)$$

since u is periodic in t with period π/ω. On the other hand,

$$\mathcal{P} = \frac{i\omega}{4\pi} (\psi_+ - \psi_-)\nabla(\psi_+ + \psi_-) \qquad (5.4.5)$$

and the time average $\overline{\mathcal{P}}$ is

$$\overline{\mathcal{P}} = \frac{i\omega \phi}{4\pi} \nabla\phi* + \text{c.c.}, \qquad (5.4.6)$$

so that energy conservation takes the form

$$\frac{4\pi}{\omega} \int \overline{\mathcal{P}} \cdot d\mathbf{S} = -i \int \phi*\nabla\phi \cdot d\mathbf{S} + \text{c.c.} = 0. \qquad (5.4.7)$$

Equation (5.4.7) is equivalent to the optical theorem. To proceed, we note that in the surface integral above, terms of the form

$$\int_{-1}^{1} e^{-ikrw} \, dw \, g(r, w),$$

following an integration by parts acquire an extra power of $1/r$. Here, $w = \cos \theta$, and g is assumed to be free of singularities in the physical region, $-1 \le w \le 1$. It follows that, as $r \to \infty$, we may write

$$\phi \to e^{i\mathbf{k} \cdot \mathbf{x}} + \frac{e^{ikr}}{r} f + \mathcal{O}\left(\frac{1}{r^2}\right) \tag{5.4.8}$$

and ignore the $1/r^2$ term. Inserting (5.4.8) into (5.4.7) gives, as $r \to \infty$, accurate to order 1,

$$-\int d\mathbf{S} \cdot i\left(e^{-i\mathbf{k} \cdot \mathbf{x}} + \frac{e^{-ikr}}{r} f^*\right)\left(i\mathbf{k} \, e^{i\mathbf{k} \cdot \mathbf{x}} + i\hat{r} \frac{k \, e^{ikr}}{r} f\right) + \text{c.c.} = 0. \tag{5.4.9}$$

With $d\mathbf{S} = r^2 \hat{r} \, d\Omega$ (we are, of course, integrating over a sphere), (5.4.9) becomes, accurate to order $1/r^2$,

$$\int d\Omega \left\{\hat{\mathbf{r}} \cdot \mathbf{k} + \frac{k}{r} f \, e^{i(kr - \mathbf{k} \cdot \mathbf{x})} + \frac{f^*}{r} \hat{\mathbf{r}} \cdot \mathbf{k} \, e^{i(\mathbf{k} \cdot \mathbf{x} - kr)} + \frac{k|f|^2}{r^2} + \text{c.c.}\right\} = 0. \tag{5.4.10}$$

The first integral is zero. In the second integral, we integrate by parts:

$$\int d\Omega \, e^{-i\mathbf{k} \cdot \mathbf{x}} f = \int_0^{2\pi} d\varphi \int_{-1}^{1} dw \, e^{-ikrw} f(w, \varphi) \tag{5.4.11}$$

where $w = \cos \theta$.

Let $e^{-ikrw} \, dw = dv$; $u = f$. Thus, integrating by parts, we obtain

$$v = -\frac{1}{ikr} e^{-ikrw},$$

and to leading order in $1/r$

$$\int_{-1}^{1} dw \, e^{-ikrw} f(w, \varphi) = -\frac{1}{ikr} e^{-ikrw} f(w, \phi)\Big|_{w=-1}^{w=1} + \mathcal{O}\left(\frac{1}{r^2}\right)$$

$$= -\left[\frac{e^{-ikr}}{ikr} f(1, \phi) - \frac{e^{ikr}}{ikr} f(-1, \phi)\right]. \quad (5.4.12)$$

Since at $\theta = 0$ or π ($w = 1$ or -1), φ dependence must disappear, we have for (5.4.11)

$$\int d\Omega \, e^{-i\mathbf{k}\cdot\mathbf{x}} f = -\frac{2\pi}{ikr} (e^{-ikr} f(\theta = 0) - e^{ikr} f(\theta = \pi)) + \mathcal{O}\left(\frac{1}{r^2}\right). \quad (5.4.13)$$

Similarly,

$$\int d\Omega \, \hat{\mathbf{r}} \cdot \mathbf{k} \, f^* \, e^{i\mathbf{k}\cdot\mathbf{x}} = \frac{2\pi}{ir} (e^{ikr} f^*(\theta = 0) + e^{-ikr} f^*(\theta = \pi)) + \mathcal{O}\left(\frac{1}{r^2}\right). \quad (5.4.14)$$

Inserting (5.4.13) and (5.4.14) into (5.4.10), we have

$$-\frac{2\pi f(\theta = 0)}{ir^2} + \frac{2\pi e^{2ikr} f(\theta = \pi)}{ir^2} + \frac{2\pi}{ir^2} f^*(\theta = 0)$$

$$+ \frac{2\pi}{ir^2} f^*(\theta = \pi) e^{-2ikr} + \frac{k}{r^2} \int |f|^2 \, d\Omega + \text{c.c.} = 0. \quad (5.4.15)$$

The contribution from $\theta = \pi$ is imaginary. This leaves, after we add the complex conjugate,

$$\text{Im} \, f(\theta = 0) = \frac{k}{4\pi} \sigma_e \quad (5.4.16)$$

as expected, with σ_e the elastic cross section,

$$\sigma_e = \int \frac{d\sigma}{d\Omega} \, d\Omega = \int |f|^2 \, d\Omega. \quad (5.4.17)$$

5.5. DIGRESSION ON RADIAL WAVE FUNCTIONS

We are about to take up the description of scattering in a system possessing spherical symmetry. For such a system, the angular dependence of the wave functions and scattering amplitudes can be expanded in a series of spherical harmonics. The coefficients of the spherical harmonics are called partial wave amplitudes. For a wave outside of the region of interaction, these partial wave amplitudes involve a specific set of radial functions called spherical Bessel functions. Since these functions appear in a large class of applications, we treat them in a separate section.

We first study the Green's function

$$\Delta_R(\mathbf{r} - \mathbf{r}') = \frac{e^{ik|\mathbf{r}-\mathbf{r}'|}}{|\mathbf{r} - \mathbf{r}'|} \tag{5.5.1}$$

for $r > r'$. As $r \to \infty$, we know the leading term is

$$\Delta_R(\mathbf{r} - \mathbf{r}') \to \frac{e^{ikr}}{r} e^{-i\mathbf{k}\cdot\mathbf{r}'} \qquad \text{with} \qquad \mathbf{k} = k\hat{r}. \tag{5.5.2}$$

We note that for arbitrary $\mathbf{r}$, $\mathbf{r}'$, but $r \neq r'$, Δ_R satisfies the homogeneous wave equation

$$(\nabla^2 + k^2) \Delta_R = 0 \tag{5.5.3}$$

as well as

$$(\nabla'^2 + k^2) \Delta_R = 0. \tag{5.5.4}$$

We expand $\Delta_R(r, r', w = \cos\theta)$ in Legendre polynomials:

$$\Delta_R(\mathbf{r} - \mathbf{r}') = \sum_{\ell=0}^{\infty} P_\ell(w) g_\ell(r, r') \tag{5.5.5}$$

where now from (5.5.3)

$$\left[\frac{\partial^2}{\partial r^2} + \frac{2}{r} \frac{\partial}{\partial r} + k^2 - \frac{\ell(\ell + 1)}{r^2} \right] g_\ell = 0 \tag{5.5.6}$$

and from (5.5.4)

$$\left[\frac{\partial^2}{\partial r'^2} + \frac{2}{r'} \frac{\partial}{\partial r'} + k^2 - \frac{\ell(\ell+1)}{r'^2} \right] g_\ell = 0 \tag{5.5.7}$$

but where

$$g_\ell \to \frac{e^{ikr}}{r} f_\ell(r') \tag{5.5.8}$$

as $r \to \infty$, and

$$f_\ell(r') = \int_{-1}^{1} dw \, P_\ell(w) \, e^{-ir'kw} \left(\frac{2\ell+1}{2} \right). \tag{5.5.9}$$

The solution of (5.5.6) subject to the boundary condition that it approach $(1/i^{\ell+1})(e^{ikr}/kr)$ for large kr is called $h_\ell(kr)$ (spherical Bessel function of the third kind). Hence,

$$g_\ell = ki^{\ell+1} h_\ell(kr) \cdot \frac{2\ell+1}{2} \int dw \, P_\ell(w) \, e^{-ir'kw} \tag{5.5.10}$$

The dw integral, for large r', can be estimated by integrating by parts: $e^{-ir'kw} dw = dv$, $P_\ell(w) = u$, so

$$\int dw \, P_\ell(w) \, e^{-ir'kw} = P_\ell(w) \frac{e^{-ir'kw}}{-ikr'} \Big|_{-1}^{1} \, O\left(\frac{1}{r'^2} \right)$$

$$= \frac{2(-i)^\ell}{kr'} \sin\left(kr' - \frac{\ell\pi}{2} \right) + O\left(\frac{1}{r'^2} \right).$$

The solution of (5.5.7) that approaches $\{\sin[kr' - (\ell\pi/2)]/kr'\}$ as $kr' \to \infty$ is called $j_\ell(kr')$ (spherical Bessel function of the first kind). Thus,

$$g_\ell = i(2\ell+1) \, kh_\ell(kr) \, j_\ell(kr') \tag{5.5.11}$$

and, for $r > r'$,

$$\frac{e^{ik|\mathbf{r}-\mathbf{r}'|}}{|\mathbf{r}-\mathbf{r}'|} = ik \sum_\ell (2\ell+1) \, P_\ell(w) \, h_\ell(kr) \, j_\ell(kr') \tag{5.5.12}$$

$$= 4\pi ik \sum_{\ell,m} Y_{\ell,m}(\Omega) \, Y^*_{\ell,m}(\Omega') \, h_\ell(kr) \, j_\ell(kr'). \tag{5.5.13}$$

Also, by taking the limit $r \to \infty$ on both sides of (5.5.12), we find

$$e^{i\mathbf{k}\cdot\mathbf{r}'} = \sum_\ell (2\ell + 1) i^\ell P_\ell(w) j_\ell(kr'). \tag{5.5.14}$$

The functions h_ℓ and j_ℓ are interesting and quite easy to study using methods similar to the standard quantum mechanical treatment of the harmonic oscillator and angular momentum. Recall the definitions:

$$\left\{ -\frac{d^2}{dr^2} - \frac{2}{r}\frac{d}{dr} + \frac{\ell(\ell + 1)}{r^2} - k^2 \right\} \begin{Bmatrix} h_\ell \\ j_\ell \end{Bmatrix} = 0 \tag{5.5.15}$$

and, as $kr \to \infty$,

$$\begin{Bmatrix} h_\ell \\ j_\ell \end{Bmatrix} \to \begin{Bmatrix} \dfrac{1}{i^{\ell+1}}\dfrac{e^{ikr}}{kr} \\[2mm] \dfrac{\sin\left(kr - \dfrac{\ell\pi}{2}\right)}{kr} \end{Bmatrix}. \tag{5.5.16}$$

The more convenient functions are

$$u_\ell(kr) = krj_\ell(kr) \qquad \text{and} \qquad w_\ell(kr) = krh_\ell(kr). \tag{5.5.17}$$

Note that this notation (u_ℓ and w_ℓ) is not standard.

$u_\ell(x)$ and $w_\ell(x)$ satisfy the equation

$$\left[-\frac{d^2}{dx^2} + \frac{\ell(\ell + 1)}{x^2} - 1 \right] \begin{Bmatrix} w_\ell \\ u_\ell \end{Bmatrix} = 0, \tag{5.5.18}$$

the boundary condition $u_\ell(0) = 0$, $w_\ell(x \to \infty) \sim e^{ix}$, and the normalization determined by

$$\begin{pmatrix} w_\ell \\ u_\ell \end{pmatrix} \to \begin{Bmatrix} \dfrac{e^{ix}}{i^{\ell+1}} \\[2mm] \sin\left(x - \dfrac{\ell\pi}{2}\right) \end{Bmatrix}$$

as $x \to \infty$. Here, $x = kr$.

We solve (5.5.18) recursively by factoring. Call the operator $-(d^2/dx^2) + [\ell(\ell + 1)/x^2] = H_\ell$,

$$A_\ell^+ = -\frac{d}{dx} + \frac{\ell}{x} \quad \text{and} \quad A_\ell^- = \frac{d}{dx} + \frac{\ell}{x}. \tag{5.5.19}$$

Then

$$A_\ell^+ A_\ell^- = -\left(\frac{d}{dx} - \frac{\ell}{x}\right)\left(\frac{d}{dx} + \frac{\ell}{x}\right) = H_\ell \tag{5.5.20}$$

and

$$A_\ell^- A_\ell^+ = H_{\ell-1} \tag{5.5.21}$$

so that, if $H_\ell \psi_\ell = \psi_\ell$, thus satisfying (5.5.18), we have

$$A_\ell^+ A_\ell^- \psi_\ell = \psi_\ell$$

and

$$A_\ell^- A_\ell^+ A_\ell^- \psi_\ell = A_\ell^- \psi_\ell$$

and hence

$$H_{\ell-1}(A_\ell^- \psi_\ell) = (A_\ell^- \psi_\ell) \tag{5.5.22}$$

so that A_ℓ^- is a lowering operator, that is, it takes ψ_ℓ into $\psi_{\ell-1}$.
Similarly, $A_{\ell+1}^- A_{\ell+1}^+ = H_\ell$, so that

$$A_{\ell+1}^- A_{\ell+1}^+ \psi_\ell = \psi_\ell$$

and

$$A_{\ell+1}^+ A_{\ell+1}^- A_{\ell+1}^+ \psi_\ell = A_{\ell+1}^+ \psi_\ell$$

or

$$H_{\ell+1}(A_{\ell+1}^+ \psi_\ell) = (A_{\ell+1}^+ \psi_\ell)$$

so that $A_{\ell+1}^+$ is a raising operator, that is, it takes ψ_ℓ into $\psi_{\ell+1}$.
As $x \to \infty$, $A^+ \to -d/dx$, $A^- \to d/dx$; acting on the asymptotic forms for u_ℓ and w_ℓ, we see that

$$A^+ w_\ell \to -\frac{d}{dx}\frac{e^{ix}}{i^{\ell+1}} = \frac{e^{ix}}{i^{\ell+2}},$$

$$A^- w_\ell \to \frac{d}{dx}\frac{e^{ix}}{i^{\ell+1}} = \frac{e^{ix}}{i^\ell}$$

$$A^+ u_\ell \to -\frac{d}{dx}\sin\left(x - \frac{\ell\pi}{2}\right) = -\cos\left(x - \frac{\ell\pi}{2}\right) = \sin\left(x - \frac{\ell+1}{2}\pi\right)$$

and

$$A^- u_\ell \to \frac{d}{dx}\sin\left(x - \frac{\ell\pi}{2}\right) = \cos\left(x - \frac{\ell\pi}{2}\right) = \sin\left(x - \frac{(\ell-1)\pi}{2}\right)$$

so that the raising and lowering operators maintain the correct asymptotic limits and therefore the correct normalizations for the functions w_ℓ and u_ℓ.

We now simply construct the functions starting from the solutions for $\ell = 0$:

$$w_0(x) = \frac{e^{ix}}{i} \tag{5.5.23}$$

and

$$u_0(x) = \sin x. \tag{5.5.24}$$

Thus,

$$w_1(x) = A_1^+ w_0(x) = \left(-\frac{d}{dx} + \frac{1}{x}\right)\frac{e^{ix}}{i} = -e^{ix}\left(1 - \frac{1}{ix}\right)$$

$$u_1(x) = A_1^+ u_0(x) = \left(-\frac{d}{dx} + \frac{1}{x}\right)\sin x = -\cos x + \frac{\sin x}{x}.$$

We note general properties of w_ℓ: $w_\ell = e^{ix}/i^{\ell+1}$ (1+ ascending powers of $1/ix$, the last being $1/x^\ell$). $u_\ell = \sin x$ and $\cos x$ times ascending powers of $1/x$, down to $1/x^\ell$, with u_ℓ odd or even in x according to ℓ even or odd, and going like $x^{\ell+1}$ as $x \to 0$. This last result can be proved by induction, using $A_{\ell+1}^+$ to raise ℓ; however, it is also evident from (5.5.9) using the orthogonality properties of the Legendre polynomials.

The coefficient of $x^{\ell+1}$ for small x in u_ℓ can be calculated by induction:

$$A_\ell^- \psi_\ell = \left(\frac{d}{dx} + \frac{\ell}{x}\right)\psi_\ell = \psi_{\ell-1}$$

so that if $\psi_\ell \to C_\ell x^{\ell+1}$ as $x \to 0$, $\psi_{\ell-1} \to C_\ell(2\ell + 1)x^\ell$ as $x \to 0$ so

$C_{\ell-1}/(2\ell + 1) = C_\ell$ that, together with $C_0 = 1$, yields

$$C_\ell = \frac{1}{(2\ell + 1)!!}$$

and

$$u_\ell \rightarrow \frac{x^{\ell+1}}{(2\ell + 1)!!} \qquad \text{as} \qquad x \rightarrow 0. \qquad (5.5.25)$$

Here, $(2\ell + 1)!! = 1 \cdot 3 \cdot 5 \ldots (2\ell + 1)$.

We can also calculate the coefficient of $1/x^\ell$ for small x in w_ℓ by using the raising operator $A_{\ell+1}^+$ on w_ℓ. We find, with $w_\ell \rightarrow B_\ell e^{ix}/x^\ell$ as $x \rightarrow 0$,

$$\frac{B_{\ell+1} e^{ix}}{x^{\ell+1}} = \frac{(2\ell + 1)}{x^{\ell+1}} e^{ix} B_\ell$$

or

$$B_{\ell+1} = (2\ell + 1) B_\ell.$$

Thus,

$$B_1 = B_0 = \frac{1}{i}, \qquad B_2 = \frac{3}{i}, \qquad \ldots, \qquad B_\ell = \frac{(2\ell - 1)!!}{i},$$

and

$$w_\ell \rightarrow \frac{(2\ell - 1)!!}{i} \frac{e^{ix}}{x^\ell} \qquad \text{as} \qquad x \rightarrow 0. \qquad (5.5.26)$$

Of course, $(2\ell - 1)!! \equiv 1$ for $\ell = 0$.

5.6. PARTIAL WAVES AND PHASE SHIFTS

Given a source S in the form $U\psi$, the resultant Schrödinger type equation is generally hard to solve for the scattering amplitude. There are two exceptions. The first is valid when the interaction is weak, in which case one can apply perturbative methods to the problem. We shall see some examples of this in Section 5.8.

The second requires that the interaction possess spherical symmetry. In that case, one can use spherical harmonic expansions to reduce the three-dimensional problem to a set of one-dimensional problems—one for each ℓ value. We turn to that case now.

We have

$$S(\mathbf{x}, t) = U\psi \tag{5.6.1}$$

where U is a real,[2] spherically symmetric linear operator, which becomes small rapidly away from the source.[3] Since U is spherically symmetric, it is useful to expand the wave function ψ and scattering amplitude f in spherical harmonics. That gives, for large r,

$$\psi \to e^{i\mathbf{k}\cdot\mathbf{x}} + \frac{e^{ikr}}{r} f(\hat{r}, k) + \mathcal{O}\left(\frac{1}{r^2}\right) \tag{5.6.2}$$

or

$$\psi \to \sum_\ell (2\ell + 1) i^\ell \frac{\sin\left(kr - \frac{\ell\pi}{2}\right)}{kr} P_\ell(\hat{r}\cdot\hat{k})$$

$$+ \frac{e^{ikr}}{r} \sum_\ell f_\ell(2\ell + 1) P_\ell(\hat{r}\cdot\hat{k}) + \mathcal{O}\left(\frac{1}{r^2}\right) \tag{5.6.3}$$

where

$$f(\hat{r}, k) = \sum_\ell f_\ell(2\ell + 1) P_\ell(\hat{r}\cdot\hat{k}) \tag{5.6.4}$$

defines f_ℓ and where the Legendre polynomial expansion of $e^{i\mathbf{k}\cdot\mathbf{x}}$ makes use of (5.5.14).

On the other hand, with

$$\psi = \sum_\ell A_\ell \frac{v_\ell(r)}{r} P_\ell(\omega)(2\ell + 1), \tag{5.6.5}$$

as $r \to \infty$, v_ℓ is real and must approach

$$v_\ell \to \sin\left(kr - \frac{\ell\pi}{2} + \delta_\ell\right), \tag{5.6.6}$$

thereby defining the phase shift δ_ℓ. The $\ell\pi/2$ is inserted to make $\delta_\ell = 0$ in the absence of interaction. Each partial wave function can have an arbitrary constant coefficient A_ℓ since the wave equation is homogeneous.

---

[2]See Problem 5.2 for a discussion of scattering by a complex potential.
[3]For example, $U\psi \equiv \int U(\mathbf{x}, \mathbf{y})\psi(\mathbf{y})\, d\mathbf{y}$, with U a real function of $|\mathbf{x}|$, $|\mathbf{y}|$ and $\mathbf{x}\cdot\mathbf{y}$.

That coefficient is determined by the incoming wave and retarded scattered wave boundary conditions implicit in (5.6.3). We find

$$f_\ell = \frac{e^{2i\delta_\ell} - 1}{2ik} = \frac{e^{i\delta_\ell} \sin \delta_\ell}{k} \tag{5.6.7}$$

and

$$f(\hat{r}, k) = \frac{1}{k} \sum_\ell (2\ell + 1) P_\ell(\cos \theta) e^{i\delta_\ell} \sin \delta_\ell. \tag{5.6.8}$$

The scattering amplitude f satisfies the optical theorem. That is, using the result

$$\int P_\ell(w) P_{\ell'}(w) d\Omega = \frac{4\pi}{2\ell + 1} \delta_{\ell\ell'}, \tag{5.6.9}$$

we have

$$\int |f(\hat{r}, k)|^2 d\Omega = \frac{4\pi}{k^2} \sum_\ell (2\ell + 1) \sin^2 \delta_\ell = \frac{4\pi}{k} \operatorname{Im} f(\theta = 0, k). \tag{5.6.10}$$

Referring back to (5.6.3), we see that once f_ℓ is known, the wave function in an interaction free region can be extrapolated back to the point where the interaction becomes significant simply by replacing e^{ikr}/kr by $h_\ell(kr) i^{\ell+1}$ and $[\sin(kr - \ell\pi)/kr]w$ by $j_\ell(kr)$.

We may calculate the behavior of δ_ℓ for small k, assuming a source that is strongly confined to the neighborhood of $r = 0$. The scattering amplitude f will in scalar scattering normally depend on $\cos \theta$ through the dot product $\mathbf{k}_i \cdot \mathbf{k}_f$. Therefore, for small k we can expand f in powers of $\mathbf{k}_i \cdot \mathbf{k}_f = k^2 \cos \theta$. This will lead to a series of the form

$$a_0 + a_1 k^2 \cos \theta + a_2 k^4 \cos^2 \theta + \cdots$$

$$= a_0 P_0 + a_1 k^2 P_1 + a_2 k^4 \left(\frac{2P_2}{3} + \frac{1}{3} P_0 \right) + \cdots.$$

Clearly, the coefficient of P_ℓ is a power series in k, starting with $k^{2\ell}$. Since

$$f_\ell = \frac{e^{i\delta_\ell} \sin \delta_\ell}{k}$$

goes like $k^{2\ell}$ for small k, δ_ℓ must go like $k^{2\ell+1}$.

A more general way of looking at the low wave number behavior of the phase shift is to introduce the logarithmic derivative ξ of the wave function at a point r_0, outside of, but close to, the interaction region. The parameter ξ is then matched to the logarithmic derivative of the wave function v_ℓ that, when $kr \gg \ell$, goes over into

$$\sin\left(kr - \frac{\ell\pi}{2} + \delta_\ell\right),$$

that is,

$$v_\ell = u_\ell(kr)\cos\delta_\ell + q_\ell(kr)\sin\delta_\ell \tag{5.6.11}$$

for r outside of the interaction region. Here, we have introduced the function

$$q_\ell = \frac{w_\ell^* - w_\ell}{2i} \tag{5.6.12}$$

which goes for small x like

$$\cos x \, \frac{(2\ell - 1)!!}{x^\ell}.$$

The logarithmic derivative ξ is obtained by integrating v_ℓ out from the origin; with a short-range, energy-independent potential, ξ will have a finite limit as $k \to 0$, obtained by integrating the equation

$$\left[-\frac{d^2}{dr^2} + U + \frac{\ell(\ell + 1)}{r^2}\right]v_\ell = k^2 v_\ell \tag{5.6.13}$$

to the point r_0 where the match is to be made.

The matching equation is [with $u'_\ell = du_\ell(kr)/dr$, $q'_\ell = dq_\ell(kr)/dr$]

$$\xi = \frac{u'_\ell \cos\delta_\ell + q'_\ell \sin\delta_\ell}{u_\ell \cos\delta_\ell + q_\ell \sin\delta_\ell} \tag{5.6.14}$$

or

$$\tan\delta_\ell = -\frac{\xi u_\ell - u'_\ell}{\xi q_\ell - q'_\ell}. \tag{5.6.15}$$

The small k limit of δ_ℓ can be calculated from (5.6.15). For ξ approaching a finite limit as $k \to 0$ (which as we shall see is not the case in scattering by a dielectric), we find, using the small x expansions of u_ℓ and q_ℓ (5.5.25–5.5.26),

$$\delta_\ell \to -k^{2\ell+1} r_0^{2\ell+2} \frac{\left(\xi - \dfrac{\ell+1}{r_0}\right)}{\xi r_0 + \ell} \frac{1}{(2\ell+1)!!} \frac{1}{(2\ell-1)!!}. \qquad (5.6.16)$$

Equation (5.6.16) is particularly simple for $\ell = 0$ (which is present for a scalar field, but not for the electromagnetic field). It becomes, for $\ell = 0$,

$$\delta_0 \to -k\left(r_0 - \frac{1}{\xi}\right) \qquad (5.6.17)$$

$$= -k\left(r_0 - \frac{v_0(r_0)}{v_0'(r_0)}\bigg|_{k=0}\right). \qquad (5.6.18)$$

It is useful, following a method of Fermi, to parametrize $v_0(r)$ for $k = 0$ in the neighborhood of r_0:

$$v_0 = b(r + a)$$
$$v_0' = b \qquad (5.6.19)$$

and

$$\frac{1}{\xi} = \frac{v_0}{v_0'} = r_0 + a \qquad (5.6.20)$$

so that

$$\frac{\delta_0}{k} \to a. \qquad (5.6.21)$$

a is called the scattering length. The differential cross section at $k = 0$ is $|a|^2$. Note that $-a$ is the value of r at which the zero energy wave function, extrapolated from its value and slope at $v = v_0$, vanishes.

We can illustrate with three cases, all for $\ell = 0$. The equation for v_0 is, at $k = 0$,

$$-\frac{d^2 v_0}{dr^2} + U v_0 = 0.$$

If U is negative (attractive in quantum mechanics) and small, v_0 will look

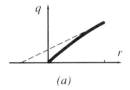 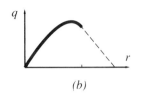 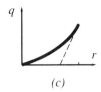

(a)   (b)   (c)

Figure 5.3.

like Figure 5.3(a), will vanish at negative values, and a will be positive. If U is not weak, as in Figure 5.3(b), it can turn the curve over. This corresponds in quantum theory to the existence of a bound state and in classical wave theory to the existence of a localized solution of the wave equation with $\omega < \mu$. In this case, the scattering length will be negative. The third possibility is U positive (in quantum mechanics a repulsion), as in Figure 5.3(c). In that case, the scattering length is again negative.

5.7. ELECTROMAGNETIC FIELD SCATTERING

As with the scalar field, we construct the vector field as a superposition of a positive frequency field and its complex conjugate. That is, for the incoming field we have

$$\mathbf{E}^0 = \mathbf{E}^0_+ + \mathbf{E}^0_- = \mathbf{E}^0_+ + \mathbf{E}^{0*}_+ \qquad (5.7.1)$$

where

$$\mathbf{E}^0_+ = \hat{\mathbf{e}}^0\, e^{i(\mathbf{k}\cdot\mathbf{x} - \omega t)}, \qquad \hat{\mathbf{e}}^{0*}\cdot\hat{\mathbf{e}}^0 = 1, \qquad k = \omega \qquad (5.7.2)$$

and

$$\mathbf{B}^0 = \mathbf{B}^0_+ + \mathbf{B}^{0*}_+, \qquad (5.7.3)$$

with

$$i\omega\mathbf{B}^0_+ = i\mathbf{k} \times \mathbf{E}^0_+ . \qquad (5.7.4)$$

The outgoing scattered wave for the $\mathbf{E}$ field will be

$$\mathbf{E}_{sc} = \frac{e^{ikr}}{r}\,\mathbf{f} + \mathcal{O}\!\left(\frac{1}{r^2}\right) \qquad (5.7.5)$$

and

$$i\omega \mathbf{B}_{sc} = ik\hat{\mathbf{r}} \times \mathbf{E}_{sc} + \mathcal{O}\left(\frac{1}{r^2}\right). \tag{5.7.6}$$

Of course, $\mathbf{E}^0$ and $\mathbf{f}$ will be transverse. That is,

$$\mathbf{k} \cdot \mathbf{E}^0 = \hat{\mathbf{r}} \cdot \mathbf{f} = 0. \tag{5.7.7}$$

The incoming time-averaged Poynting flux will be

$$\overline{\mathcal{P}}^0 = \frac{1}{2\pi} \mathbf{E}_+^* \times \mathbf{B}_+$$

$$= \frac{1}{2\pi} \mathbf{E}_+^* \times \left(\frac{\mathbf{k}}{\omega} \times \mathbf{E}_+\right)$$

or

$$\overline{\mathcal{P}}^0 = \frac{1}{2\pi} \hat{\mathbf{k}} \mathbf{E}_+^* \cdot \mathbf{E}_+ = \frac{1}{2\pi} \hat{\mathbf{k}}. \tag{5.7.8}$$

The time-averaged scattered flux per unit area will be

$$\overline{\mathcal{P}}^{sc} = \frac{1}{2\pi} \mathbf{f}^* \times (\hat{\mathbf{r}} \times \mathbf{f}) \frac{1}{r^2} \tag{5.7.9}$$

$$= \frac{1}{2\pi} \frac{\hat{\mathbf{r}}}{r^2} \mathbf{f}^* \cdot \mathbf{f}, \tag{5.7.10}$$

giving a scattered flux per unit solid angle

$$\overline{\mathcal{P}}^{sc} \cdot \hat{\mathbf{r}} \, r^2 = \frac{1}{2\pi} \mathbf{f}^* \cdot \mathbf{f}$$

and a differential scattering cross section[4]

$$\frac{d\sigma}{d\Omega} = \mathbf{f}^* \cdot \mathbf{f}. \tag{5.7.11}$$

---

[4]The wave packet discussion given earlier for scalar scattering evidently goes through equally well for the vector field we are considering now, with the final result being the justification of (5.7.11).

The polarization of the scattered wave is described by the vector scattering amplitude $\mathbf{f}(\mathbf{k}, \hat{\mathbf{r}}, \hat{\mathbf{e}}^0)$, and is a linear function of $\hat{\mathbf{e}}^0$. The function $\mathbf{f}$ can be resolved into any complete pair of polarization vectors $\hat{\mathbf{e}}^\lambda$, where we would normally choose $\hat{\mathbf{e}}^{\lambda*} \cdot \hat{\mathbf{e}}^{\lambda'} = \delta_{\lambda\lambda'}$. Thus,

$$\mathbf{f} = \sum_{\lambda=1}^{2} \hat{\mathbf{e}}^\lambda \hat{\mathbf{e}}^{\lambda*} \cdot \mathbf{f} \qquad (5.7.12)$$

and $\hat{\mathbf{e}}^{\lambda*} \cdot \mathbf{f}$ gives the scattering that would be measured by a detector detecting only the polarization state λ.

5.8. THE OPTICAL THEOREM FOR LIGHT

As in our discussion of the scalar field, we make use of the surface integral of the Poynting vector, which satisfies the equation

$$\int_S \mathscr{P} \cdot d\mathbf{S} = -\frac{d}{dt} \int_V d\mathbf{r}\, u \qquad (5.8.1)$$

so that for a monochromatic wave, the time average

$$\int_S \overline{\mathscr{P}} \cdot d\mathbf{S} = \frac{1}{2\pi} \mathrm{Re} \int_S \mathbf{E}_+^* \times \mathbf{B}_+ \cdot d\mathbf{S} = 0, \qquad (5.8.2)$$

where we will integrate over a distant sphere. For $r \to \infty$, we know

$$\mathbf{E}_+ = \left(\hat{\mathbf{e}}^0\, e^{i\mathbf{k}\cdot\mathbf{x}} + \mathbf{f}\frac{e^{ikr}}{r} \right) e^{-i\omega t} + \mathcal{O}\!\left(\frac{1}{r^2}\right) \qquad (5.8.3)$$

and

$$\mathbf{B}_+ = \left(\hat{\mathbf{k}} \times \hat{\mathbf{e}}^0\, e^{i\mathbf{k}\cdot\mathbf{x}} + \frac{\hat{\mathbf{r}} \times \mathbf{f}\, e^{ikr}}{r} \right) e^{-i\omega t} + \mathcal{O}\!\left(\frac{1}{r^2}\right). \qquad (5.8.4)$$

Equations (5.8.2), (5.8.3), and (5.8.4) then tell us that

$$0 = \mathrm{Re} \int d\Omega r\, \hat{\mathbf{r}} \cdot \left(\hat{\mathbf{e}}^{0*}\, e^{-i\mathbf{k}\cdot\mathbf{x}} + \frac{\mathbf{f}^*\, e^{-ikr}}{r} \right) \times \left(\hat{\mathbf{k}} \times \hat{\mathbf{e}}^0\, e^{i\mathbf{k}\cdot\mathbf{x}} + \frac{\hat{\mathbf{r}} \times \mathbf{f}}{r}\, e^{ikr} \right).$$

$$(5.8.5)$$

We proceed as in Section 5.4, noting that

$$\int\limits_{-1}^{1} dw\, e^{-ikrw} F(w) = -\frac{1}{ikr}\left(e^{-ikr}F(1) - e^{ikr}F(-1)\right) + \mathcal{O}\left(\frac{1}{r^2}\right). \qquad (5.8.6)$$

We have, with $w = \hat{r} \cdot \hat{k}$,

$$\mathrm{Re}\left\{\frac{2\pi}{ik}[\hat{\mathbf{r}} \cdot \mathbf{f}^* \times (\hat{\mathbf{k}} \times \hat{e}^0)|_{w=1} - \hat{\mathbf{r}} \cdot \mathbf{f}^* \times (\hat{\mathbf{k}} \times \hat{e}^0)|_{w=-1}\, e^{-2ikr}]\right.$$

$$-\frac{2\pi}{ik}[\hat{\mathbf{r}} \cdot \hat{e}^{0*} \times (\hat{\mathbf{r}} \times \mathbf{f})|_{w=1} - \hat{\mathbf{r}} \cdot \hat{e}^{0*} \times (\hat{\mathbf{r}} \times \mathbf{f})|_{w=-1}\, e^{2ikr}]$$

$$\left. + \int d\Omega\, \mathbf{f}^* \cdot \mathbf{f}\right\} = 0. \qquad (5.8.7)$$

Combining terms and noting that $\hat{\mathbf{r}} \cdot \hat{\mathbf{k}} = \pm 1$ at $w = \pm 1$, we obtain

$$\mathrm{Re}\left[\frac{2\pi}{ik}(\mathbf{f}^* \cdot \hat{e}^0 - \hat{e}^{0*} \cdot \mathbf{f})\right|_{w=1} - \frac{2\pi}{ik}(\mathbf{f}^* \cdot \hat{e}^0\, e^{-2ikr}$$

$$\left. + \hat{e}^{0*} \cdot \mathbf{f}\, e^{2ikr})\right|_{w=-1} + \int d\Omega\, \mathbf{f}^* \cdot \mathbf{f}\right] = 0 \qquad (5.8.8)$$

or finally,

$$\frac{4\pi}{k} \mathrm{Im}\, \hat{e}^{0*} \cdot \mathbf{f}\Big|_{w=1} = \sigma_{el} \qquad (5.8.9)$$

where, as before, σ_{el} is the total elastic cross section, and $\hat{e}^{0*} \cdot \mathbf{f}$ is the forward, polarization-preserving scattering amplitude.

5.9. PERTURBATION THEORY OF SCATTERING

We consider a situation where the source-field coupling is sufficiently weak so that we may calculate the charge and current distribution of the source induced by the field without taking into account the reaction of the source on itself. We illustrate with two examples: scattering by a damped oscillator and scattering by a dielectric with a dielectric constant near 1.

5.9.1. Scattering by a Damped Oscillator and Radiation Reaction

We write the equation of the charged oscillator in a weak electromagnetic field:

$$m\ddot{\mathbf{x}} + R\dot{\mathbf{x}} + k\mathbf{x} = q(\mathbf{E}(\mathbf{x}, t) + \dot{\mathbf{x}} \times \mathbf{B}(\mathbf{x}, t)). \qquad (5.9.1)$$

Here, R is a damping constant that we will adjust to give overall energy conservation via the optical theorem. m and k are, respectively, the mass and force constant of the oscillator.

The displacement $\mathbf{x}$ and velocity $\dot{\mathbf{x}}$ in steady-state motion will be linear in the field strength; therefore, since the field is weak, we may neglect the $\mathbf{x}$ and $\dot{\mathbf{x}}$ dependence on the right-hand side of (5.9.1). In this linear approximation, the scattering is independent of the strength of the field, which we normalize to unity. The incoming field $\mathbf{E}_0$ is then

$$\mathbf{E}_0 = \hat{\mathbf{e}}^0 \, e^{-i\omega t} \qquad (5.9.2)$$

and the magnetic field is

$$\mathbf{B}_0 = \hat{\mathbf{k}} \times \hat{\mathbf{e}}^0 \, e^{-i\omega t}.$$

The incoming average energy flux per unit area is, with $\hat{\mathbf{e}}^{0*} \cdot \hat{\mathbf{e}}^0 = 1$,

$$\overline{\mathcal{P}}_0 = \frac{1}{2\pi} \hat{\mathbf{k}}. \qquad (5.9.3)$$

The steady-state motion of the oscillator is given (in our linear approximation) by

$$\mathbf{x} = \mathbf{x}_0 \, e^{-i\omega t} \qquad (5.9.4)$$

with

$$\mathbf{x}_0 = \frac{\dfrac{q\hat{\mathbf{e}}^0}{m}}{\dfrac{\omega_0^2 - \omega^2 - i\omega R}{m}}. \qquad (5.9.5)$$

and $\omega_0^2 = k/m$.

The field radiated by the oscillator is then

$$\mathbf{E}^{sc} = -q \frac{\ddot{\mathbf{x}}_T(t_R)}{r} \tag{5.9.6}$$

and the scattering amplitude $\mathbf{f}$ is given by

$$\mathbf{f} = \frac{q^2 \omega^2 \hat{\mathbf{e}}_T^0}{m \left[\omega_0^2 - \omega^2 - \dfrac{i\omega R}{m} \right]}. \tag{5.9.7}$$

The total elastic cross section is

$$\sigma_{el} = \frac{q^4 \omega^4}{m^2} \frac{1}{\left| \omega_0^2 - \omega^2 - i \dfrac{R\omega}{m} \right|^2} \int |\hat{\mathbf{e}}_T^0|^2 \, d\Omega$$

$$= \frac{8\pi}{3} \frac{q^4 \omega^4}{m^2} \cdot \frac{1}{\left| \omega_0^2 - \omega^2 - i \dfrac{R\omega}{m} \right|^2}. \tag{5.9.8}$$

Before discussing the result (5.9.8), we use the optical theorem to determine $R(\omega)$. From (5.9.7),

$$\text{Im} \, \hat{\mathbf{e}}^0 \cdot \mathbf{f}(\theta = 0) = \frac{q^2 \omega^2}{m^2} \frac{\omega R}{\left| \omega_0^2 - \omega^2 - i \dfrac{R\omega}{m} \right|^2} \tag{5.9.9}$$

Since the optical theorem requires

$$\text{Im} \, \hat{\mathbf{e}}^0 \cdot \mathbf{f}(\theta = 0) = \frac{\omega}{4\pi} \sigma_{el},$$

we find

$$\frac{\omega}{4\pi} \frac{8\pi}{3} \frac{q^4 \omega^4}{m^2} \frac{1}{\left| \omega_0^2 - \omega^2 - i \dfrac{R\omega}{m} \right|^2} = \frac{q^2 \omega^2}{m^2} \frac{\omega R}{\left| \omega_0^2 - \omega^2 - i \dfrac{R\omega}{m} \right|^2}$$

or

$$R = \frac{2}{3} q^2 \omega^2 \tag{5.9.10}$$

so that the damping term in (5.9.1) could be written as $R \cdot \dot{\mathbf{x}} = -\frac{2}{3}q^2\dddot{\mathbf{x}}$, which is a more usual form for the so-called radiation reaction force, $\mathbf{f}_r = \frac{2}{3}q^2\dddot{\mathbf{x}}$. We see (as in Problem 4.8) that inserting the force $\mathbf{f}_r$ into the equation of motion gives a correct overall energy balance for this situation. In general, for confined motion this will be the case. It is clear however that in the absence of a confining potential, the force $\mathbf{f}_r$ gives nonsensical results. Thus, with the force constant k and the incident field $\mathbf{E}_0$ set to zero, (5.9.1) would be

$$m\ddot{\mathbf{x}} = \frac{2}{3}q^2\dddot{\mathbf{x}} \qquad (5.9.11)$$

which has the general solution

$$\mathbf{x} = \mathbf{x}_0 - \frac{\mathbf{a}_0}{\gamma^2} + \left(\mathbf{v}_0 - \frac{\mathbf{a}_0}{\gamma}\right)t + \frac{\mathbf{a}_0}{\gamma^2}e^{\gamma t} \qquad (5.9.12)$$

with $\mathbf{x}_0$, $\mathbf{v}_0$, and $\mathbf{a}_0$ arbitrary and

$$\gamma = \left(\frac{2}{3}\frac{q^2}{m}\right)^{-1}.$$

Thus, unless $\mathbf{a}_0 = 0$, the motion explodes exponentially, and the formula makes no sense. Note that this problem is not resolved by a harmonic binding force.

We return to (5.9.8) for the cross section. We see that the energy denominator $1/(\omega_0^2 - \omega^2)$ has been damped by the imaginary term $-\frac{2}{3}i(\omega^3/m)q^2$. Note that the scattering amplitude $\mathbf{f}$ at the resonant $\omega = \omega_0$ takes on the imaginary value

$$\mathbf{f}_{\omega=\omega_0} = \frac{3}{2}\frac{i\hat{\mathbf{e}}_{\lambda T}}{\omega}$$

independent of q and m. This is a characteristic of resonant behavior: The resonant scattering amplitude is given by the wavelength multiplied by a kinematically determined constant (here, $\frac{3}{2}$).

The damping constant determines the width of the resonant curve. The cross section as a function of ω is

$$\sigma_{el} = \frac{C\omega^4}{(\omega^2 - \omega_0^2)^2 + \dfrac{\omega^2 R^2(\omega)}{m^2}}$$

with C a constant. For small R, the maximum is at $\omega = \omega_0$:

$$\sigma_{max} = \frac{C\omega_0^2 m^2}{R^2(\omega_0)}. \tag{5.9.13}$$

The cross section takes on half this value at

$$(\omega^2 - \omega_0^2)^2 \cong \frac{\omega_0^2 R^2}{m^2} \qquad \text{or} \qquad \omega - \omega_0 \cong \frac{1}{2}\frac{R}{m};$$

thus, the full width at half maximum is $\Gamma = R/m$.

On the other hand, the possible decay constants of the isolated oscillator are given by the roots of the equation

$$\omega_0^2 - \omega^2 - i\frac{\omega R(\omega)}{m} = 0 \tag{5.9.14}$$

or

$$\omega \cong \pm \omega_0 - \frac{iR(\omega_0)}{2m}$$

giving a time dependence to the oscillation amplitude

$$x = x(t = 0)\, e^{\pm i\omega_0 t - (R/2m)t},$$

and an energy decay given by $e^{-\Gamma t} = e^{-(R/m)t}$. Thus, the decay constant is, in fact, equal to the width of the resonance.

There is unfortunately a third unwanted root of (5.9.14) produced at high ω by taking the ω^2 dependence of R into account; this root is closely related to the unphysical runaway solutions of the free particle equation found earlier. Note that this root is in the upper half ω plane and, hence, produces acausal behavior in the scattering. We evade all these problems (without justification) by taking $R(\omega)$ to be a constant $R = R(\omega_0)$. Furthermore, the energy balance now only works near the resonance. A justification of this procedure cannot be given within the framework of classical field theory. The contradiction between energy conservation and causality is a genuine difficulty of classical electromagnetic theory describing the interaction of electromagnetic fields with point particles. This problem (the unwanted root) does not appear in relativistic quantum electrodynamics;

however, other new problems do. For further discussion of this problem, see Section 7.5.

We finally note the low-frequency limit of (5.9.7) for the scattering amplitude by a free particle,

$$\mathbf{f} = -\frac{q^2}{m}\mathbf{e}_{\lambda T},$$

(5.9.15)

and of (5.9.8) for the elastic cross section,

$$\sigma_{el} = \frac{8\pi}{3}\left(\frac{q^2}{m}\right)^2.$$

(5.9.16)

The expression (5.9.16) is called the Thompson cross section. This result was first used by Thompson to measure the number of electrons in carbon by X-ray scattering.

Note also that the scattering amplitude of a bound electron goes to zero for small ω like ω^2, accounting for the dominance of short wavelengths in the scattering of visible light by air molecules and therefore the blue color of the sky.

Scattering by a Dielectric with a Dielectric Constant Near 1

Our second example is scattering by a dielectric with permeability $\mu = 1$ and a dielectric constant ϵ near 1: $\epsilon - 1 = 4\pi\chi \ll 1$.

As always, we return to Maxwell for guidance:

$$\mathbf{B} = \frac{\nabla \times \mathbf{E}}{i\omega} \quad \text{and} \quad \nabla \times \mathbf{B} = -i\omega\mathbf{D} = -i\omega\epsilon\mathbf{E}$$

giving a single equation for $\mathbf{E}$:

$$\nabla \times \left(\frac{\nabla \times \mathbf{E}}{i\omega}\right) = -i\omega\epsilon\mathbf{E}$$

(5.9.17)

or

$$\nabla \times (\nabla \times \mathbf{E}) = \omega^2\epsilon\mathbf{E}.$$

We expand about $\mathbf{E} = \mathbf{E}_0$, the incident field, and $\epsilon = 1$:

$$\nabla \times (\nabla \times (\mathbf{E}_0 + \mathbf{E}_1)) = \omega^2(1 + (\epsilon - 1))(\mathbf{E}_0 + \mathbf{E}_1)$$

where $\nabla \cdot \mathbf{E}_0 = 0$ and $-\nabla^2\mathbf{E}_0 = \omega^2\mathbf{E}_0$, so that $\mathbf{E}_0 = \hat{\mathbf{e}}_\lambda\,e^{i(\mathbf{k}\cdot\mathbf{x}-\omega t)}$. There follows

$$\nabla \times (\nabla \times \mathbf{E}_1) = \omega^2[\mathbf{E}_1 + [(\epsilon - 1)\,\mathbf{E}_0]] \tag{5.9.18}$$

from which we deduce

$$\nabla \cdot \mathbf{E}_1 = -\nabla \cdot [(\epsilon - 1)\,\mathbf{E}_0] \tag{5.9.19}$$

and

$$-\nabla^2\mathbf{E}_1 + \nabla(\nabla \cdot \mathbf{E}_1) = \omega^2[\mathbf{E}_1 + (\epsilon - 1)\,\mathbf{E}_0]. \tag{5.9.20}$$

Substituting (5.9.19) into (5.9.20) yields

$$(-\nabla^2 - \omega^2)\,\mathbf{E}_1 = \omega^2(\epsilon - 1)\,\mathbf{E}_0 + \nabla(\nabla \cdot (\epsilon - 1)\,\mathbf{E}_0). \tag{5.9.21}$$

We solve (5.9.21) with the usual retarded Green's function. As $r \to \infty$.

$$\mathbf{E}_1 \to \frac{e^{ikr}}{r}\int dx\, e^{-i\mathbf{k}_f\cdot\mathbf{x}'}\left[\omega^2\left(\frac{\epsilon - 1}{4\pi}\right)\mathbf{E}_0(\mathbf{x}')\right.$$
$$\left. + \nabla'\left(\nabla' \cdot \left(\frac{\epsilon - 1}{4\pi}\right)\mathbf{E}_0(\mathbf{x}')\right)\right] \tag{5.9.22}$$

or

$$\mathbf{E}_1 = \omega^2\frac{e^{ikr}}{r}\int dx\, e^{-i\mathbf{k}_f\cdot\mathbf{x}'}\left(\frac{\epsilon - 1}{4\pi}\right)e^{i\mathbf{k}\cdot\mathbf{x}'}(\hat{\mathbf{e}}_\lambda - \hat{\mathbf{k}}_f\hat{\mathbf{k}}_f \cdot \hat{\mathbf{e}}_\lambda), \tag{5.9.23}$$

yielding a scattering amplitude

$$\mathbf{f} = \omega^2\int dx'\, e^{i(\mathbf{k}-\mathbf{k}_f)\cdot\mathbf{x}'}\chi(\mathbf{x}')\,\hat{\mathbf{e}}_{\lambda T}, \tag{5.9.24}$$

where $\chi(\mathbf{x}')$ is the dielectric susceptibility at the point $\mathbf{x}'$, $\chi = (\epsilon - 1)/4\pi$.

5.10. VECTOR MULTIPOLES

We come next to the partial wave expansion for a vector field. As remarked earlier, the method is essential for a large class of problems. We therefore take it up, even though the required algebra is quite complicated (but elementary). The prerequisite groundwork was done in Appendix B and in Section 5.5.

The mathematics is based on a nontrivial operator dyadic identity in three dimensions:

$$\overset{\leftrightarrow}{\mathbf{1}} = \nabla \frac{1}{\nabla^2} \nabla + \mathbf{L} \frac{1}{\mathbf{L}^2} \mathbf{L} + \frac{\nabla \times \mathbf{L}}{i} \frac{1}{\nabla^2 \mathbf{L}^2} \frac{\mathbf{L} \times \nabla}{i}$$

where $\overset{\leftrightarrow}{\mathbf{1}}$ is the unit dyadic and $\mathbf{L}$ the quantum mechanical angular momentum operator:

$$\mathbf{L} = \mathbf{r} \times \frac{\nabla}{i}.$$

A vector field can be expanded as

$$\mathbf{V} = \nabla \psi_1 + \mathbf{L}\psi_2 + \frac{\nabla \times \mathbf{L}}{i} \psi_3 \qquad (5.10.1)$$

To see that this is correct, we note that we can always expand

$$\mathbf{V} = \nabla \psi_1 + \nabla \times \mathbf{Q} \qquad (5.10.2)$$

for some ψ_1 and $\mathbf{Q}$. Equation (5.10.1) prescribes

$$\mathbf{Q} = i\mathbf{r}\psi_2 - i\mathbf{L}\psi_3 - \nabla\psi_4 \qquad (5.10.3)$$

where ψ_4 does not affect $\mathbf{V}$. We now show that any vector function $\mathbf{Q}$ can be expanded in the form (5.10.3). We introduce spherical coordinates r, θ, and φ and orthogonal unit vectors $\hat{\mathbf{r}}$, $\hat{\boldsymbol{\theta}}$, $\hat{\boldsymbol{\varphi}}$ such that $\hat{\mathbf{r}} \times \hat{\boldsymbol{\theta}} = \hat{\boldsymbol{\varphi}}$, $\hat{\boldsymbol{\theta}} \times \hat{\boldsymbol{\varphi}} = \hat{\mathbf{r}}$, and $\hat{\boldsymbol{\varphi}} \times \hat{\mathbf{r}} = \hat{\boldsymbol{\theta}}$. The gradient operator is

$$\nabla = \hat{\mathbf{r}} \frac{\partial}{\partial r} + \frac{\hat{\boldsymbol{\theta}}}{r} \frac{\partial}{\partial \theta} + \frac{\hat{\boldsymbol{\varphi}}}{r \sin \theta} \frac{\partial}{\partial \varphi} \qquad (5.10.4)$$

and

$$\mathbf{L} = \frac{1}{i} \mathbf{r} \times \nabla = \frac{1}{i} \left(\hat{\boldsymbol{\varphi}} \frac{\partial}{\partial \theta} - \frac{\hat{\boldsymbol{\theta}}}{\sin \theta} \frac{\partial}{\partial \varphi} \right). \qquad (5.10.5)$$

The expansion in question is then, with $\mathbf{Q} = \hat{\mathbf{r}}Q_r + \hat{\boldsymbol{\theta}}Q_\theta + \hat{\boldsymbol{\varphi}}Q_\varphi$ and the Q_i

arbitrary single-valued functions,

$$Q_r = ir\psi_2 - \frac{\partial \psi_4}{\partial r} \tag{5.10.6}$$

$$Q_\theta = \frac{1}{\sin \theta} \frac{\partial \psi_3}{\partial \varphi} - \frac{1}{r} \frac{\partial \psi_4}{\partial \theta}, \tag{5.10.7}$$

and

$$Q_\phi = - \frac{\partial \psi_3}{\partial \theta} - \frac{1}{r} \frac{1}{\sin \theta} \frac{\partial \psi_4}{\partial \varphi}. \tag{5.10.8}$$

Clearly, one must choose ψ_3 and ψ_4 to satisfy (5.10.7) and (5.10.8). ψ_2 is then chosen to satisfy (5.10.6). So, our problem is to show that ψ_3 and ψ_4 can be found.

We multiply (5.10.7) and (5.10.8) by $\sin \theta$. There results

$$\sin \theta \frac{\partial \bar{\psi}_4}{\partial \theta} - \frac{\partial \psi_3}{\partial \varphi} = -P_\theta \tag{5.10.9}$$

$$\frac{\partial \bar{\psi}_4}{\partial \varphi} + \sin \theta \frac{\partial \psi_3}{\partial \theta} = -P_\varphi \tag{5.10.10}$$

where

$$\bar{\psi}_4 = \frac{\psi_4}{r}$$

and where the $P_i = \sin \theta \, Q_i$ are still arbitrary and single-valued. Since the operators $\sin \theta (\partial / \partial \theta)$ and $\partial / \partial \varphi$ commute with each other, (5.10.9) and (5.10.10) can be solved algebraically:

$$\bar{\psi}_4 = \frac{1}{\sin \theta \dfrac{\partial}{\partial \theta} \sin \theta \dfrac{\partial}{\partial \theta} + \dfrac{\partial^2}{\partial \varphi^2}} \left(\sin \theta \frac{\partial P_\theta}{\partial \theta} + \frac{\partial P_\varphi}{\partial \varphi} \right) \tag{5.10.11}$$

and

$$\psi_3 = - \frac{1}{\sin \theta \dfrac{\partial}{\partial \theta} \sin \theta \dfrac{\partial}{\partial \theta} + \dfrac{\partial^2}{\partial \varphi^2}} \left(\sin \theta \frac{\partial}{\partial \theta} P_\varphi - \frac{\partial P_\theta}{\partial \varphi} \right). \tag{5.10.12}$$

We factor out $\sin^2 \theta$:

$$-\frac{1}{\sin\theta\,\dfrac{\partial}{\partial\theta}\sin\theta\,\dfrac{\partial}{\partial\theta}+\dfrac{\partial^2}{\partial\varphi^2}}=-\frac{1}{\left[\dfrac{1}{\sin\theta\,\partial\theta}\dfrac{\partial}{\partial\theta}\sin\theta\,\dfrac{\partial}{\partial\theta}+\dfrac{1}{\sin^2\theta}\dfrac{\partial^2}{\partial\varphi^2}\right]}\cdot\frac{1}{\sin^2\theta}$$

so that

$$\overline{\psi}_4 = \frac{1}{\mathbf{L}^2}\left(\frac{1}{\sin\theta}\frac{\partial P_\theta}{\partial\theta}+\frac{1}{\sin^2\theta}\frac{\partial P_\varphi}{\partial\varphi}\right) \tag{5.10.13}$$

and

$$\psi_3 = \frac{1}{\mathbf{L}^2}\left(\frac{1}{\sin\theta}\frac{\partial}{\partial\theta}P_\varphi-\frac{1}{\sin^2\theta}\frac{\partial P_\theta}{\partial\varphi}\right), \tag{5.10.14}$$

where we recognize the form

$$-\left[\frac{1}{\sin\theta\,\partial\theta}\frac{\partial}{\partial\theta}\sin\theta\,\frac{\partial}{\partial\theta}+\frac{1}{\sin^2\theta}\frac{\partial^2}{\partial\varphi^2}\right]$$

as the operator

$$\mathbf{L}^2 = r^2\left(-\nabla^2+\frac{\partial^2}{\partial r^2}+\frac{2}{r}\frac{\partial}{\partial r}\right).$$

Of course, one here obtains ψ_3 and ψ_4 as a sum of spherical harmonics. Note that the operator $1/\mathbf{L}^2 = 1/\ell(\ell + 1)$ does not become singular since the two functions in parentheses in (5.10.13) and (5.10.14) have no $\ell = 0$ projection, and P_θ and P_φ vanish at $\theta = 0$ and π.

We return now to the electric field. We consider first radiation by a prescribed current. The radiation equations for $\mathbf{E}$ may be obtained directly from Maxwell's equations:

$$\nabla \times \mathbf{E} = i\omega\mathbf{B}, \qquad \nabla \times \mathbf{B} = -i\omega\mathbf{E} + 4\pi\mathbf{j}, \qquad \text{and} \qquad \nabla \cdot \mathbf{E} = 4\pi\rho$$

so

$$\nabla \times \left(\frac{\nabla \times \mathbf{E}}{i\omega}\right) = -i\omega\mathbf{E} + 4\pi\mathbf{j}$$

and

$$\nabla^2\mathbf{E} + \omega^2\mathbf{E} = -4\pi[i\omega\mathbf{j} - \nabla\rho], \tag{5.10.15}$$

together with the continuity equation, $\nabla \cdot \mathbf{j} = i\omega\rho$.

We solve directly for the retarded field:

$$\mathbf{E} = \int d\mathbf{x}' \frac{e^{ik|\mathbf{x}-\mathbf{x}'|}}{|\mathbf{x}-\mathbf{x}'|} [i\omega\mathbf{j}(\mathbf{x}') - \nabla'\rho(\mathbf{x}')], \qquad (5.10.16)$$

with the large r form

$$\mathbf{E} = \frac{e^{ikr}}{r} \int d\mathbf{x}' \, e^{-i\mathbf{k}\cdot\mathbf{x}'} [i\omega\mathbf{j}(\mathbf{x}') - \nabla'\rho(\mathbf{x}')] + \mathcal{O}\!\left(\frac{1}{r^2}\right). \qquad (5.10.17)$$

We now apply the expansion (5.10.1) to (5.10.16):

$$\mathbf{E} = \nabla\psi_L + \mathbf{L}\psi_M + \frac{\nabla}{i} \times \frac{\mathbf{L}}{\omega} \psi_E \qquad (5.10.18)$$

where L stands for longitudinal, M for magnetic, and E for electric. The normalization factor ω is inserted for convenience.

We solve for the ψ's by three orthogonal projections. First,

$$\nabla^2\psi_L = \nabla \cdot \mathbf{E}. \qquad (5.10.19)$$

Since $\nabla \cdot \mathbf{E} = 4\pi\rho$, we can solve (5.10.19) for ψ_L:

$$\psi_L = -\int d\mathbf{r}' \frac{\rho(\mathbf{r}')}{|\mathbf{r}-\mathbf{r}'|} \qquad (5.10.20)$$

which has inverse power law behavior at large r and is instantaneously related to the charge density ρ. Of course, the electric field itself does not have such acausal behavior since the factor e^{ikr} in (5.10.17) guarantees proper retardation. This power behavior reflects the use of the potentials ψ and must disappear (by cancellation) in a calculation of the field itself.

We solve for ψ_M by projecting with $\mathbf{L}$. We note that

$$\mathbf{L} \cdot \nabla = \frac{\mathbf{r}}{i} \times \nabla \cdot \nabla = \frac{\mathbf{r}}{i} \cdot \nabla \times \nabla = 0$$

and since $\nabla \times \mathbf{L} = -\mathbf{L} \times \nabla - (2\nabla/i)$,

$$\mathbf{L} \cdot \nabla \times \mathbf{L} = -\mathbf{L} \times \nabla = -\mathbf{L} \times \mathbf{L} \cdot \nabla = -i\mathbf{L} \cdot \nabla = 0,$$

and

$$\mathbf{L}^2 \psi_M = \mathbf{L} \cdot \mathbf{E}. \tag{5.10.21}$$

Thus,

$$\psi_M = \frac{1}{\mathbf{L}^2} \mathbf{L} \cdot \mathbf{E} \tag{5.10.22}$$

provides an expansion in spherical harmonics for ψ_M. Note again that the factor $\mathbf{L}$ in $\mathbf{L} \cdot \mathbf{E}$ precludes any problem with $\ell = 0$.

Finally, ψ_E is projected with $\dfrac{\mathbf{L} \times \nabla}{i\omega}$ so that

$$\frac{\mathbf{L} \times \nabla \cdot \nabla \times \mathbf{L}}{\omega^2} \psi_E = \frac{\mathbf{L} \times \nabla}{i\omega} \cdot \mathbf{E} \quad \text{or} \quad \frac{\nabla^2 \mathbf{L}^2}{\omega^2} \psi_E = \frac{\mathbf{L} \times \nabla}{i\omega} \cdot \mathbf{E}. \tag{5.10.23}$$

The magnetic field is given by a formula very similar to (5.10.18):

$$\mathbf{B} = \frac{\nabla \times \mathbf{E}}{i\omega} = \frac{\nabla \times \mathbf{L}}{i\omega} = \frac{\nabla \times \mathbf{L}}{i\omega} \psi_M + \frac{\nabla \times (\nabla \times \mathbf{L})}{-\omega^2} \psi_E \tag{5.10.24}$$

$$= \frac{\nabla \times \mathbf{L}}{i\omega} \psi_M + \frac{\nabla^2}{\omega^2} \mathbf{L} \psi_E \tag{5.10.25}$$

Outside the charge distribution,

$$\nabla^2 = -\omega^2 \quad \text{and} \quad \mathbf{B} = \frac{\nabla \times \mathbf{L}}{i\omega} \psi_M - \mathbf{L} \psi_E. \tag{5.10.26}$$

Note the generalization of the relation (4.5.20) between electric and magnetic dipole radiation. We interchange electric and magnetic radiation by the transformation $\psi'_E = \psi_M$ and $\psi'_M = -\psi_E$, which produces $\mathbf{E}' = \mathbf{B}$ and $\mathbf{B}' = -\mathbf{E}$.

We next calculate the potentials from (5.10.19), (5.10.22), and (5.10.23). We already know from (5.10.20) that

$$\psi_L = -\int d\mathbf{x}' \frac{\rho(\mathbf{x}')}{|\mathbf{x} - \mathbf{x}'|}. \tag{5.10.20}$$

For ψ_M, we have from (5.10.22)

$$\psi_M = \frac{1}{\mathbf{L}^2} \mathbf{L} \cdot \mathbf{E} \tag{5.10.22}$$

$$= \frac{\omega}{L^2} \int d\mathbf{r}' \frac{e^{ik|\mathbf{r}-\mathbf{r}'|}}{|\mathbf{r}-\mathbf{r}'|} \mathbf{r}' \times \nabla' \cdot \mathbf{j}(\mathbf{r}') \qquad (5.10.27)$$

$$= \frac{\omega}{L^2} \int d\mathbf{r}' \frac{e^{ik|\mathbf{r}-\mathbf{r}'|}}{|\mathbf{r}-\mathbf{r}'|} (-\nabla') \cdot [\mathbf{r}' \times \mathbf{j}(\mathbf{r}')]. \qquad (5.10.28)$$

We recognize the source of magnetic radiation as $-\nabla \cdot \mathbf{M}$, where $\mathbf{M}$ is the magnetic moment per unit volume. If we call $-\nabla' \cdot [\mathbf{r}' \times \mathbf{j}(\mathbf{r}')]/2 = \rho_M(\mathbf{r}')$, we have

$$\psi_M = \frac{2\omega}{L^2} \int d\mathbf{r}' \frac{e^{ik|\mathbf{r}-\mathbf{r}'|}}{|\mathbf{r}-\mathbf{r}'|} \rho_M(\mathbf{r}') \qquad (5.10.29)$$

so that a $Y_{\ell,m}$ expansion will give us ψ_M for all r outside the source, once we know the asymptotic form. We use (5.5.13), which gives

$$\psi_M = 8\pi i \omega^2 \sum_{\ell,m} h_\ell(kr) \frac{Y_{\ell,m}(\Omega)}{\ell(\ell+1)} \int d\mathbf{r}' Y^*_{\ell,m}(\Omega') j_\ell(kr') \rho_M(\mathbf{r}') \qquad (5.10.30)$$

for r outside the source.

For ψ_E, we have from (5.10.23)

$$\psi_E = \frac{\omega}{iL^2\nabla^2} \mathbf{L} \times \nabla \cdot \mathbf{E} \qquad (5.10.23)$$

$$= \frac{1}{\nabla^2} \frac{\omega}{iL^2} \int d\mathbf{r}' \frac{e^{ik|\mathbf{r}-\mathbf{r}'|}}{|\mathbf{r}-\mathbf{r}'|} \mathbf{L}' \times \nabla' \cdot (i\omega\mathbf{j}(\mathbf{r}') - \nabla'\rho). \qquad (5.10.31)$$

Here again, we find a long-range instantaneous interaction arising from the Coulomb operator $1/\nabla^2$ in (5.10.31). Since (5.10.16) shows that such terms are absent in the electric and magnetic fields outside the source, we may set $1/\nabla^2 = -1/\omega^2$ in (5.10.31), confident that the residual Coulomb-like term will cancel against the field generated from ψ_L (5.10.20). The cancellation clearly depends on the identity that justifies our calculation of ψ_L, ψ_M, and ψ_E. That is that the dyadic

$$\nabla \frac{1}{\nabla^2} \nabla + \mathbf{L} \frac{1}{L^2} \mathbf{L} - \nabla \times \mathbf{L} \frac{1}{L^2\nabla^2} \mathbf{L} \times \nabla = \overset{\leftrightarrow}{\mathbf{1}} \qquad (5.10.32)$$

where $\overset{\leftrightarrow}{\mathbf{1}}$ is the unit dyadic. This identity is not at all obvious and difficult to prove directly, but must be true in view of the completeness and uniqueness of the representation (5.10.1), subject to the usual boundary conditions.

We continue to work on (5.10.31). Since

$$\frac{1}{\nabla^2} = -\frac{1}{\omega^2} + \frac{\omega^2 + \nabla^2}{\omega^2 \nabla^2} \tag{5.10.33}$$

and

$$(\omega^2 + \nabla^2)\frac{e^{ikr}}{r} = -4\pi\delta(\mathbf{r}) \tag{5.10.34}$$

(5.10.31) becomes

$$\psi_E = \psi_{E_1} + \psi_{E_2} \tag{5.10.35}$$

where

$$\psi_{E_1} = -\frac{1}{\mathbf{L}^2}\int d\mathbf{r}'\,\frac{e^{ik|\mathbf{r}-\mathbf{r}'|}}{|\mathbf{r}-\mathbf{r}'|}\mathbf{L}' \times \nabla' \cdot \mathbf{j}(\mathbf{r}') \tag{5.10.36}$$

and

$$\psi_{E_2} = -\frac{4\pi}{\nabla^2}\frac{1}{\mathbf{L}^2}\mathbf{L} \times \nabla \cdot \mathbf{j}. \tag{5.10.37}$$

Our result is contained in (5.10.36), the retarded field. The electric field generated by ψ_{E_2} must cancel that of ψ_L outside the source. That is, we must have

$$\nabla\psi_L(\mathbf{r}) + \frac{\nabla \times \mathbf{L}}{i\omega}\psi_{E_2}(\mathbf{r}) = 0 \tag{5.10.38}$$

for $\mathbf{r}$ outside of the source, with ψ_L and ψ_{E_2} given by (5.10.20) and (5.10.37), respectively.

We recall, for a localized function f,

$$\frac{4\pi}{\nabla^2}f(\mathbf{r}) = -\int \frac{d\mathbf{r}'\,f(\mathbf{r}')}{|\mathbf{r}'-\mathbf{r}|} \tag{5.10.39}$$

so that

$$\psi_{E_2} = \int \frac{d\mathbf{r}'}{|\mathbf{r}'-\mathbf{r}|}\frac{1}{\mathbf{L}'^2}\mathbf{L}' \times \nabla' \cdot \mathbf{j}(\mathbf{r}') \tag{5.10.40}$$

and for $r > r'$

$$\psi_{E_2} = \sum_{\ell,m} \frac{Y_{\ell m}(\Omega)}{r^{\ell+1}} \frac{1}{\ell(\ell+1)} \int d\mathbf{r}' \, r'^{\ell} Y^*_{\ell m}(\Omega') \, \mathbf{L}' \times \nabla' \cdot \mathbf{j}(\mathbf{r}'). \quad (5.10.41)$$

We expand

$$\mathbf{L}' \times \nabla' \cdot \mathbf{j}' = \frac{1}{i} (\mathbf{r}' \times \nabla') \times \nabla' \cdot \mathbf{j}(\mathbf{r}) = \frac{1}{i} [-\nabla'^2 \mathbf{r}' + 2\nabla' + \mathbf{r}' \cdot \nabla'\nabla'] \cdot \mathbf{j}(\mathbf{r}')$$

$$(5.10.42)$$

We substitute $\nabla' \cdot \mathbf{j}(\mathbf{r}') = i\omega\rho(\mathbf{r}')$, integrate (5.10.41) by parts, and remember that $\nabla'^2 r'^{\ell} Y^*_{\ell m}(\Omega') = 0$. There results

$$\psi_{E_2} = -\sum_{\ell,m} \frac{Y_{\ell m}(\Omega)}{r^{\ell+1}} \frac{\omega}{\ell(\ell+1)} \int d\mathbf{r}' \, \rho(\mathbf{r})(\nabla' \cdot \mathbf{r}' - 2) \, r'^{\ell} Y^*_{\ell m}(\Omega')$$

$$= -\sum \frac{Y_{\ell m}(\Omega)}{r^{\ell+1}} \frac{\omega}{\ell} \int d\mathbf{r}' \, r'^{\ell} Y^*_{\ell m}(\Omega') \, \rho(\mathbf{r}'). \quad (5.10.43)$$

Note here that the coefficient of $1/\ell$ vanishes at $\ell = 0$, so there is no singularity at $\ell = 0$. Since the electric and magnetic fields are generated from ψ_{E_2} by acting with $(\nabla \times \mathbf{L})/i\omega$ and $\nabla^2 \mathbf{L}$, respectively, the $\ell = 0$ component of ψ_{E_2} makes no contribution to either. Note also that the magnetic field outside of the source generated by ψ_{E_2} vanishes, since $\nabla^2 \psi_{E_2} = 0$ there. We turn finally to (5.10.38):

$$\nabla\psi_L + \frac{\nabla \times \mathbf{L}}{i\omega} \psi_{E_2} = \nabla\psi_L + \nabla \frac{1}{\omega}\left(1 + r\frac{\partial}{\partial r}\right)\psi_{E_2} = 0 \quad (5.10.44)$$

where we have again made use of the equation $\nabla^2 \psi_{E_2} = 0$ for r outside of the source. We find thus that the fields generated by ψ_L and ψ_{E_2} cancel outside of the source; only ψ_{E_1} and ψ_M contribute. This is not the case in the source, as we shall see in Section 5.12.

Our final result for ψ_{E_1} is then

$$\psi_{E_1} = -\frac{1}{\mathbf{L}^2} \int d\mathbf{r}' \, \frac{e^{ik|\mathbf{r}-\mathbf{r}'|}}{|\mathbf{r} - \mathbf{r}'|} \mathbf{L}' \times \nabla' \cdot \mathbf{j}(\mathbf{r}'); \quad (5.10.36)$$

expanded in spherical harmonics, (5.10.36) becomes

$$\psi_{E_1} = 4\pi ik \sum_{\ell,m} \frac{h_\ell(kr)}{\ell(\ell+1)} Y_{\ell,m}(\Omega) \int d\mathbf{r}' \, j_\ell(kr') \, Y^*_{\ell,m}(\Omega') \, \mathbf{L}' \times \nabla' \cdot \mathbf{j}(\mathbf{r}').$$

$$(5.10.45)$$

It is again the case that the asymptotic e^{ikr}/r coefficient of each $Y_{\ell,m}$ produces a known r dependence, $h_\ell(kr)$, as long as we stay outside of the source.

We can see why ψ_E is called electric by considering the small kr' behavior of the integrand in (5.10.45). The algebra is almost identical to that leading to (5.10.43). The source function is

$$\mathbf{L}' \times \nabla' \cdot \mathbf{j}(\mathbf{r}') = \frac{1}{i} (\mathbf{r} \times \nabla') \times \nabla' \cdot \mathbf{j}$$

$$= \frac{1}{i} (\mathbf{r} \cdot \nabla'\nabla' - \mathbf{r}'\nabla'^2) \cdot \mathbf{j}$$

$$= \frac{1}{i} (\mathbf{r}' \cdot \nabla'\nabla' - \nabla'^2\mathbf{r} + 2\nabla') \cdot \mathbf{j}$$

$$= \frac{1}{i} [(\mathbf{r}' \cdot \nabla' + 2)\nabla' \cdot \mathbf{j} - \nabla'^2\mathbf{r}' \cdot \mathbf{j}];$$

Finally,

$$\mathbf{L}' \times \nabla' \cdot \mathbf{j}(\mathbf{r}') = \omega[(\mathbf{r}' \cdot \nabla' + 2)\rho] - \frac{\nabla'^2\mathbf{r}' \cdot \mathbf{j}}{i}. \qquad (5.10.46)$$

Now integrate (5.10.45) by parts, keeping only the $(kr)^\ell$ term in $j_\ell(kr)$. Since $\nabla'^2 r'^\ell Y_{\ell,m}(\Omega') = 0$, the last term in (5.10.46) makes no contribution. The first term gives

$$\left(-3 + 2 - r'\frac{\partial}{\partial r'}\right)r^\ell = -(1 + \ell)r^\ell$$

so that the ℓ, m amplitude is given in this approximation by

$$\int d\mathbf{r}' \, \rho(\mathbf{r}') \, r'^\ell Y_{\ell,m}(\Omega'),$$

appropriate to an electric multipole. Of course, the labels electric and magnetic are only labels. The importance of the scalar functions ψ_E and ψ_M is that they permit a spherical harmonic expansion of the vector field radiation to be made. We shall see, just as in the scalar case, that the ℓ, mth multipole has simple angular momentum properties.

5.11. ENERGY AND ANGULAR MOMENTUM

We define the multipole amplitudes as

$$\psi_M \to \frac{e^{ikr}}{r} \sum_{\ell,m} Y_{\ell,m} a_{\ell,m}^M \quad \text{and} \quad \psi_E \to \frac{e^{ikr}}{r} \sum_{\ell,m} Y_{\ell,m} a_{\ell,m}^E$$

The average flux of energy is the Poynting vector,

$$\overline{\mathcal{P}} = \text{Re}\, \frac{\mathbf{E}^* \times \mathbf{B}}{2\pi}.$$

With

$$\mathbf{E} = \mathbf{L}\psi_M + \frac{\nabla}{i\omega} \times \mathbf{L}\psi_E \tag{5.11.1}$$

$$\mathbf{B} = \frac{\nabla \times \mathbf{L}}{i\omega} \psi_M - \mathbf{L}\psi_E \tag{5.11.2}$$

the radiated power is then

$$\frac{dW}{dt} = \int \overline{\mathcal{P}} \cdot d\mathbf{S} = \frac{1}{2\pi} \int \left(\mathbf{L}\psi_M + \frac{\nabla \times \mathbf{L}}{i\omega} \psi_E \right)^* \times \left(\frac{\nabla \times \mathbf{L}}{i\omega} \psi_M - \mathbf{L}\psi_E \right) \cdot d\mathbf{S}. \tag{5.11.3}$$

Since we only need the constant (as $r \to \infty$) term in (5.11.3), we may set $\nabla = ik\hat{\mathbf{r}}$, so that

$$\frac{dW}{dt} = \frac{1}{2\pi} \int (\mathbf{L}\psi_M + \hat{\mathbf{r}} \times \mathbf{L}\psi_E)^* \times (\hat{\mathbf{r}} \times \mathbf{L}\psi_M - \mathbf{L}\psi_E) \cdot \hat{\mathbf{r}} r^2 \, d\Omega, \tag{5.11.4}$$

The purely magnetic term is

$$\frac{dW_M}{dt} = \frac{1}{2\pi} \int r^2 \, d\Omega (\mathbf{L}\psi_M)^* \times (\hat{\mathbf{r}} \times \mathbf{L}\psi_M) \cdot \hat{\mathbf{r}}$$

$$= \frac{1}{2\pi} \int (\mathbf{L}\psi_M)^* \cdot (\mathbf{L}\psi_M) \, r^2 \, d\Omega \tag{5.11.5}$$

so

$$\frac{dW_M}{dt} = \frac{1}{2\pi} \sum_{\ell,m} \ell(\ell + 1) \left| a^M_{\ell,m} \right|^2. \qquad (5.11.6)$$

Similarly,

$$\frac{dW_E}{dt} = \frac{1}{2\pi} \sum_{\ell,m} \ell(\ell + 1) \left| a^E_{\ell,m} \right|^2. \qquad (5.11.7)$$

The cross-terms between ψ_E and ψ_M, when integrated over Ω, give zero.

We turn next to the radiation of angular momentum. Recall from (3.8.21) that the outgoing flux of angular momentum through a surface is

$$\bar{F}_i = -\epsilon_{ijk} \int dS_\ell x_j T_{k\ell} \qquad (5.11.8)$$

where $T_{k\ell}$ is the Maxwell stress tensor

$$T_{k\ell} = \frac{1}{4\pi} \left(E_k E_\ell + B_k B_\ell - \frac{1}{2} \delta_{k\ell} \mathbf{E}^2 - \frac{1}{2} \delta_{n\ell} \mathbf{B}^2 \right). \qquad (5.11.9)$$

The terms with $\delta_{k\ell}$ are orthogonal to the spherical surface, so there remains, in vector notation,

$$\bar{\mathbf{F}} = \frac{r}{2\pi} \int d\Omega (\mathbf{r} \cdot \mathbf{E}^* \mathbf{E} \times \mathbf{r} + \mathbf{r} \cdot \mathbf{B}^* \mathbf{B} \times \mathbf{r}). \qquad (5.11.10)$$

Note that $\bar{\mathbf{F}}$ appears to grow linearly with r, since the $\mathbf{E}$ and $\mathbf{B}$ fields go inversely with r. Hence, there must be a cancellation of one power of r.

With

$$\mathbf{E} = \mathbf{L}\psi_M + \frac{\nabla}{i\omega} \times \mathbf{L}\psi_{E_1}$$

and $\qquad\qquad\qquad\qquad\qquad\qquad\qquad\qquad\qquad\qquad (5.11.11)$

$$\mathbf{B} = \frac{\nabla}{i\omega} \times \mathbf{L}\psi_M - \mathbf{L}\psi_{E_1}$$

(since ψ_{E_2} and ψ_L cancel away from the source)

$$\mathbf{r} \cdot \mathbf{E} = \frac{\mathbf{L}^2}{\omega} \psi_{E_1} \qquad (5.11.12)$$

and

$$\mathbf{r} \cdot \mathbf{B} = \frac{\mathbf{L}^2}{\omega} \psi_M \tag{5.11.13}$$

and the cancellation has occurred.

There remain $\mathbf{E} \times \mathbf{r}$ and $\mathbf{B} \times \mathbf{r}$:

$$\mathbf{E} \times \mathbf{r} = -\left(\mathbf{r} \times \mathbf{L}\psi_M + \mathbf{r} \times \left(\frac{\nabla \times \mathbf{L}}{i\omega}\right)\psi_{E_1}\right) \tag{5.11.14}$$

and

$$\mathbf{B} \times \mathbf{r} = -\left[\mathbf{r} \times \left(\frac{\nabla \times \mathbf{L}}{i\omega}\right)\psi_M - \mathbf{r} \times \mathbf{L}\psi_{E_1}\right]. \tag{5.11.15}$$

Again, the gradients must act on e^{ikr} to give $ik\hat{\mathbf{r}}$, so

$$\mathbf{E} \times \mathbf{r} = -(\mathbf{r} \times \mathbf{L}\psi_M + \mathbf{r} \times (\hat{\mathbf{r}} \times \mathbf{L})\psi_{E_1}) \tag{5.11.16}$$

and

$$\mathbf{B} \times \mathbf{r} = -(\mathbf{r} \times (\hat{\mathbf{r}} \times \mathbf{L})\psi_M - \mathbf{r} \times \mathbf{L}\psi_{E_1}). \tag{5.11.17}$$

Combining, we find for purely electric radiation an angular momentum

$$\overline{\mathbf{F}}_E = \frac{1}{2\pi\omega} \sum_{\ell,m,m'} \ell(\ell + 1)a_{\ell,m'}^{*E}\mathbf{L}_{m'm}a_{\ell,m}^E \tag{5.11.18}$$

and for purely magnetic radiation

$$\overline{\mathbf{F}}_M = \frac{1}{2\pi\omega} \sum_{\ell,m,m'} \ell(\ell + 1)a_{\ell,m'}^{*M}\mathbf{L}_{m'm}a_{\ell,m}^M \tag{5.11.19}$$

where $\mathbf{L}_{m'm}$ is the quantum mechanical matrix element of the operator $\mathbf{L}$ between the states m' and m:

$$\mathbf{L}_{m'm} = \int d\Omega Y_{\ell,m'}^*\mathbf{L}Y_{\ell,m}. \tag{5.11.20}$$

Note that different ℓ values do not interfere. Since $(L_z)_{mm'} = m\delta_{mm'}$, the z-component of angular momentum is particularly simple. For a single

ℓ, m multipole, it is

$$\bar{F}_z = \frac{1}{2\pi\omega}\ell(\ell + 1)m|a_{\ell,m}|^2 \qquad (5.11.21)$$

and the ratio of the z-component of angular momentum radiated to energy radiated is

$$\frac{\bar{F}_z}{\dfrac{dW}{dt}} = \frac{m}{\omega} \qquad (5.11.22)$$

which suggests that the wave described by $a_{\ell,m}$ would in quantum theory carry a z-component of angular momentum $\hbar m$. The suggestion is correct.

The mixed EM fluxes are not so simple. They involve integrals of the form

$$\mathbf{F}_{EM} = \int \left[\frac{\mathbf{L}^2}{\omega} \psi_M^* \mathbf{r} \times \mathbf{L}\psi_E - \frac{\mathbf{L}^2}{\omega} \psi_E^* \mathbf{r} \times \mathbf{L}\psi_M \right] d\Omega \qquad (5.11.23)$$

which connect a_ℓ^M's and $a_{\ell\pm1}^E$'s.

5.12. MULTIPOLE SCATTERING BY A DIELECTRIC

The expansion of a field amplitude in electric and magnetic multipoles of given ℓ makes it possible to reduce three-dimensional scattering problems to one radial dimension for each ℓ, provided the system is isotropic—that is, rotationally invariant.

We illustrate this by deriving the general equations for scattering by an isotropic medium and carry out the calculation for the simple case of a uniform spherical dielectric.

As usual, we start with Maxwell's equations:

$$\nabla \times \mathbf{E} = i\omega\mathbf{B}, \qquad \nabla \times \mathbf{B} = -i\omega\epsilon\mathbf{E}, \qquad \text{and} \qquad \nabla \cdot \epsilon\mathbf{E} = 0,$$

which give the equation for $\mathbf{E}$

$$\nabla \times (\nabla \times \mathbf{E}) = \omega^2\epsilon\mathbf{E}$$

or, from (5.10.18),

$$\mathbf{E} = \nabla\psi_L + \mathbf{L}\psi_M + \frac{\nabla \times \mathbf{L}}{i\omega}\psi_E$$

so

$$-\nabla^2\left(\mathbf{L}\psi_M + \frac{\nabla \times \mathbf{L}}{i\omega}\psi_E\right) = \omega^2\epsilon\left(\nabla\psi_L + \mathbf{L}\psi_M + \frac{\nabla \times \mathbf{L}}{i\omega}\psi_E\right) \quad (5.12.1)$$

and

$$\nabla \cdot \epsilon\left(\nabla\psi_L + \mathbf{L}\psi_M + \frac{\nabla \times \mathbf{L}}{i\omega}\psi_E\right) = 0. \quad (5.12.2)$$

We project the magnetic amplitude by operating on (5.12.1) with $\mathbf{L}$ dot. Since the commutator $[\mathbf{L}, \epsilon(r)] = 0$, we find

$$-\nabla^2\mathbf{L}^2\psi_M = \omega^2\epsilon\mathbf{L}^2\psi_M \quad (5.12.3)$$

or

$$-\nabla^2\psi_M = \omega^2\epsilon\psi_M \quad (5.12.4)$$

provided we leave out the undetermined $\ell = 0$ component that does not contribute to the fields.

Equation (5.12.4) shows that the magnetic amplitude is decoupled from the longitudinal and electric amplitudes, and can be found independently of the other two. Since (5.12.4) is a scalar equation, the function ψ_M can be expanded in amplitudes of definite ℓ, just as in the scalar case studied earlier. There is one important difference: The incoming field has a polarization direction that produces a significant azimuthal dependence in the scattering and requires the introduction of $Y_{\ell,m}$'s for $m \neq 0$ in the expansion. We will see this shortly when we take up scattering by a sphere.

We project the electric amplitude in (5.12.1) by the operator $\mathbf{L} \times \nabla$, which gives

$$\nabla^2\frac{\nabla^2\mathbf{L}^2}{i\omega}\psi_E = \omega^2 \cdot \mathbf{L} \times \nabla \cdot \epsilon\left(\nabla\psi_L + \mathbf{L}\psi_M + \frac{\nabla \times \mathbf{L}}{i\omega}\psi_E\right)$$

or, with $\epsilon' = d\epsilon/dr$,

$$\frac{(\nabla^2)^2\mathbf{L}^2}{i\omega}\psi_E = \omega^2\mathbf{L} \times \hat{r} \cdot \epsilon'\left(\nabla\psi_L + \mathbf{L}\psi_M + \frac{\nabla \times \mathbf{L}}{i\omega}\psi_E\right) + \omega^2\epsilon\left(-\frac{\nabla^2\mathbf{L}^2}{i\omega}\right)\psi_E$$

$$(5.12.5)$$

or

$$\frac{\nabla^4 \mathbf{L}^2}{i\omega} \psi_E - \nabla^2 \mathbf{L}^2 i\omega \epsilon \psi_E = \frac{i\epsilon' \omega^2}{r} \left(\mathbf{L}^2 \psi_L + \frac{\mathbf{L}^2 (\mathbf{r} \cdot \nabla + 1)}{\omega} \psi_E \right)$$

or

$$\nabla^4 \psi_E + \omega^2 \epsilon \nabla^2 \psi_E = -\frac{\omega^3}{r} \epsilon' \left(\psi_L + \frac{\mathbf{r} \cdot \nabla + 1}{\omega} \psi_E \right)$$

or

$$\nabla^2 (\nabla^2 + \omega^2) \psi_E = \omega^2 (1 - \epsilon) \nabla^2 \psi_E - \frac{\omega^3}{r} \epsilon' \left(\psi_L + \frac{(\mathbf{r} \cdot \nabla + 1)}{\omega} \psi_E \right).$$

$$(5.12.6)$$

The equation for ψ_L is given by (5.12.2):

$$\epsilon \nabla^2 \psi_L + \frac{\epsilon'}{r} \left(\mathbf{r} \cdot \nabla \psi_L + \frac{\mathbf{L}^2}{\omega} \psi_E \right) = 0.   (5.12.7)$$

Thus, ψ_L and ψ_E are coupled via (5.12.6) and (5.12.7); however, since both equations are spherically symmetric, different ℓ values are not coupled. The coupling of ψ_L and ψ_E results from the fact that the parity of each $Y_{l,m}$ is the same in ψ_L and ψ_E, but opposite in ψ_M. Thus, the rotational and inversion invariance of the system that requires decoupling of the different ℓ values permits coupling of ψ_L and ψ_E; since any interaction that is not forbidden is, in general, allowed, this coupling appears. Note that in especially simple cases, such as our example of a spherical dielectric, with $\epsilon' = 0$ except at the surface of the sphere, this coupling may disappear. We find again the presence of an apparent pole at $\nabla^2 = 0$ (leading to $1/r$ at infinity), which now must cancel between the actual fields produced by ψ_L and ψ_E.

Clearly, the general problem of scattering by a dielectric is a hard one, and we do not address it here. We turn instead to the special case of scattering by a uniform dielectric sphere.

The equations satisfied by the potentials follow from (5.12.1) and (5.12.2) by setting ϵ equal to a constant: $\epsilon = 1$ outside the sphere and $\epsilon = \epsilon$ inside the sphere. Thus, from (5.12.7), $\nabla^2 \psi_L = 0$, inside and outside, and we are free to set $\psi_L = 0$ and $\psi_{E_2} = 0$ (provided we can satisfy the appropriate boundary conditions at the surface without them). From (5.12.4) and (5.12.6), we find

$$(\nabla^2 + \epsilon \omega^2) \psi_M = 0 \quad \text{and} \quad (\nabla^2 + \epsilon \omega^2) \psi_E = 0.   (5.12.8)$$

The boundary condition at $r = R$ (the radius) are, as usual, E_T and B_T continuous, D_{normal} and B_{normal} continuous.

From

$$\mathbf{E} = \mathbf{L}\psi_M + \frac{\nabla \times \mathbf{L}}{i\omega}\psi_E \tag{5.12.9}$$

we look for continuity in $\mathbf{E}_T$. Since $\mathbf{E}_T$ continuous at $r = R$ implies $\mathbf{L} \cdot \mathbf{E}$ continuous, $\mathbf{L}^2 \psi_M$ must be continuous; therefore, ψ_M is continuous at the spherical boundary. The tangential component of the second term is $\hat{\mathbf{r}} \times (\nabla \times \mathbf{L})$, so we must have

$$\mathbf{r} \times (\nabla \times \mathbf{L})\psi_E \quad \text{continuous,} \tag{5.12.10}$$

or

$$\mathbf{L}(1 + \mathbf{r} \cdot \nabla)\psi_E$$

and hence,

$$(1 + \mathbf{r} \cdot \nabla)\psi_E = \left(1 + r\frac{\partial}{\partial r}\right)\psi_E \quad \text{continuous.} \tag{5.12.11}$$

From

$$\mathbf{B} = \frac{\nabla \times \mathbf{L}}{i\omega}\psi_M + \frac{\nabla^2 \mathbf{L}}{\omega^2}\psi_E \tag{5.12.12}$$

or

$$\mathbf{B} = \frac{\nabla \times \mathbf{L}}{i\omega}\psi_M - \epsilon \mathbf{L}\psi_E \tag{5.12.13}$$

we learn that

$$\epsilon \psi_E = \text{continuous} \tag{5.12.14}$$

and

$$\left(1 + r\frac{\partial}{\partial r}\right)\psi_M = \text{continuous} \tag{5.12.15}$$

It is easy to verify (as we already know from Problem 3.9) that the boundary conditions we have just found from the tangential continuity of $\mathbf{E}$ and $\mathbf{B}$ also guarantee the normal continuity of $\mathbf{B}$ and $\mathbf{D} = \epsilon \mathbf{E}$. Therefore, ψ_M and ψ_E are completely determined: They satisfy the wave equation (5.12.8) and the boundary conditions (5.12.10), (5.12.11), (5.12.14), and (5.12.15).

ψ_M and ψ_E have $Y_{\ell,m}$ expansions

$$\psi_M = \sum_{\ell,m} C^M_{\ell,m} Y_{\ell,m}(\theta, \varphi) \frac{v^M_\ell(r)}{r} \qquad (5.12.16)$$

and

$$\psi_E = \sum_{\ell,m} C^E_{\ell,m} Y_{\ell,m} \frac{v^E_\ell(r)}{r} \qquad (5.12.17)$$

where the functions v^M_ℓ and v^E_ℓ are independent of m.

The coefficients $C^M_{\ell,m}$ and $C^E_{\ell,m}$ must be determined by the incoming wave boundary condition at large r.

We easily solve for v_ℓ (electric or magnetic). For $r < R$,

$$v_\ell = u_\ell(k_\epsilon r) \qquad (5.12.18)$$

where $k_\epsilon = \sqrt{\epsilon}\omega = n\omega$. For $r > R$,

$$v_\ell = A_\ell u_\ell + B_\ell q_\ell \qquad (5.12.19)$$

where A_ℓ and B_ℓ are determined by the boundary conditions, and u_ℓ and q_ℓ are defined in (5.5.17) and (5.6.12).

Note that q_ℓ may be chosen to be real, since the equations and boundary conditions are real; then A_ℓ and B_ℓ are real, and define a phase shift via

$$\frac{B_\ell}{A_\ell} = \tan \delta_\ell.$$

Matching the boundary conditions for v^M_ℓ, we have

$$\left(1 + r \frac{\partial}{\partial r}\right) \frac{v^M_\ell}{r} \quad \text{continuous} \qquad \text{and} \qquad v^m_\ell \text{ continuous.}$$

Thus,

$$A_\ell u_\ell(nkR) = \cos \delta_\ell u_\ell(kR) + \sin \delta_\ell q_\ell(kR) \qquad (5.12.20)$$

and

$$A_\ell u_\ell(nkR) = \cos \delta_\ell u'_\ell(kR) + \sin \delta_\ell q'_\ell(kR). \qquad (5.12.21)$$

The prime in (5.12.21) and (5.12.22) stands for differentiation with respect to R, not kR. The parameter ξ_ℓ^M to go into Eq. (5.6.15) is

$$\xi_\ell^M = \frac{u_\ell'(nkR)}{u_\ell(nkR)},$$ (5.12.22)

which for small k goes like $(\ell + 1)/R$. Thus, from (5.6.16) the normal term in δ_ℓ^M of order $k^{2\ell+1}$ vanishes; the first nonvanishing term is of order $k^{2\ell+3}$, clearly a consequence of the ω^2 dependence of the interaction strength. This does *not* happen for the electric amplitude, as we now show.

The boundary conditions for ψ_E are

$$\epsilon \psi_E \text{ continuous}$$ (5.12.23)

and

$$\left(1 + r\frac{\partial}{\partial r}\right)\psi_E \text{ continuous}.$$ (5.12.24)

With

$$\psi_E = \frac{v_\ell}{r},$$ (5.12.25)

these become

$$\epsilon v_\ell^E \text{ continuous}$$ (5.12.26)

and

$$\frac{\partial v_\ell^E}{\partial r} \text{ continuous}.$$ (5.12.27)

In this case, the parameter ξ becomes for small k

$$\xi_\ell = \frac{(\ell + 1)}{\epsilon R},$$ (5.12.28)

and, from (5.6.16),

$$\delta_\ell \rightarrow -k^{2\ell+1} \frac{R^{2\ell+2}}{(2\ell + 1)!!(2\ell - 1)!!} \frac{\left(\dfrac{\ell + 1}{\epsilon R} - \dfrac{\ell + 1}{R}\right)}{\left(\dfrac{\ell + 1}{\epsilon} + \ell\right)}$$ (5.12.29)

which for $\ell = 1$ is

$$\delta_1 \rightarrow \frac{2}{3}(kR)^3 \frac{\epsilon - 1}{\epsilon + 2}. \qquad (5.12.30)$$

To relate this calculation to scattering, we must determine the incoming fields ψ_M^0 and ψ_M^0 in terms of

$$\mathbf{E}_0 = \hat{\mathbf{e}}^0\, e^{i\mathbf{k}\cdot\mathbf{r}} \qquad \text{and} \qquad \mathbf{B}_0 = \hat{k} \times \hat{\mathbf{e}}^0\, e^{i\mathbf{k}\cdot\mathbf{r}}. \qquad (5.12.31)$$

We use the expansion (5.5.14) to expand $e^{i\mathbf{k}\cdot\mathbf{r}}$ in spherical harmonics:

$$e^{i\mathbf{k}\cdot\mathbf{r}} = \sum (2\ell + 1)\, P_\ell(\hat{\mathbf{k}} \cdot \hat{r})\, j_\ell(kr)\, i^\ell, \qquad (5.12.32)$$

and

$$\mathbf{E}_0 = \frac{\hat{\mathbf{e}}^0}{kr} \sum (2\ell + 1)\, P_\ell(\hat{\mathbf{k}} \cdot \hat{\mathbf{r}})\, u_\ell\, i^\ell. \qquad (5.12.33)$$

We solve for ψ_M^0 and ψ_E^0 as usual

$$\psi_M^0 = \frac{1}{kr} \sum_\ell (2\ell + 1)\, u_\ell\, \hat{\mathbf{e}}^0 \cdot \frac{\mathbf{L}}{\mathbf{L}^2}\, P_\ell(\hat{\mathbf{k}} \cdot \hat{\mathbf{r}})\, i^\ell \qquad (5.12.34)$$

and

$$\psi_E^0 = \frac{\omega}{i} \frac{1}{\nabla^2 \mathbf{L}^2} \mathbf{L} \times \nabla \cdot \hat{\mathbf{e}}^0\, e^{i\mathbf{k}\cdot\mathbf{r}} = -\frac{1}{\mathbf{L}^2} \mathbf{L} \times \frac{\mathbf{k}}{\omega} \cdot \hat{\mathbf{e}}^0\, e^{i\mathbf{k}\cdot\mathbf{r}}$$

$$= +\frac{1}{\mathbf{L}^2} \hat{\mathbf{e}}^0 \cdot \hat{\mathbf{k}} \times \mathbf{L}\, e^{i\mathbf{k}\cdot\mathbf{r}} \qquad (5.12.35)$$

$$= \sum (2\ell + 1)\, i^\ell j_\ell(kr) \frac{1}{\mathbf{L}^2} \hat{\mathbf{e}}^0 \cdot \hat{\mathbf{k}} \times \mathbf{L} P_\ell(\hat{\mathbf{k}} \cdot \hat{\mathbf{r}}). \qquad (5.12.36)$$

We have to deal with two spherical harmonics of order ℓ:

$$Y_\ell^M = \frac{\hat{\mathbf{e}}^0 \cdot \mathbf{L}}{\mathbf{L}^2}\, P_\ell(\hat{\mathbf{k}} \cdot \hat{\mathbf{r}}) \qquad (5.12.37)$$

and

$$Y_\ell^E = \frac{\hat{\mathbf{e}}^0 \times \hat{\mathbf{k}} \cdot \mathbf{L}}{\mathbf{L}^2}\, P_\ell(\hat{\mathbf{k}} \cdot \hat{\mathbf{r}}). \qquad (5.12.38)$$

Note that both $\hat{\mathbf{e}}^0 = \mathbf{a}_1$ and $\hat{\mathbf{e}}^0 \times \hat{\mathbf{k}} = \mathbf{a}_2$ are transverse vectors to the wave vector $\mathbf{k}$. For either of them,

$$Y^j_\ell = \mathbf{a}_j \cdot \frac{\mathbf{L}}{L^2} P_\ell(\hat{\mathbf{k}} \cdot \hat{\mathbf{r}})$$

$$= \mathbf{a}_j \cdot \frac{\mathbf{r}}{\ell(\ell+1)} \times \frac{1}{i} P'_\ell \nabla \hat{\mathbf{k}} \cdot \hat{\mathbf{r}}$$

$$= \frac{\mathbf{a}_j \cdot \hat{\mathbf{r}}}{\ell(\ell+1)} \times \frac{\hat{\mathbf{k}}}{i} P'_\ell. \qquad (5.12.39)$$

Now choose $\hat{\mathbf{k}}$ in the z-direction. Since $\hat{\mathbf{r}} = \hat{\mathbf{k}} \cos\theta + \hat{\imath} \sin\theta \cos\varphi + \hat{\jmath} \sin\theta \sin\varphi$, $\hat{\mathbf{r}} \times \hat{\mathbf{k}}$ is

$$\hat{\mathbf{r}} \times \hat{\mathbf{k}} = -\hat{\jmath} \sin\theta \cos\varphi + \hat{\imath} \sin\theta \sin\varphi \qquad (5.12.40)$$

and

$$Y^j_\ell = \frac{1}{i} (a_{jx} \sin\varphi - a_{jy} \cos\varphi) \sin\theta P'_\ell(\cos\theta), \qquad (5.12.41)$$

which is a spherical harmonic of order ℓ, linear in $e^{\pm i\varphi}$, and hence a combination of $Y_{\ell,1}(\theta, \varphi)$ and $Y_{\ell,-1}(\theta, \varphi)$. In expanding the scattering amplitude $\mathbf{f}(\theta, \varphi)$, we must accordingly introduce the spherical functions Y^M_ℓ and Y^E_ℓ defined in (5.12.37) and (5.12.38).

As in the scalar case studied earlier, we require that the ψ_ℓ amplitudes consist of the known incoming wave ψ_0, plus an outgoing wave with an unknown coefficient. That coefficient is then fixed by the requirement that the resulting wave function for each (ℓ, m) be a multiple of the known radial solution v_ℓ.

The incoming magnetic field ψ^0_M is

$$\psi^0_M = \frac{1}{kr} \sum_\ell i^\ell (2\ell+1) u_\ell Y^M_\ell(\theta, \varphi), \qquad (5.12.42)$$

and the incoming electric field ψ^0_E is

$$\psi^0_E = \frac{1}{kr} \sum_\ell i^\ell (2\ell+1) u_\ell Y^E_\ell(\theta, \varphi). \qquad (5.12.43)$$

The scattered field is, in each case,

$$\psi^{sc} = \frac{1}{r} \sum_\ell i^{\ell+1} h_\ell (2\ell + 1) Y_\ell (\theta, \varphi) f_\ell \qquad (5.12.44)$$

where f_ℓ is determined by the requirement that as $r \to \infty$,

$$\psi^0 + \psi^{sc} \to \frac{1}{kr} \sum_\ell (2\ell + 1) Y_\ell C_\ell \sin\left(kr - \frac{\ell \pi}{2} + \delta_\ell \right) \qquad (5.12.45)$$

for some C_ℓ. Equation (5.12.45), in turn, requires

$$f_\ell = \frac{e^{2i\delta_\ell} - 1}{2ik} \qquad (5.12.46)$$

for the magnetic or electric amplitude.
The scattered electric field is now

$$\mathbf{E}^{sc} = \mathbf{L}\psi_M^{sc} + \frac{\nabla}{i\omega} \times \mathbf{L}\psi_E^{sc} \qquad (5.12.47)$$

or, asymptotically,

$$\mathbf{E}^{sc} \to \mathbf{L}\psi_M^{sc} + \hat{\mathbf{r}} \times \mathbf{L}\psi_E^{sc},$$

and the scattering cross section is the square of the vector coefficient of e^{ikr}/r in ($\mathbf{E}^{sc}$ (for $\hat{\mathbf{e}}^{0*} \cdot \hat{\mathbf{e}}^0 = 1$), as defined in (5.7.5); for each partial wave, that coefficient is

$$\mathbf{f}_\ell = (2\ell + 1)(\mathbf{L}Y_\ell^M f_\ell^M + \hat{\mathbf{r}} \times \mathbf{L}Y_\ell^E f_\ell^E). \qquad (5.12.48)$$

We calculate separately the noninterfering total magnetic and electric cross section for each ℓ:

$$\sigma_\ell^M = (2\ell + 1)^2 |f_\ell^M|^2 \int d\Omega (\mathbf{L}Y_\ell^M)^* \cdot \mathbf{L}Y_\ell^M$$

$$= \ell(\ell + 1)(2\ell + 1)^2 |f_\ell^M|^2 \int d\Omega \, Y_\ell^{M*} Y_\ell^M. \qquad (5.12.49)$$

But

$$Y_\ell^M = \frac{\hat{\mathbf{e}}^0 \cdot \mathbf{L}}{\ell(\ell + 1)} P_\ell$$

and

$$\int d\Omega \, Y_\ell^{M*} \, Y_\ell^M = \int d\Omega \, P_\ell \, \frac{\hat{\mathbf{e}}^{0*} \cdot \mathbf{L} \hat{\mathbf{e}}^0 \cdot \mathbf{L}}{[\ell(\ell + 1)]^2} P_\ell$$

$$= \frac{4\pi}{2(2\ell + 1)(\ell(\ell + 1))} \tag{5.12.50}$$

so that

$$\sigma_\ell^M = \frac{4\pi(2\ell + 1)}{2k^2} \left| \frac{e^{2i\delta_\ell^M} - 1}{2i} \right|^2. \tag{5.12.51}$$

σ_ℓ^E is a little harder:

$$\sigma_\ell^E = (2\ell + 1)^2 |f_\ell^E|^2 \int d\Omega (\hat{\mathbf{r}} \times \mathbf{L} Y_\ell^E)^* (\hat{\mathbf{r}} \times \mathbf{L} Y_\ell^E) \tag{5.12.52}$$

$$= (2\ell + 1)^2 |f_\ell^E|^2 \int d\Omega \, Y_\ell^{E*} \hat{\mathbf{r}} \times \mathbf{L} \cdot \hat{\mathbf{r}} \times \mathbf{L} Y_\ell^E + \mathcal{O}\left(\frac{1}{r}\right) \tag{5.12.53}$$

$$= (2\ell + 1) |f_\ell^E|^2 \ell(\ell + 1) \int d\Omega \, Y_\ell^{E*} \, Y_\ell^E.$$

But

$$Y_\ell^E = \frac{(\hat{\mathbf{e}} \times \hat{\mathbf{k}}) \cdot \mathbf{L}}{\mathbf{L}^2} P_\ell,$$

so again,

$$\sigma_\ell^E = \frac{4\pi(2\ell + 1)}{2k^2} \left| \frac{e^{2i\delta_\ell^E} - 1}{2i} \right|^2. \tag{5.12.54}$$

Substituting the result for δ_ℓ^E for $\ell = 1$, we find for small k,

$$\sigma_1^E \rightarrow \frac{8\pi}{3} k^4 \left(\frac{R^3(\epsilon - 1)}{\epsilon + 2} \right)^2. \tag{5.12.55}$$

We recover, for small $\epsilon - 1$, the result given by (5.9.24), which as $k \rightarrow 0$

gives

$$\mathbf{f} = \omega^2 \cdot \frac{(\epsilon - 1)}{4\pi} \cdot \frac{4\pi}{3} R^3 \mathbf{e}_T \qquad (5.12.56)$$

and

$$\sigma_T = \omega^4 \left(\frac{\epsilon - 1}{3}\right)^2 R^6 \cdot 4\pi \cdot \frac{2}{3}. \qquad (5.12.57)$$

The result (5.12.55) shows that the electrostatic polarizability $\alpha = (\epsilon - 1)/(\epsilon + 2) R^3$ of a dielectric sphere dominates the low-frequency scattering cross section.

CHAPTER 5 PROBLEMS

5.1. From the reality of the interaction applied to the scattering of a scalar wave, we found the scattering amplitude to go from wave number $\mathbf{k}_1$ to $\mathbf{k}_2$ (with $|k_1| = |k_2| = k$), $f(\mathbf{k}_2, \mathbf{k}_1)$, was given by

$$f(\mathbf{k}_2, \mathbf{k}_1) = \sum (2\ell + 1) P_\ell(\hat{\mathbf{k}}_1 \cdot \hat{\mathbf{k}}_2) f_\ell(k) \qquad (1)$$

where

$$f_\ell(k) = \frac{e^{2i\delta_\ell} - 1}{2ik} = \frac{e^{i\delta_\ell} \sin \delta_\ell}{k}, \qquad (2)$$

and δ real. f_ℓ therefore satisfies a partial wave optical theorem:

$$\operatorname{Im} f_\ell = k |f_\ell|^2.$$

From (1) and (2), prove the generalized optical theorem for $|\mathbf{k}_2| = |\mathbf{k}_1| = k$:

$$\operatorname{Im} f(\mathbf{k}_2, \mathbf{k}_1) = k \int \frac{d\Omega_\mathbf{k}}{4\pi} f^*(\mathbf{k}_2, \mathbf{k}) f(\mathbf{k}_1, \mathbf{k}).$$

5.2. For a scalar field scattered by a complex potential, the wave function cannot be chosen real, and the asymptotic form for a given ℓ will not approach $q_\ell \to \sin[kr - (\ell\pi/2) + \delta_\ell]$ as $r \to \infty$ with δ_ℓ real. The asymptotic wave function q_ℓ will approach $q_\ell \to \alpha \sin[kr - (\ell\pi/2)] + \beta \cos[kr - (\ell\pi/2)]$, where α and β are complex functions of k.

(a) Show that one can always uniquely (to within $\pm n\pi$) find a δ_ℓ such that $q_\ell \to A \sin[kr - (\ell\pi/2) + \delta_\ell]$ with complex δ_ℓ and that the scattering amplitude for that case is $f_\ell = (e^{2i\delta_\ell} - 1)/2ik$.

(b) Suppose energy is absorbed by the system. Apply the energy calculation used to derive the optical theorem to a single partial wave to decide what the sign of Im δ_ℓ must be if the target absorbs energy.

(c) Give an example of a (nonlocal) potential that will scatter only an ℓ wave.

*5.3. Consider the scattering of light by a perfectly conducting sphere of radius R. The complex electric field outside the sphere satisfies the wave equation

$$(\nabla^2 + \omega^2)\mathbf{E} = 0 \tag{3}$$

and the divergence condition

$$\nabla \cdot \mathbf{E} = 0. \tag{4}$$

(a) From (3) and (4) and the multipole expansion

$$\mathbf{E} = \nabla \psi_L + \frac{\nabla \times \mathbf{L}}{i\omega}\psi_E + \mathbf{L}\psi_M, \tag{5}$$

verify that outside the sphere, for $\ell \neq 0$,

$$\nabla^2 \psi_L = 0, \quad (\nabla^2 + \omega^2)\psi_M = 0, \quad \text{and} \quad \nabla^2(\nabla^2 + \omega^2)\psi_E = 0. \tag{6}$$

(b) With

$$\psi_E = \psi_{E_1} + \psi_{E_2}, \tag{7}$$

and

$$(\omega^2 + \nabla^2)\psi_{E_1} = 0, \qquad \nabla^2 \psi_{E_2} = 0, \tag{8}$$

verify from (3–8) that we must have

$$\nabla \psi_L + \frac{\nabla \times \mathbf{L}}{i\omega}\psi_{E_2} = 0$$

and that this equation can be satisfied outside the conductor for both ψ's $\neq 0$.

(c) Write the boundary condition that must be satisfied by ψ_{E_1} and ψ_M at the surface of the sphere. From these, calculate the electric and magnetic phase shifts δ_ℓ^M and δ_ℓ^E and the low-frequency limits of the partial wave cross sections σ_ℓ^E and σ_ℓ^M.

(d) In part (c), you found that δ_ℓ^E and $\delta_\ell^M \to k^{2\ell+1}$ for small k. How can this be? We found in Section 5.12 that, for any ϵ, $\delta_\ell^M \to k^{2\ell+3}$ as $k \to 0$; but we can include conductivity in ϵ by letting $\epsilon \to \epsilon + (i\sigma/\omega)$, and a perfect conductor by taking the limit $\sigma \to \infty$, or $\epsilon \to \infty$. What went wrong?

(e) From (5.6.15) and (5.12.22), calculate the first nonvanishing power of k and its coefficient in the magnetic phase shift $f_\ell^M(k)$. Repeat the calculation starting from the perturbative result (5.9.24) and show that the results agree for small $\epsilon - 1$.

(f) Calculate the $\ell = 1$ electric phase shift from (5.9.24) and show that it agrees with (5.12.30) for small $\epsilon - 1$ and $k \to 0$.

*5.4. Calculate the $\ell = 0$ scattering length for a scalar field scattered by a square well potential

$$
\begin{aligned}
U(r) &= U_0, & r < R \\
U &= 0, & r > R
\end{aligned}.
$$

The equation for q_ℓ is the Schrödinger equation (for $\ell = 0$):

$$
-\frac{d^2 q}{dr^2} + U(r)\, q = k^2 q.
$$

The boundary conditions are $q(r = 0) = 0$, and q and dq/dr continuous at $r = R$. Consider separately the three cases (a) $U_0 > 0$, (b) $U_0 < 0$ and $\sqrt{-U_0}R \ll 1$, and, (c) $U_0 < 0$ and $\sqrt{-U_0}R$ close to $\pi/2$.

*5.5. The Schrödinger equation (5.3.2) has the integral form (with a slight redefinition of U)

$$
\phi(\mathbf{r}) = e^{i\mathbf{k}_0 \cdot \mathbf{r}} - \frac{1}{4\pi} \int \frac{e^{ik_0|\mathbf{r} - \mathbf{r}'|}}{|\mathbf{r} - \mathbf{r}'|} U(\mathbf{r}')\, d\mathbf{r}'\, \phi(\mathbf{r}').
$$

It is sometimes useful to Fourier-transform the integral equation. To do this, show first that

(a)

$$
\frac{e^{ik_0 r}}{r} = \lim_{\epsilon \to 0+} \frac{4\pi}{(2\pi)^3} \int \frac{d\mathbf{k}\, e^{i\mathbf{k} \cdot \mathbf{r}}}{k^2 - k_0^2 - i\epsilon}.
$$

(b) Then show that the integral equation for $\chi(\mathbf{k}) = \int d\mathbf{r}'\, e^{-i\mathbf{k} \cdot \mathbf{r}} \phi(\mathbf{r})$ is

$$
\chi(\mathbf{k}) = (2\pi)^3\, \delta^3(\mathbf{k} - \mathbf{k}_0) - \frac{4\pi}{k^2 - k_0^2 - i\epsilon} \int d\mathbf{k}'\, U(\mathbf{k}, \mathbf{k}')\chi(\mathbf{k}')
$$

where $U(\mathbf{k}, \mathbf{k}') = 1/(2\pi)^3 \int e^{-i\mathbf{k} \cdot \mathbf{r}} U(\mathbf{r})\, e^{i\mathbf{k}' \cdot \mathbf{r}} d\mathbf{r}$.

(c) Show that the scattering amplitude f is

$$
f = -\frac{1}{4\pi} \int d\mathbf{k}'\, U(\mathbf{k}_f, \mathbf{k}')\chi(\mathbf{k}'),
$$

where $\mathbf{k}_f = k_0 \hat{\mathbf{r}}$, with $\hat{\mathbf{r}}$ the direction of observation, and therefore

$$f = \frac{1}{4\pi}(k^2 - k_0^2)(\chi(\mathbf{k}) - (2\pi)^3 \, \delta^3(\mathbf{k} - \mathbf{k}_0)).$$

(d) Check that for small U (to be defined) and $k_0 \to 0$, the results of (c) above agree with what you found in Problem 5.4.

(e) Repeat (a–c) above for U, a nonlocal potential, defined by

$$U\phi(\mathbf{r}) = U(\mathbf{r}, \mathbf{r}') \, \phi(\mathbf{r}') \, d\mathbf{r}'.$$

Show that the only change is that $U(\mathbf{k}, \mathbf{k}')$ becomes

$$U(\mathbf{k}, \mathbf{k}') = \frac{1}{(2\pi)^3} \int e^{-i\mathbf{k}\cdot\mathbf{r}} d\mathbf{r} \, U(\mathbf{r}, \mathbf{r}') \, e^{i\mathbf{k}'\cdot\mathbf{r}'} \, d\mathbf{r}'.$$

5.6. A nonlocal potential for which the scattering equation can be solved exactly (i.e., in terms of integrals) is a product potential:

$$U(\mathbf{r}, \mathbf{r}') = g(\mathbf{r}) \, g^*(\mathbf{r}').$$

Find the scattering amplitude in terms of the Fourier transform of g:

$$h(\mathbf{k}) = \int e^{-i\mathbf{k}\cdot\mathbf{r}} g(\mathbf{r}) \, d\mathbf{r}.$$

5.7. Apply the factorization techniques of Section 5.5 to the Schrödinger equation for the one-dimensional harmonic oscillator:

$$H\psi_n = \left(-\frac{1}{2}\frac{d^2}{dx^2} + \frac{1}{2}x^2\right)\psi_n = E_n\psi_n, \qquad -\infty < x < \infty.$$

The factors are $a_+ = -(d/dx) + x$ and $a_- = (d/dx) + x$.

(a) Show that $a_+a_- = 2H - 1$ and $a_-a_+ = 2H + 1$.

(b) Show that if $H\psi = E\psi$, then $Ha_+\psi = (E + 1)a_+\psi$ so that a_+ is a raising operator, and that $Ha_-\psi = (E - 1)\psi$ so that a_- is a lowering operator.

(c) The positivity of H shows that there must be an eigenfunction of H, ψ_0, that cannot be lowered. Use this to find the lowest eigenvalue $(E = \frac{1}{2})$ and corresponding normalized eigenfunction $(\psi_0 = e^{-x^2/2}/\pi^{1/4})$.

(d) The nth eigenfunction, with energy $E_n = n + \frac{1}{2}$, is given by

$$\psi_n = A(a_+)^n \psi_0.$$

Calculate the value of A required to normalize ψ_n.

*5.8. Apply the factorization techniques of Section 5.5 to the Schrödinger equation for the isotropic three-dimensional harmonic oscillator, with angular momentum quantum number ℓ:

$$H_\ell v_\ell = \left(-\frac{1}{2}\frac{d^2}{dr^2} + \frac{\ell(\ell+1)}{2r^2} + \frac{1}{2}r^2 \right) v_\ell$$

$$= E_\ell v_\ell, \qquad 0 \le r < \infty \quad \text{and} \quad v_\ell(0) = 0.$$

The factors are $A_\ell^+ = -(d/dr) + (\ell/r) + r$ and $A_\ell^- = (d/dr) + (\ell/r) + r$.

(a) Show that A_ℓ^+ raises ℓ and E_ℓ by one unit each, and that A_ℓ^- lowers ℓ and E_ℓ by one unit each.

(b) Finding the spectrum here is more subtle. The lowest value of ℓ is zero, for which H_ℓ becomes H of the preceding problem, with the difference that in this case the wave function must vanish at $r = 0$. Thus, only odd harmonic oscillator eigenfunctions are allowed.

 Using all this, give the eigenvalues $E_\ell(n_r)$ as explicit functions of ℓ and n_r (the harmonic oscillator quantum number). Give the allowed values of ℓ and n_r.

(c) Show how the $\ell = 0$ wave functions fail to be lowered by A_ℓ^-.

(d) Using the separability of the Schrödinger equation,

$$\left(-\frac{1}{2}\left(\frac{d^2}{dx^2} + \frac{d^2}{dy^2} + \frac{d^2}{dz^2} \right) + \frac{1}{2}(x^2 + y^2 + z^2) \right)\psi = E\psi,$$

find the eigenvalues E as sums and the eigenfunction ψ_{n_x,n_y,n_z} as products of the eigenvalues and eigenfunctions of the three one dimensional oscillators.

 Write explicitly all the eigenfunctions for which $n_x + n_y + n_z \le 2$ and show how these appear when characterized by ℓ and n_r.

5.9. Show that the next term in the expansion of $h_\ell(kr)$ at large kr is given by

$$h_\ell(kr) = \frac{e^{ikr}}{kr\, i^{\ell+1}}\left(1 + \frac{\ell(\ell+1)\,i}{2kr} + \cdots \right).$$

Invariance and Special Relativity

6.1. INVARIANCE

When we say the laws of nature are invariant under a transformation, we mean that it is impossible to determine whether that transformation has taken place. To illustrate, consider invariance under time translation. The physical meaning of this invariance is that were we to go to sleep and wake up some time later, no experiments we perform before and after our nap (not counting looking at a clock) could tell us how long we had been asleep. That is, the laws of Nature do not change with time; they are the same now as they were then.

The mathematical expression of this invariance is that the equations governing the system we are describing are invariant under the transformation $t' = t + \Delta$. For example, Newton's law of gravitation,

$$m_i \frac{d^2\mathbf{r}_i}{dt^2} = -\sum_{j \neq i} G \frac{(\mathbf{r}_i - \mathbf{r}_j)\, m_i m_j}{|\mathbf{r}_i - \mathbf{r}_j|^3},$$

can be expressed in terms of t':

$$m_i \frac{d^2\mathbf{r}_i}{dt'^2} = -\sum_{j \neq i} G \frac{(\mathbf{r}_i - \mathbf{r}_j)\, m_i m_j}{|\mathbf{r}_i - \mathbf{r}_j|^3}.$$

Since the two laws are the same, the phenomena they describe (gravitational motions) cannot tell us whether we are using t or t' as a clock. (See Appendix A.2 for further elementary discussion.)

Similarly, translational invariance makes it impossible to tell whether our laboratory has been picked up and moved to a new location.

245

Rotational invariance makes it impossible to tell whether our laboratory has been turned around.

Thus, translation and rotation invariance together tell us that space is homogeneous and isotropic.

Invariance with respect to a constant velocity transformation makes it impossible to tell if our laboratory has been gently accelerated to a new velocity. All these invariance principles are believed to hold exactly.[1]

There is one other class of invariances believed to be exactly true: these are gauge invariances analogous to the gauge invariance of electrodynamics.

There are two more space-time connected invariance principles that are approximately true: inversion invariance (called P), which forbids knowing whether you are looking in a mirror or at the real world; P is broken by the weak β decay interactions. An invariance that holds much more accurately, called CP, forbids knowing whether you are looking at the reflection of a particle in a mirror, or at an antiparticle in the real world. CP is known to be violated, but *very* weakly.

The second approximate space-time invariance, which is also weakly broken, is time reversal invariance, called T. This invariance tells us that every motion has a reversed motion that is equally possible, with the same coordinates and accelerations, but opposite velocities, occurring in backward order in time. The classical laws of mechanics and electromagnetism are invariant under P and T.

In relativistic quantum field theory, it is shown that CPT must hold exactly. That is, T must be violated just enough to compensate in CPT for the violation of CP. CPT says: You cannot tell whether you are looking at a certain motion of particles in the real world, or looking in a mirror at antiparticles undergoing the time-reversed motion.

We may express the space-time invariances in terms of infinitesimal transformation of coordinates. Thus, for translation in space or time:

$$t' = t + \Delta t, \qquad x' = x + \Delta x, \qquad y' = y + \Delta y, \qquad z' = z + \Delta z.$$
$$(6.1.1)$$

We have three rotations:

---

[1]Of course, real space in our universe is filled with masses that generate gravitational fields. It is our common experience on Earth that space above our planet is neither isotropic nor homogeneous. What does appear to be true is that for phenomena occurring on a distance and time scale that is small compared to the length and timescale of the gravitational field, there exists a class of observers for whom these space-time invariances hold. This follows from the equivalence principle, which tells us that an observer in free fall in a gravitational field will have no local way of determining that there is a gravitational field in the neighborhood. Of course, the falling elevator must not be so large that the convergence of the field lines toward their source can be detected.

$$x' = x + y\Delta\theta, \qquad y' = y - x\Delta\theta, \qquad z' = z. \qquad (6.1.2)$$

and two more infinitesimal rotations about x and y.

One can express finite translations and rotations by iterating infinitesimal ones. For translation, this is trivial. For rotations, note that (6.1.2) is equivalent to

$$\begin{pmatrix} x' \\ y' \end{pmatrix} = (1 + i\Delta\theta\sigma_y)\begin{pmatrix} x \\ y \end{pmatrix} \qquad (6.1.3)$$

where $\sigma_y = \begin{pmatrix} 0 & -i \\ i & 0 \end{pmatrix}$. Repeated N times, with $N\Delta\theta = \theta$, the finite rotation is

$$\begin{pmatrix} x' \\ y' \end{pmatrix} = \left(1 + \frac{i\theta\sigma_y}{N}\right)^N \begin{pmatrix} x \\ y \end{pmatrix} \qquad (6.1.4)$$

or, as $N \to \infty$, with θ remaining finite,

$$\begin{pmatrix} x' \\ y' \end{pmatrix} = e^{i\sigma_y\theta}\begin{pmatrix} x \\ y \end{pmatrix} = \begin{pmatrix} x\cos\theta + y\sin\theta \\ -x\sin\theta + y\cos\theta \end{pmatrix}. \qquad (6.1.5)$$

An inversion cannot be generated by a succession of rotations, since the determinant of the rotation matrix a_{ij} in $x_i' = a_{ij}x_j$ (summation convention) is 1, whereas an inversion $x_i' = -x_i$ has determinant -1.

Finally, we come to the transformation between observers moving with constant but different velocities. Newtonian mechanics has an invariance of this kind, called Galilean invariance, with the transformation given by

$$x_i' = x_i - v_i t, \qquad i = 1, 2, 3. \qquad (6.1.6)$$

The Newtonian equations

$$m_a \frac{d^2 x_i^a}{dt^2} = \sum_b F_i^{ab}(x_j^a - x_j^b) \qquad (6.1.7)$$

are clearly invariant under the transformation (6.1.6) for a force law F_i that is independent of velocity. Thus, the Newtonian world would have velocity invariance (called Galilean relativity) were distance measurements to transform according to (6.1.7)—that is, were the distance between simultaneous events the same for all observers.

Another consequence of (6.1.6) is the addition rule for velocities:

$$\frac{dx_i'}{dt} = \frac{dx_i}{dt} - v_i. \tag{6.1.8}$$

Thus, if observer O observes a velocity v_i^O, an observer O' will observe the velocity

$$v_i^{O'} = v_i^O - v_i. \tag{6.1.9}$$

Evidently, Maxwell's equations cannot be invariant to the transformation (6.1.6), since they predict a unique light velocity c, independent of the velocity of the observer or the source.

Thus, if there is a principle of relativity (invariance under constant velocity transformation), then either the transformation (6.1.6) is wrong, and Newton's laws must be modified; or Maxwell's equations are wrong, and the modified equations must somehow contain the possibility of the velocity addition law *without* containing the observer's velocity as a parameter.

No such modification has been found, although it must have been diligently sought. Einstein took the first view: There is an invariance principle, Maxwell's equations are correct, and they are invariant to the correct, transformation law (to be found) replacing (6.1.6).

6.2. THE LORENTZ TRANSFORMATION

We are thus led to study (first in one dimension) the kinds of transformation that could hold between space-time measurements made by different observers moving with constant velocity with respect to each other. It will turn out that we are almost uniquely led to a Lorentz transformation, with a velocity parameter c. Galilean relativity results from $c \to \infty$. Einstein's relativity results from $c =$ velocity of light.

The most general transformation that leaves space homogeneous must be linear:

$$x' = \gamma(v)(x - vt). \tag{6.2.1}$$

Thus, the trajectory $x = vt$ corresponds to $x' = 0$. That is, the origin of the O' coordinate system moves with velocity v to the right in the O system.

Let us assume that the O system moves (to the left) with velocity $-v$ with respect to the O' system. Then, of course, assuming $t' = t$, we must have

$$x = \gamma(-v)(x' + vt) \tag{6.2.2}$$

which together with (6.2.1) implies $\gamma = 1$; back to Galileo.

It was the remarkable insight of Einstein here that relations between space-time measurements made by different observers cannot be derived by logic, but are empirical. There is therefore no reason to insist that time intervals between events be invariant. Einstein wrote, instead,

$$x' = \gamma(v)(x - vt) \quad \text{and} \quad t' = \delta(v)(t - \beta(v)\,x). \tag{6.2.3}$$

The transformation from $-x$ to $-x'$ should have a velocity $-v$. Then $-x' = \gamma(v)(-x - vt)$ should be equivalent to $x' = \gamma(-v)(x + vt)$, or $\gamma(v) = \gamma(-v)$. Similarly,

$$t' = \delta(v)(t + \beta(v)\,x)$$

should be equivalent to

$$t' = \delta(-v)(t - \beta(-v)\,x)$$

so that $\delta(v) = \delta(-v)$ and $\beta(-v) = -\beta(v)$.

Now solve for x, t as functions of x', t'. From (6.2.3),

$$t = \frac{t'}{\delta} + \beta x \tag{6.2.4}$$

so that

$$x' = \gamma(v)\left(x - v\left(\frac{t'}{\delta} + \beta\right)\right) = \gamma(v)\left[x(1 - \beta v) - \frac{vt'}{\delta}\right]$$

or

$$x\gamma(1 - \beta v) = x' + v\frac{\gamma}{\delta}t' \tag{6.2.5}$$

and

$$\frac{\delta}{\gamma} = 1. \tag{6.2.6}$$

Since we have here

$$x = \gamma(x' + vt')$$

it follows that

$$\gamma^2(1 - \beta v) = 1. \tag{6.2.7}$$

Define

$$\beta = \frac{v}{c^2(v)} \tag{6.2.8}$$

where $c(v)$ has the dimension of velocity.

We now show that $c(v)$ must be a constant by requiring that two successive transformations of the form (6.2.1) and (6.2.3) must again be of the same form. That is, we set

$$x'' = \gamma(v')(x' - v't') \tag{6.2.9}$$

and

$$t'' = \gamma(v')\left(t' - \frac{v'}{c'^2}x'\right) \tag{6.2.10}$$

or

$$x'' = \gamma(v')\,\gamma(v)\left[x\left(1 + \frac{vv'}{c^2}\right) - t(v + v')\right] \tag{6.2.11}$$

and

$$t'' = \gamma(v)\,\gamma(v')\left[t\left(1 + \frac{vv'}{c'^2}\right) - x\left(\frac{v}{c^2} + \frac{v}{c'^2}\right)\right] \tag{6.2.12}$$

so that c' must equal c; thus, c is a constant. The rule for adding velocities is

$$u = \frac{v + v'}{1 + \frac{vv'}{c^2}} \tag{6.2.13}$$

and

$$\gamma(v) = \frac{1}{\sqrt{1 - \frac{v^2}{c^2}}}. \tag{6.2.14}$$

We note that the addition law for v and c is

$$u = \frac{v + c}{1 + \dfrac{vc}{c^2}} = c \tag{6.2.15}$$

so that the speed c is a limit, seen to be the same for all observers. Thus, the identification of c with the velocity of light will guarantee the constancy of that velocity for all observers, consistent with the straightforward interpretation of Maxwell's equations.

Clearly, the simplest three-dimensional expression of this transformation is

$$x' = \gamma(x - vt), \qquad t' = \gamma\left(t - \frac{v}{c^2}x\right), \qquad y' = y, \qquad z' = z. \tag{6.2.16}$$

A few remarks on (6.2.16) are in order. First, these equations must relate space-time *intervals*. Evidently, the origins of the O and O' space-time coordinate system could be different, in which case we would have

$$x' = \gamma(x - vt) + x_0, \qquad t' = \gamma\left(t - \frac{vx}{c^2}\right) + t_0$$

with x_0 and to the same for all events instead of (6.2.16). Equation (6.2.16) holds when the origins of the two coordinate systems pass each other at a time that both observers call zero. If there is any confusion about this point, rewrite (6.2.16) as

$$\Delta x' = \gamma(\Delta x - v\Delta t), \qquad \Delta t' = \gamma\left(\Delta t - \frac{v\Delta x}{c^2}\right) \tag{6.2.17}$$

where Δx and Δt are the space and time intervals between two events; $\Delta x'$ and $\Delta t'$ are those intervals as measured by the primed observer.

Second, let us consider the behavior of clocks. Suppose we have a clock at rest in O, and the time between ticks (say, birth and death of a μ meson) is T. Then the time between these events seen by O' is, since they occur at the same position in the O frame,

$$T' = \gamma T \quad \text{(i.e., a longer time).} \tag{6.2.18}$$

Third, moving rods appear shorter. Suppose a rod is at rest in O' with length L, that is, $x_1' - x_2' = L$. Then, a measurement of the distance

$x_1 - x_2$ at the same O time would give

$$L = \gamma(x_1 - x_2) \quad \text{or} \quad (x_1 - x_2) = \frac{L}{\gamma} = L\sqrt{1 - \frac{v^2}{c^2}}, \quad (6.2.19)$$

the Lorentz–Fitzgerald contraction.

The simplest way to characterize the Lorentz transformation (6.2.17) is by the invariance of the quantity

$$(\Delta\tau)^2 = (\Delta t)^2 - \frac{(\Delta x)^2}{c^2} \tag{6.2.20}$$

or, in three dimensions,

$$(\Delta\tau)^2 = (\Delta t)^2 - \frac{(\Delta \mathbf{x})^2}{c^2}. \tag{6.2.21}$$

Thus, the transformation between space-time measurements is (remember the summation convention)

$$x'^{\mu} = a^{\mu} + \Lambda^{\mu}{}_{\nu}x^{\nu} \tag{6.2.22}$$

where $\mu = 0, 1, 2, 3$, $x^0 = ct$, and the invariant interval[2] between two events is

$$(\Delta\tau)^2 = \Delta x'^{\mu}\eta_{\mu\nu}\Delta x'^{\nu} = \Delta x^{\mu}\eta_{\mu\nu}\Delta x^{\nu} \tag{6.2.23}$$

with

$$\eta_{00} = 1, \qquad \eta_{ij} = -\delta_{ij}, \qquad \eta_{0i} = \eta_{i0} = 0, \qquad i = 1, 2, 3.$$

In our one-dimensional example, with $x = x_1$, etc.,

$$\Lambda^1{}_1 = \gamma, \qquad \Lambda^1{}_0 = -\frac{v}{c}\gamma, \qquad \Lambda^0{}_0 = \gamma, \qquad \Lambda^0{}_1 = -\frac{v}{c}\gamma. \tag{6.2.24}$$

The transformations (6.2.22) form the Poincaré group. The subgroup with $a^{\mu} = 0$ is the Lorentz group. The Lorentz group has six parameters: three velocity transformations (boosts) and three rotations. The Poincaré group has four more: the four space-time translations.

---

[2]The reader is warned that there is no general agreement on the sign of η. Our choice makes $(\Delta\tau)^2 > 0$ for a timelike interval, that is, one with $|\Delta t| > |\Delta\mathbf{x}|/c$.

From (6.2.23) we learn that

$$\Delta x'^{\mu} \eta_{\mu\nu} \Delta x'^{\nu} = \Lambda^{\mu}_{\ \sigma} \Delta x^{\sigma} \eta_{\mu\nu} \Lambda^{\nu}_{\ \lambda} \Delta x^{\lambda} = \Delta x^{\sigma} \eta_{\sigma\lambda} \Delta x^{\lambda} \qquad (6.2.25)$$

or

$$\Lambda^{\mu}_{\ \sigma} \eta_{\mu\nu} \Lambda^{\nu}_{\ \lambda} = \eta_{\sigma\lambda}. \qquad (6.2.26)$$

So, if we define

$$\Lambda^{\nu}_{\ \lambda} = A_{\nu\lambda} \quad \text{(a matrix)} \qquad \text{and} \qquad \Lambda^{\mu}_{\ \sigma} = A_{\mu\sigma} \equiv A^{T}_{\sigma\mu}, \quad (6.2.27)$$

Eq. (6.2.26) requires that

$$A^{T} \eta A = \eta \qquad (6.2.28)$$

as a matrix identity defining the most general Lorentz transformation.
Just as in the case of rotations, we find from (6.2.28),

$$\det \eta = \det(A^{T} \eta A) = (\det A)^{2} \det \eta$$

so $\det A = \pm 1$.

Transformations with determinant $+1$ can be reached by a succession of infinitesimal Lorentz transformations.[3] Thus, analogous to (6.1.2), we can consider

$$x' = x - \Delta v t, \qquad t' = t - \Delta v x$$

(where from now on we use units in which $c = 1$). With $\xi = N\Delta v$,

$$\begin{pmatrix} x' \\ t' \end{pmatrix} = \left(1 - \frac{\xi \sigma_x}{N}\right)^{N} \begin{pmatrix} x \\ t \end{pmatrix}, \qquad \text{where } \sigma_x = \begin{pmatrix} 0 & 1 \\ 1 & 0 \end{pmatrix}$$

As $N \to \infty$,

$$= e^{-\xi \sigma_x} \begin{pmatrix} x \\ t \end{pmatrix} = \cosh \xi - \sigma_x \sinh \xi) \begin{pmatrix} x \\ t \end{pmatrix}$$

so that

$$\begin{pmatrix} x' \\ t' \end{pmatrix} = \begin{pmatrix} x \cosh \xi - t \sinh \xi \\ -x \sinh \xi + t \cosh \xi \end{pmatrix}. \qquad (6.2.29)$$

---

[3]This does not include $x'^{\mu} = -x^{\mu}$.

The transformation velocity is clearly

$$v = \tanh \xi. \tag{6.2.30}$$

The most general infinitesimal Lorentz transformation is specified by a 4×4 antisymmetric matrix that we call $\epsilon_{\sigma\lambda}$. Thus, we set $\Lambda^\mu{}_\sigma = \delta^\mu{}_\sigma + \epsilon^\mu{}_\sigma$, and with $\epsilon_{\sigma\lambda} = \eta_{\sigma\nu}\epsilon^\nu{}_\lambda$, (6.2.26) shows that

$$\epsilon_{\sigma\lambda} + \epsilon_{\lambda\sigma} = 0. \tag{6.2.31}$$

Returning to the invariant $(d\tau)^2$, we remark that the notation is deceptive, since $(d\tau)^2$ can be negative or positive. For two simultaneous events, $(\Delta t)^2 = 0$, $(\Delta\tau)^2 = -(\Delta \mathbf{x})^2$; the interval is called spacelike. For two events at the same spatial location, $(\Delta \mathbf{x})^2 = 0$ and $(\Delta t)^2 = (\Delta\tau)^2$; the interval is called timelike. In general, if the interval is spacelike, there is a coordinate system within which the events are simultaneous. In one dimension, $\Delta t' = (\Delta t - v\Delta x)/\sqrt{1 - v^2}$, so setting $v = \Delta t/\Delta x$, with $|\Delta t/\Delta x| < 1$ for a spacelike interval, we find $\Delta t' = 0$. Clearly, slightly smaller or greater velocities than v can make $\Delta t'$ either positive or negative; time order for a spacelike interval is not invariant.

Similarly, for a timelike interval, there is always an observer for whom the two events happen at the same place. Also, for a timelike interval, the sign of the time difference is clearly invariant.

We finally consider the consequence of an object moving with a super-luminal velocity, that is, a velocity greater than 1. Suppose the object goes from x_1, t_1 to x_2, t_2, where $x_2 - x_1 = v(t_2 - t_1)$, with $v > 1$. Then the $(1, 2)$ interval is spacelike, and the time order t_1 and t_2 will be different for different observers. That is, it will be impossible to say whether the object went from 1 to 2 or from 2 to 1. Thus, no causal relationship between events 1 and 2 can be established.

How can we write equations that are covariant with respect to the Lorentz transformation—that is, equations which are the same for all Lorentz observers, but which do not contain explicitly the velocity of the observer with respect to a given coordinate system?

We have a clue in our treatment of rotational invariance. There, we required that all equations be tensor equations—the tensors being defined by a specific set of transformations: rotations. Thus, the basic tensor in classical mechanics is a vector x_i, or better, dx_i, the (infinitesimal) displacement between two points. Since dt is a rotational invariant, dx_i/dt, d^2x_i/dt^2, etc. are all vectors, so linear equations joining d^2x_i/dt^2 to other vectors, for example, $(x_i - y_i)/|\mathbf{x} - \mathbf{y}|^3$, will be vector equations and hence invariant under rotations.

We can do the same for Lorentz transformations. We define a contra-variant vector under Lorentz transformations as an object V^μ such that

$$V'^{\mu} = \Lambda^{\mu}{}_{\nu} V^{\nu}.$$

An immediate example of such a vector is the coordinate differential,

$$dx^{\mu} = (dx^0, dx^1, dx^2, dx^3),$$

since $dx'^{\mu} = \Lambda^{\mu}{}_{\nu} dx^{\nu}$ precisely describes the way coordinate differentials (space-time intervals between events) transform.

A more compact and convenient way of characterizing the transformation properties of dx^{μ} is

$$dx'^{\mu} = \frac{\partial x'^{\mu}}{\partial x^{\nu}} dx^{\nu} \tag{6.2.33}$$

or, for a general contravariant vector,

$$V'^{\mu} = \frac{\partial x'^{\mu}}{\partial x^{\nu}} V^{\nu}.$$

How do we construct a finite Lorentz (or four-) vector? We clearly need an analogue to the use of dt described above for rotational invariance. We recall the invariant

$$(d\tau)^2 = (dt)^2 - (d\mathbf{x})^2 = dx^{\mu} \, \eta_{\mu\nu} \, dx^{\nu}.$$

If the two events are separated by a timelike interval (and for two locations of a particle moving with $v < 1$, they will be), $(d\tau)^2$ is positive, and $d\tau = dt\sqrt{1 - v^2}$ will be invariant; thus,

$$u^{\mu} = \frac{dx^{\mu}}{d\tau} = \left(\frac{1}{\sqrt{1 - v^2}}, \frac{\mathbf{v}}{\sqrt{1 - v^2}} \right) \tag{6.2.34}$$

will be a four-vector, sometimes called the four-velocity.

We can now write a covariant equation for momentum conservation. Let us call the four-vector p^{μ} (where $mu^{\mu} = p^{\mu}$) the four-momentum of the particle with mass m. Then a covariant conservation law for a two-body interaction, with a and b incoming, possibly different c and d outgoing, would be

$$P^{\mu} = p_a^{\mu} + p_b^{\mu} = p_c^{\mu} + p_d^{\mu}. \tag{6.2.35}$$

The covariance of the momentum conservation law requires a fourth conservation law: energy. The energy, of course, includes rest energy:

The four-vector

$$p^\mu = (p^0, \mathbf{p}) \tag{6.2.36}$$

has

$$\mathbf{p} = \frac{m\upsilon}{\sqrt{1 - \upsilon^2}} \tag{6.2.37}$$

and

$$p^0 = \frac{m}{\sqrt{1 - \upsilon^2}}. \tag{6.2.38}$$

For $\upsilon \ll 1$, $\mathbf{p} \sim m\upsilon$ and $p^0 \sim m + m\upsilon^2/2$, the nonrelativistic forms, but with the rest energy (mc^2) added.

The individual momenta are timelike with positive time components, so their sum P^μ is timelike. Therefore, there is a coordinate system in which the space component P^i is zero; it is called the rest system and is usually the most convenient place to carry out calculations. We consider the two-body example. First, we express the energy in terms of the momentum of each particle:

$$\mathbf{p} = \frac{m\mathbf{v}}{\sqrt{1 - \upsilon^2}} \quad \text{or} \quad \mathbf{v} = \frac{\mathbf{p}}{\sqrt{\mathbf{p}^2 + m^2}} \tag{6.2.39}$$

and

$$p^0 = \sqrt{\mathbf{p}^2 + m^2}. \tag{6.2.40}$$

Therefore, in the rest system of a and b, $\mathbf{p}_a = -\mathbf{p}_b = \mathbf{p}$ and

$$P^0 = p_a^0 + p_b^0 = \sqrt{m_a^2 + \mathbf{p}^2} + \sqrt{m_b^2 + \mathbf{p}^2}. \tag{6.2.41}$$

Since momentum is conserved, $\mathbf{p}_c = -\mathbf{p}_d = \mathbf{p}'$ and energy is conserved,

$$P^0 = p_c^0 + p_d^0 = \sqrt{m_c^2 + p'^2} + \sqrt{m_d^2 + p'^2}. \tag{6.2.42}$$

The system will be exothermic or endothermic according to whether $m_a + m_b$ is greater than or less than $m_c + m_d$.

6.3. LORENTZ TENSORS

Since, for any contravariant vector, $V^\mu \eta_{\mu\nu} V^\nu$ is invariant, so is

$$(V^\mu + U^\mu)\, \eta_{\mu\nu}(V^\nu + U^\nu)$$

and hence so is $U^\mu \eta_{\mu\nu} V^\nu$. Thus, given a contravariant vector U^μ, there exists an object

$$U_\mu = \eta_{\mu\nu} U^\nu \tag{6.3.1}$$

such that $U_\mu V^\mu$ is invariant for any vectors $U\mu$ and V^μ.

U_μ is called a covariant vector. From (6.3.1), with $U^\mu = (U^0, U^i)$, we have[4]

$$U_\mu = (U^0, -U^i) = (U_0, U_i). \tag{6.3.2}$$

Tensors under rotation have the property that their covariant and contravariant representations are the same. However, we note that if we had used a nonorthogonal basis system, say, $\mathbf{e}_\lambda$, as the basis vectors in three-dimensional space, then any vector $\mathbf{V}$ could be expanded as

$$\mathbf{V} = \sum_\lambda \mathbf{e}_\lambda V^\lambda \tag{6.3.3}$$

and would be invariant under rotation. In particular, for the vector

$$d\mathbf{x} = \sum_\lambda \mathbf{e}_\lambda\, dx^\lambda, \tag{6.3.4}$$

the dx^λ's would transform contravariantly, and since $d\mathbf{x}$ is invariant, the $\mathbf{e}_\lambda$'s transform covariantly.

The transformation law for covariant Lorentz vectors is

$$U'_\mu = \frac{\partial x'}{\partial x'^\mu}\, U_\nu \tag{6.3.5}$$

since with

$$V'^\mu = \frac{\partial x'^\mu}{\partial x^\sigma}\, V^\sigma,$$

---

[4]We usually represent four-vectors and tensors with Greek indices, $\mu, \nu, \ldots$, and three-dimensional space vectors with latin indices, $i, j, k, \ldots$.

we have

$$U'_\mu V'^\mu = \frac{\partial x^\nu}{\partial x'^\mu} \frac{\partial x'^\mu}{\partial x^\sigma} U_\nu V^\sigma = U_\nu \delta^\nu_\sigma V^\sigma = U_\sigma V^\sigma \qquad (6.3.6)$$

where δ^ν_σ is the usual Kronecker delta.

Conversely, if $U_\mu V^\mu$ is invariant and V^μ is an arbitrary contravariant vector, then U_μ is a covariant vector, since

$$U'_\mu V'^\mu = U'_\mu \frac{\partial x'^\mu}{\partial x^\sigma} V^\sigma = U_\sigma V^\sigma$$

from which

$$U'_\mu \frac{\partial x'^\mu}{\partial x^\sigma} = U_\sigma.$$

Lorentz tensors are constructed in the same way as rotational tensors, except that they can be covariant, contravariant, or mixed. Thus, we have a second-rank contravariant tensor

$$T'^{\mu\nu} = \frac{\partial x'^\mu}{\partial x^\sigma} \frac{\partial x'^\nu}{\partial x^\lambda} T^{\sigma\lambda}, \qquad (6.3.7)$$

a second-rank covariant tensor

$$T'_{\mu\nu} = \frac{\partial x^\sigma}{\partial x'^\mu} \frac{\partial x^\lambda}{\partial x'^\nu} T_{\sigma\lambda}, \qquad (6.3.8)$$

and a second-rank mixed tensor

$$T'^\mu_{\ \nu} = \frac{\partial x'^\mu}{\partial x^\sigma} \frac{\partial x^\lambda}{\partial x'^\nu} T^\sigma_{\ \lambda}. \qquad (6.3.9)$$

Symmetry properties are invariant: For example, let $T_{\nu\mu} = \pm T_{\mu\nu}$. Then

$$T'_{\mu\nu} = \frac{\partial x_\sigma}{\partial x'^\mu} \frac{\partial x_\lambda}{\partial x'^\nu} T_{\sigma\lambda} \quad \text{and} \quad T'_{\nu\mu} = \frac{\partial x_\sigma}{\partial x'^\nu} \frac{\partial x_\lambda}{\partial x'^\mu} T_{\sigma\lambda};$$

interchanging the dummy indices σ and λ gives

$$T'_{\nu\mu} = \frac{\partial x_\lambda}{\partial x'^\nu} \frac{\partial x_\sigma}{\partial x'^\mu} T_{\lambda\sigma}$$

$$= \pm \frac{\partial x_\lambda}{\partial x'^\nu} \frac{\partial x_\sigma}{\partial x'^\mu} T_{\sigma\lambda} = \pm T'_{\mu\nu}.$$

There are several universal tensors. δ^μ_ν is a second-rank mixed tensor:

$$\delta'^\mu_\nu = \frac{\partial x^\sigma}{\partial x'^\nu} \frac{\partial x'^\mu}{\partial x^\lambda} \sigma^\lambda_\sigma = \frac{\partial x'^\mu}{\partial x^\lambda} \frac{\partial x^\lambda}{\partial x'^\nu} = \delta^\mu_\nu. \tag{6.3.10}$$

$\eta_{\mu\nu}$ is a second-rank covariant tensor: With x^μ and y^ν as contravariant vectors, we see, by the definition of the Lorentz transformation, that

$$x'^\mu \eta'_{\mu\nu} y'^\nu = x^\lambda \eta_{\lambda\sigma} y^\sigma$$

is invariant, where $\eta'_{\mu\nu}$ is the same function of its indices as $\eta_{\lambda\sigma}$. Since

$$x'^\mu = \frac{\partial x'^\mu}{\partial x^\lambda} x^\lambda, \qquad y'^\nu = \frac{\partial x'^\nu}{\partial x^\sigma} y^\sigma,$$

we have

$$x^\lambda y^\sigma \eta_{\lambda\sigma} = x^\lambda y^\sigma \frac{\partial x'^\mu}{\partial x^\lambda} \frac{\partial x'^\nu}{\partial x^\sigma} \eta'_{\mu\nu}$$

so that

$$\eta_{\lambda\sigma} = \frac{\partial x'^\mu}{\partial x^\lambda} \frac{\partial x'^\nu}{\partial x^\sigma} \eta'_{\mu\nu}. \tag{6.3.11}$$

Of course, (6.3.11) is equivalent to the original equations that determined the Lorentz transformation matrix. The result could also have been deduced from the fact that $\eta_{\mu\nu}V^\nu$ transforms as a covariant vector for any V^ν, since if $T_{\mu\nu}V^\nu$ transforms like a covariant vector for any contravariant V^ν, then $T_{\mu\nu}$ is a tensor. The proof is identical to the one given above for a covariant vector. Clearly, these rules apply to tensors of arbitrary rank: Any covariant index can, with summation, cancel a contravariant index. Thus, with A and B tensors, $A_{\mu\nu\lambda\eta}B^{\eta\sigma} = T^\sigma_{\mu\nu\lambda}$ is a mixed tensor, as indicated. Single indices can be lowered with $\eta_{\mu\nu}$. However, note that $\eta_{\lambda\mu}\eta_{\mu\nu} = \delta^\lambda_\nu$ and is therefore a mixed tensor; it follows that $\eta_{\lambda\mu} = \eta^{\lambda\mu}$ is a second-rank contravariant tensor as well as a second-rank covariant tensor. Thus, $\eta^{\mu\nu} = \eta_{\mu\nu}$ can also be used to raise indices and to lower the rank of a tensor by tracing:

$$Tr\,T_{\mu\nu} = \eta^{\nu\mu} T_{\mu\nu},$$

or with

$$T^{\mu}{}_{\nu} = \eta^{\mu\sigma} T_{\sigma\nu}, \qquad Tr\,T_{\mu\nu} = T^{\mu}{}_{\mu}. \tag{6.3.12}$$

One more universal tensor; $\epsilon_{\mu\nu\lambda\eta}$, is defined to be totally antisymmetric in $\mu\nu\lambda\eta$; hence, all four indices must be different, so $\epsilon_{\mu\nu\lambda\eta} = \pm\epsilon_{0123}$, the plus sign for $\mu\nu\lambda\eta$, an even permutation of 0123, the minus sign for an odd permutation. We define $\epsilon_{0123} = 1$. (Beware of applying the three-dimensional rule that cyclic permutations are even. They are not. In four dimensions they are odd.)

Is ϵ a tensor? We calculate

$$\epsilon'_{\mu\nu\lambda\eta} = \frac{\partial x^{\sigma}}{\partial x'^{\mu}}\frac{\partial x^{\lambda}}{\partial x'^{\nu}}\frac{\partial x^{\alpha}}{\partial x'^{\lambda}}\frac{\partial x^{\beta}}{\partial x'^{\eta}}\,\epsilon_{\sigma\lambda\alpha\beta}.$$

Since ϵ' has the total antisymmetry of ϵ, we need only calculate

$$\epsilon'_{0123} = \frac{\partial x^{\sigma}}{\partial x'^{0}}\frac{\partial x^{\lambda}}{\partial x'^{1}}\frac{\partial x^{\alpha}}{\partial x'^{2}}\frac{\partial x^{\beta}}{\partial x'^{3}}\,\epsilon_{\sigma\lambda\alpha\beta}$$

$$= \det\!\left(\frac{\partial x^{\sigma}}{\partial x'^{\mu}}\right). \tag{6.3.13}$$

So, $\epsilon_{\mu\nu\lambda\eta}$ transforms like a tensor for proper Lorentz transformations, but like a pseudotensor for inversions.

6.4. TENSOR FIELDS: COVARIANT ELECTRODYNAMICS

Tensor fields are functions of **x** and t that transform according to tensor laws. Thus, a tensor of rank zero is a scalar, and a scalar field is an invariant, say, $\psi(x)$.[5]

The invariance of the field means that two Lorentz observers observing the field will measure the same value of ψ at the same space-time point, so $\psi'(x'(x)) = \psi(x)$. That is,

---

[5]From now on, we will frequently use symbols like x and y to stand for a four-vector, x^{μ} and y^{μ}. Thus, a scalar field $\psi(x) = \psi(x^{0}, \mathbf{x})$.

$$\psi'(x') = \psi(x) \tag{6.4.1}$$

where x' and x represent the same space-time point.

The prototype covariant vector field is the gradient of a scalar field:

$$V_\mu = \frac{\partial \psi}{\partial x^\mu}, \tag{6.4.2}$$

since

$$V'_\mu = \frac{\partial \psi'(x')}{\partial x'^\mu} = \frac{\partial \psi(x)}{\partial x^\sigma} \frac{\partial x^\sigma}{\partial x'^\mu} = \frac{\partial x^\sigma}{\partial x'^\mu} V_\sigma. \tag{6.4.3}$$

Derivatives of tensor fields generate higher tensors, for example,

$$T_{\mu\nu} = \frac{\partial}{\partial x^\mu} V_\nu. \tag{6.4.4}$$

Derivatives can also be used to lower rank by tracing:

$$\frac{\partial}{\partial x^\mu} V^\mu = V, \tag{6.4.5}$$

a scalar.

The first example of a physical scalar field is the scalar density associated with a point particle, moving along a given trajectory

$$x^i_p = x^i_p(t) \tag{6.4.6}$$

or more covariantly,

$$x^\mu_p = x^\mu_p(\tau) \tag{6.4.7}$$

where $d\tau$ is the proper time along the trajectory:

$$d\tau = dx^0 \sqrt{1 - \mathbf{v}^2}. \tag{6.4.8}$$

We obtain (6.4.6) from (6.4.7) by solving $x^0 = x^0(\tau)$ for τ as a function of x^0 and substituting it into $x^i(\tau)$.

We wish to construct a scalar density[6] that, for a point particle, must

---

[6]We here give the word "density" its physics meaning. The expressions scalar (and vector and tensor) density are sometimes used to designate certain transformation properties under general coordinate transformations [as in (7.6.15)]. That is not the case here.

be proportional to a three-dimensional delta function

$$d(x^0, \mathbf{x}) = A\delta^3(\mathbf{x} - \mathbf{x}_p(x^0)) \tag{6.4.9}$$

where A may depend on the velocity of the particle.

To show that an A exists, we note that the four-dimensional volume element $d^4x = dx^0 d^3x$ is invariant, since the Jacobian from x to x' is just the absolute value of the determinant

$$\left| \det\left(\frac{\partial x^\mu}{\partial x'^\nu} \right) \right|,$$

which is equal to 1. Therefore, the four-dimensional delta function

$$\delta^4(x - x_p(\tau)) = \delta(x^0 - x_p^0(\tau))\, \delta^3(\mathbf{x} - \mathbf{x}_p(\tau)) \tag{6.4.10}$$

is invariant, as is its integral over the invariant τ [which is held to a unique value $\tau(x^0)$ by the first delta function in (6.4.10)]. There results

$$d(x) = \int d\tau\, \delta(x^0 - x_p^0(\tau))\, \delta^3(\mathbf{x} - \mathbf{x}_p(\tau))$$

$$= \frac{d\tau}{dx^0}\, \delta^3(\mathbf{x} - \mathbf{x}_p(\tau))$$

at $x_p^0(\tau) = x^0$, so

$$d(x) = \sqrt{1 - v^2}\, \delta^3(\mathbf{x} - \mathbf{x}_p(x^0)) \tag{6.4.11}$$

where

$$\mathbf{v} = \frac{d\mathbf{x}_p(x^0)}{dx^0}.$$

Similarly, a four-vector density for a point particle can be defined by

$$J^\mu = \frac{dx_p^\mu}{d\tau}\, d(x),$$

that is,

$$J^\mu = \int d\tau \frac{dx_p^\mu(\tau)}{d\tau}\, \delta^4(x - x_p(\tau))$$

$$= \frac{dx_P^\mu}{d\tau} \frac{d\tau}{dx^0} \delta^3(\mathbf{x} - \mathbf{x}_p(x_0)) \tag{6.4.12}$$

or

$$J^0 = \delta^3(\mathbf{x} - \mathbf{x}_p(x^0)) \quad \text{and} \quad \mathbf{J} = \mathbf{v}\delta^3(\mathbf{x} - \mathbf{x}_p(x^0)). \tag{6.4.13}$$

We recognize in (6.4.13) the electromagnetic charge and current density of a point-particle-carrying unit charge. For a particle of charge q, we would have

$$j^\mu = qJ^\mu. \tag{6.4.14}$$

Finally, we construct a second-rank tensor density for a point particle:

$$T^{\mu\nu} = \frac{dx_P^\mu}{d\tau} \frac{dx_P^\nu}{d\tau} \frac{d\tau}{dx^0} \delta^3(\mathbf{x} - \mathbf{x}_p(x^0)). \tag{6.4.15}$$

Covariant densities can be constructed by using the tensor $\eta_{\mu\nu}$ to lower contravariant indices. Remember that this operation is actually very simple: Leave the 0th component alone and change the sign of the three space components.

A useful clue to help construct a relativistic electrodynamics is furnished by (6.4.13), which tells us that the charge and current densities of a point particle form a contravariant four-vector; so therefore does a sum of these objects over different particles. Therefore, a general charge and current density form a contravariant vector field $j^\mu(x)$. The local charge conservation law is a scalar equation

$$0 = \frac{\partial j^0}{\partial x^0} + \nabla \cdot \mathbf{j} = \frac{\partial}{\partial x^\mu} j^\mu, \tag{6.4.16}$$

which we know is satisfied by (6.4.13). (See Problem 2.1.)

Recall now the equation for the vector and scalar potentials in the Lorentz gauge:

$$\left(\nabla^2 - \frac{\partial^2}{(\partial x^0)^2}\right)\mathbf{A} = -4\pi\mathbf{j} \quad \text{and} \quad \left(\nabla^2 - \frac{\partial^2}{(\partial x^0)^2}\right)\phi = -4\pi\rho. \tag{6.4.17}$$

Since

$$\nabla^2 - \frac{\partial^2}{(\partial x^0)^2} = -\frac{\partial}{\partial x^\mu} \eta^{\mu\nu} \frac{\partial}{\partial x^\nu}$$

is a scalar operator, it is natural to treat **A** and ϕ as a contravariant four-vector. In general, of course, there can be no required relativistic transformation properties for **A** and ϕ, since **A** and ϕ are only determined to within a gauge transformation, and gauge transformations have no special covariance properties. Note however that the Lorentz gauge allows **A** and ϕ to be given vector transformation properties, since the Lorentz condition

$$\nabla \cdot \mathbf{A} + \frac{\partial \phi}{\partial x^0} = 0,$$

expressed covariantly becomes

$$\frac{\partial}{\partial x^\mu} A^\mu = 0. \tag{6.4.18}$$

A gauge transformation takes

$$A_\mu \quad \text{into} \quad A_\mu + \frac{\partial \chi}{\partial x^\mu}. \tag{6.4.19}$$

None of these considerations chooses a sign for the contravariant vector A^μ. It is actually most convenient to deal with the covariant vector A_μ, which we take to be given in terms of the vector and scalar potentials **A** and ψ by

$$A_\mu = (-\phi, \mathbf{A}). \tag{6.4.20}$$

A second clue tells us how to obtain the fields. Remember $\mathbf{B} = \nabla \times \mathbf{A}$, or in tensor language[7] $B_{ij} = \partial_i A_j - \partial_j A_i$, so that $B_{12} = B_3$, etc. The obvious covariant generalization is to define a field tensor[8]:

$$F_{\mu\nu} = \partial_\mu A_\nu - \partial_\nu A_\mu. \tag{6.4.21}$$

The antisymmetric tensor $F_{\mu\nu}$ is gauge invariant (under $A_\mu \to A_\mu + \partial\chi/\partial x^\mu$) and has the right number of components to generate the electromagnetic field: F_{12}, F_{23}, F_{31} (which we have defined to generate B_{ij}) and F_{10}, F_{20}, F_{30} (which must generate E_i). Let us check:

---

[7]We introduce here another convention: $\partial_\mu \equiv \partial/\partial x^\mu$ and $\partial^\mu = \partial/\partial x_\mu$.

[8]Here again, the reader is warned that there is no general consensus on the sign of $F_{\mu\nu}$. Our choice makes $F_{i0} = E_i$.

$$F_{ij} = (\partial_i A_j - \partial_j A_i) = B_{ij} \tag{6.4.22}$$

and

$$F_{i0} = \partial_i A_0 - \partial_0 A_i = E_i. \tag{6.4.23}$$

Finally, we must be able to write Maxwell's equations in covariant form. First, the inhomogeneous equations involve derivatives of F on the left and current components on the right. The covariant expression of this must be $\partial_\mu F^{\mu\nu}$ on the left and j^ν on the right. Thus,

$$\partial_\mu F^{\mu\nu} = \kappa j^\nu \tag{6.4.24}$$

with κ to be determined. We write out (6.4.24):

$$\partial_i F^{i\nu} + \partial_0 F^{0\nu} = \kappa j^\nu. \tag{6.4.25}$$

Let $\nu = 0$: $\partial_i F^{i0} = \kappa j^0$, or using (6.4.23), we obtain

$$\nabla \cdot \mathbf{E} = -\kappa j^0 \tag{6.4.26}$$

so $\kappa = -4\pi$. Now let $\nu = \ell$: We find

$$\partial_i F^{i\ell} + \partial_0 F^{0\ell} = \kappa j^\ell. \tag{6.4.27}$$

Since the form of (6.4.27) guarantees rotational invariance, it is sufficient to consider one value of ℓ, say, $\ell = 1$. Then (6.4.27) becomes

$$\partial_i F_{i1} + \partial_0 F^{01} = \kappa j^1$$

or

$$\partial_i F_{i1} + \partial_0 F_{10} = \kappa j^1$$

or

$$\partial_2 F_{21} + \partial_3 F_{31} + \partial_0 E_1 = \kappa j^1$$

or

$$(\partial_3 B_2 - \partial_2 B_3) + \partial_0 E_1 = \kappa j^1; \tag{6.4.28}$$

with $\kappa = -4\pi$, (6.4.28) gives

$$-\left(\nabla \times \mathbf{B}\right) + \partial_0 \mathbf{E} = -4\pi \mathbf{j}$$

which is correct. Therefore, the four inhomogeneous equations are put together into one four-vector equation:

$$\partial_\mu F^{\mu\nu} = -4\pi j^\nu. \tag{6.4.29}$$

The homogeneous equations are also four in number, suggesting a four-vector equation. However, we know that the homogeneous equations involve axial quantities: $\partial \mathbf{B}/\partial t$, $\nabla \times \mathbf{E}$, and $\nabla \cdot \mathbf{B}$. The unique axial vector consisting of first derivatives of F is $\partial_\mu \tilde{F}^{\mu\nu}$, where $\tilde{F}^{\mu\nu}$ is the dual of F:

$$2\tilde{F}^{\mu\nu} \equiv \epsilon^{\mu\nu\lambda\sigma} F_{\lambda\sigma}. \tag{6.4.30}$$

$\epsilon^{\mu\nu\lambda\sigma}$ is, of course, $-\epsilon_{\mu\nu\lambda\sigma}$, since one time and three space indices are raised to go from $\epsilon_{\mu\nu\lambda\sigma}$ to $\epsilon^{\mu\nu\lambda\sigma}$. The dual $\tilde{F}^{\mu\nu}$ of $F_{\mu\nu}$ simply interchanges **E** and **B**. That is,

$$\tilde{F}^{ij} = \frac{1}{2}\epsilon^{ij\mu\nu} F_{\mu\nu} = \epsilon^{ijk0} F_{k0}$$

$$= \epsilon_{ijk} E_k$$

and

$$\tilde{F}^{j0} = \frac{1}{2}\epsilon^{j0k\ell} F_{k\ell} = \frac{1}{2}\epsilon_{jk\ell} F_{k\ell} = B_j.$$

Thus, we conjecture that the homogeneous equations are equivalent to

$$\partial_\nu \epsilon^{\mu\nu\lambda\sigma} F_{\lambda\sigma} = 0. \tag{6.4.31}$$

Written out, (6.4.31) is for $\mu = 0$:

$$\epsilon^{0ijk} \partial_i F_{jk} = -\epsilon_{ijk} \partial_i F_{jk} = 0 \quad \text{or} \quad \nabla \cdot \mathbf{B} = 0.$$

For $\mu = j$,

$$\epsilon^{j\nu\lambda\sigma} \partial_\nu F_{\lambda\sigma} = \epsilon^{j0k\ell} \partial_0 F_{k\ell} + \epsilon^{jk0\ell} \partial_k F_{0\ell} + \epsilon^{jk\ell0} \partial_k F_{\ell0}$$

$$= \epsilon_{jk\ell} \partial_0 F_{k\ell} - 2\epsilon_{jk\ell} \partial_k F_{0\ell}$$

$$= 2\partial_0 B_j + 2\epsilon_{jk\ell} \partial_k E_\ell = 0$$

or

$$\frac{\partial \mathbf{B}}{\partial t} + \nabla \times \mathbf{E} = 0.$$

A more familiar form for (6.4.31) is

$$\partial_\mu F_{\nu\lambda} + \partial_\lambda F_{\mu\nu} + \partial_\nu F_{\lambda\mu} = 0. \tag{6.4.32}$$

To see the equivalence, note that (6.4.32) is antisymmetric in any interchange. For example, interchanging μ and ν gives

$$\partial_\nu F_{\mu\lambda} + \partial_\lambda F_{\nu\mu} + \partial_\mu F_{\lambda\nu}$$

which is the negative of the original expression. Therefore, we rewrite (6.4.31):

$$0 = 3\epsilon^{\mu\nu\lambda\sigma} \partial_\nu F_{\lambda\sigma} = \epsilon^{\mu\nu\lambda\sigma} \partial_\nu F_{\lambda\sigma} + \epsilon^{\mu\sigma\nu\lambda} \partial_\sigma F_{\nu\lambda} + \epsilon^{\mu\lambda\sigma\nu} \partial_\lambda F_{\sigma\nu}$$

$$= \epsilon^{\mu\nu\lambda\sigma}(\partial_\nu F_{\lambda\sigma} + \partial_\sigma F_{\nu\lambda} + \partial_\lambda F_{\sigma\nu}),$$

for each μ. The sum over $\nu\lambda\sigma$ simply multiplies the result by 6 (the number of permutations of three objects). Therefore, the quantity in parenthesis is 0.

We close this section by studying the transformation laws for the **E** and **B** fields. We have, since $F_{\mu\nu}$ is a tensor,

$$F'_{\mu\nu} = \frac{\partial x^\sigma}{\partial x'^\mu} \frac{\partial x^\lambda}{\partial x'^\nu} F_{\sigma\lambda} \tag{6.4.33}$$

Let the primed observer be moving with velocity v in the x direction. Then

$$x^0 = \frac{x'^0 + vx'}{\sqrt{1 - v^2}}, \qquad x = \frac{x' + vx'^0}{\sqrt{1 - v^2}}, \tag{6.4.34}$$

$y = y'$ and $z = z'$, from which we can calculate $\partial x^\sigma / \partial x'^\mu$.

We first calculate the longitudinal fields $B_x = F_{yz}$ and $E_x = F_{x0}$:

$$F'_{yz} = \frac{\partial x^\sigma}{\partial y'} \frac{\partial x^\lambda}{\partial z'} F_{\sigma\delta} = F_{yz} \tag{6.4.35}$$

and B_x is invariant.

$$F'_{x0} = \frac{\partial x^\sigma}{\partial x'} \frac{\partial x^\lambda}{\partial x'^0} F_{\sigma\lambda}$$

$$= \frac{\partial x}{\partial x'} \frac{\partial x^0}{\partial x'^0} F_{x0} + \frac{\partial x^0}{\partial x'} \frac{\partial x}{\partial x'^0} F_{0x}$$

$$= \left(\frac{1}{1 - v^2} - \frac{v^2}{1 - v^2} \right) F_{x0} = F_{x0} \tag{6.4.36}$$

and E_x is also invariant.

Consider next the transverse components

$$B'_y = F'_{zx} = \frac{\partial x^\sigma}{\partial z'} \frac{\partial x^\lambda}{\partial x'} F_{\sigma\lambda}$$

$$= \frac{\partial x^\lambda}{\partial x'} F_{z\lambda}$$

$$= \frac{\partial x}{\partial x'} F_{zx} + \frac{\partial x^0}{\partial x'} F_{z0}$$

$$= \frac{1}{\sqrt{1-v^2}} [B_y + vE_z]$$

or

$$\mathbf{B}'_T = \frac{1}{\sqrt{1-v^2}} \left(\mathbf{B}_T - \mathbf{v} \times \mathbf{E}_T \right) \qquad (6.4.37)$$

since

$$\left(\mathbf{v} \times \mathbf{E}_T \right)_y = (v\hat{e}_x \times (\hat{e}_y E_y + \hat{e}_z E_z))_y = -vE_z.$$

Finally,

$$E'_y = F'_{y0} = \frac{\partial x^\sigma}{\partial y'} \frac{\partial x^\lambda}{\partial x'^0} F_{\sigma\lambda}$$

$$= \frac{\partial x^\lambda}{\partial x'^0} F_{y\lambda}$$

$$= \frac{\partial x}{\partial x'^0} F_{yx} + \frac{\partial x^0}{\partial x'^0} F_{y0}$$

$$= \frac{1}{\sqrt{1-v^2}} (E_y - vB_z)$$

or

$$\mathbf{E}'_T = \frac{1}{\sqrt{1-v^2}} \left(\mathbf{E}_T + \mathbf{v} \times \mathbf{B}_T \right). \qquad (6.4.38)$$

Suppose there is a magnetic field but no electric field in the O system. Then an object at rest in the O' system would have velocity $\mathbf{v}$ in the O system, but would only experience the electric force in the O' system, which is, for small $\mathbf{v}$, $e\mathbf{v} \times \mathbf{B}_T = e\mathbf{v} \times \mathbf{B}$, as it should be.

We observe that we can construct two invariants from the $F_{\mu\nu}$ field. The first is

$$I_1 = -\frac{1}{2} F_{\mu\nu} F^{\mu\nu} = \mathbf{E}^2 - \mathbf{B}^2. \tag{6.4.39}$$

The second is, in fact, a pseudoscalar:

$$I_2 = \frac{1}{4} F_{\mu\nu} \tilde{F}^{\mu\pi} = \mathbf{E} \cdot \mathbf{B}. \tag{6.4.40}$$

One verifies easily that I_1 and I_2 are left invariant by the above transformation rules for $\mathbf{E}$ and $\mathbf{B}$.

Another invariant is the phase of $\mathbf{E}$ or $\mathbf{B}$; thus, for a monochromatic wave,

$$k_\mu x^\mu = k \cdot x = k' \cdot x' = k'_\mu x'^\mu \tag{6.4.41}$$

defines k^μ to transform like x^μ. For a wave vector in the v direction,

$$\omega' = \frac{\omega - vk}{\sqrt{1 - v^2}}$$

and since $\omega = k$, $\tag{6.4.42}$

$$\omega' = \omega \sqrt{\frac{1 - v}{1 + v}}.$$

This is the relativistic Doppler shift. For other angles, the formula is more complicated but straightforward, although some care must be taken to define the observation when θ is near 90°. (See Problem 6.9.)

6.5. EQUATIONS OF MOTION FOR A POINT CHARGE IN AN ELECTROMAGNETIC FIELD

We can guess various covariant equations of motion for a charged point particle. We narrow the list of suspects by requiring that they (the equations) be linear in the $F_{\mu\nu}$ field and contain no derivatives.[9]

We start from the four-velocity $u^\mu = dx^\mu/d\tau$. A four-vector acceleration is

$$a^\mu = \frac{d^2 x^\mu}{d\tau^2} = \frac{du^\mu}{d\tau};$$

---

[9]The presence of first derivatives would, for example, result from an internal structure of the particle, such as a magnetic moment.

a covariant equation would have a four-vector on the right-hand side, linear in $F_{\mu\nu}$. The only possibility is $F_{\mu\nu}u^\nu$. We conjecture[10]

$$m\frac{du_\mu}{d\tau} = \kappa q F_{\mu\nu}u^\nu \tag{6.5.1}$$

where q is the charge and the constant κ will be adjusted to agree with the nonrelativistic limit. Consider first $\mu = i$:

$$-m\frac{d}{d\tau}u^i = \kappa q(F_{ij}u^j + F_{i0}u^0) \tag{6.5.2}$$

or

$$= \frac{\kappa q}{\sqrt{1-v^2}}\left(E_i + (\mathbf{v} \times \mathbf{B})_i\right)$$

so that $\kappa = -1$. The particle equation is

$$\frac{d}{dt}\mathbf{p} = q(\mathbf{E} + \mathbf{v} \times \mathbf{B}) \tag{6.5.3}$$

where $\mathbf{p}$ is the particle momentum:

$$\mathbf{p} = \frac{m\mathbf{v}}{\sqrt{1-v^2}}.$$

The fourth component of (6.5.1) not independent, since

$$u^\mu\frac{du_\mu}{d\tau} = \frac{\kappa q}{m}u^\mu F_{\mu\nu}u^\nu = 0,$$

or

$$\frac{d}{d\tau}(u^\mu u_\mu) = 0.$$

This is as it should be:

---

[10]Observe that covariance requires four equations for four variables, the u_μ. The equations, however, cannot be independent, since $u_\mu u^\mu = 1$. The existence of such constraints occurs often in relativistic theories.

$$u^{\mu}u_{\mu} = u_0^2 - \mathbf{u}^2 = \frac{1 - v^2}{1 - v^2} = 1.$$

The fourth equation is

$$m\frac{d}{d\tau}u_0 = -qF_{0i}u^i \qquad \text{or} \qquad m\frac{d}{dt}\frac{1}{\sqrt{1 - v^2}} = q\mathbf{E} \cdot \mathbf{v}. \qquad (6.5.4)$$

Equation (6.5.4) looks like an energy equation; the right-hand side reminds us of work done on the charge, the left-hand side the increase in energy $m/\sqrt{1 - v^2}$. It is, in fact, an energy conservation law, when $\mathbf{E}$ is a static field derivable from a potential $\mathbf{E} = -\nabla\phi$. It is easily seen that (6.5.4) under these circumstances implies

$$\frac{d}{dt}\left(\frac{m}{\sqrt{1 - v^2}} + q\phi\right) = 0. \qquad (6.5.5)$$

6.6. RELATIVISTIC CONSERVATION LAWS

We prove a theorem. Let Q^{μ} be a conserved vector field, that is, let

$$\partial_{\mu}Q^{\mu} = 0. \qquad (6.6.1)$$

Then

$$\frac{d}{dx^0}\int_V Q^0\,d\mathbf{r} = -\int_S \mathbf{Q} \cdot d\mathbf{S} \qquad (6.6.2)$$

and there is a globally conserved quantity

$$Q = \int_V d\mathbf{r}\,Q^0$$

provided there is no outgoing flux:

$$\int_S \mathbf{Q} \cdot d\mathbf{S} = 0.$$

The theorem is that Q is an invariant.[11] We already know an example: the charge of a particle.

Proof: Consider

$$Q(t = 0) = \int d^4x\, j^\mu(x)\, \partial_\mu \theta(n \cdot x) \tag{6.6.3}$$

with

$$\theta(t) = \begin{cases} 1, & t > 0 \\ 0, & t < 0 \end{cases}, \tag{6.6.4}$$

$n \cdot x = n_\lambda x^\lambda$, and n_λ a unit vector in the time direction, so that $n \cdot x = x^0$. Then

$$Q = Q(0) = \int d^4x\, j^0(x)\, \frac{\partial}{\partial x^0}\, \theta(x^0)$$

$$= \int d^4x\, j^0(x)\, \delta(x^0) \tag{6.6.5}$$

and

$$Q = \int d^3x\, j^0(\mathbf{x}, 0).$$

Consider a different observer O'. He calculates Q':

$$Q' = \int d^4x'\, j'^\mu(x')\, \frac{\partial}{\partial x'^\mu}\, \theta(n \cdot x') = \int d^4x\, \frac{\partial x'^\mu}{\partial x^\sigma}\, j^\sigma(x)\, \frac{\partial x^\tau}{\partial x'^\mu}\, \frac{\partial}{\partial x^\tau}\, \theta(n \cdot x')$$

$$= \int d^4x\, j^\mu(x)\, \partial_\mu \theta(n \cdot x').$$

But

$$n_\lambda \cdot x'^\lambda = n_\lambda\, \frac{\partial x'^\lambda}{\partial x^\sigma}\, x^\sigma = n'_\sigma x^\sigma$$

and

---

[11]We consider a vector Q^μ for simplicity. The theorem also holds for a conserved tensor $Q^{\mu\nu}$, with $\partial_\mu Q^{\mu\nu} = 0$. Then $Q^\nu = \int_V d\mathbf{r}\, Q^{0\nu}$ will be globally conserved and a four-vector.

$$Q - Q' = \int d^4x\, j^\mu(x)\, \partial_\mu[\theta(n \cdot x) - \theta(n' \cdot x)].$$

Since $\partial_\mu j^\mu = 0$, ∂_μ can be placed to the left of j^μ in the integral:

$$Q - Q' = \int d^4x\, \partial_\mu\Big[j^\mu(x)\big(\theta(n \cdot x) - \theta(n' \cdot x)\big)\Big]. \qquad (6.6.6)$$

Integrating each derivative, we see that the surface terms all vanish, since for large $\mathbf{x}$, j^μ vanishes and for large x_0, both $n \cdot x$ and $n' \cdot x$ are positive or negative, so that the θ functions cancel.

Are there other conserved integrals of functions of the field? Evidently, yes—we already know about energy, momentum, and angular momentum. Consider first energy and momentum. We have seen that in particle mechanics they together form a conserved four-vector; we also know that in electromagnetic field theory, they both can be expressed as integrals of quadratic functions of the fields. It is natural to suppose, therefore, that we can write

$$P_F^\mu = \int d^3x\, T_F^{0\mu} \qquad (6.6.7)$$

where P_F^0 is the energy, P_F^i the momentum of the field, and $T_F^{\nu\mu}$ a conserved tensor, that is,

$$\partial_\nu T^{\nu\mu} = 0 \qquad (6.6.8)$$

in the absence of sources.

This is, in fact, the case; $T_F^{\nu\mu}$ is called the electromagnetic stress-energy (sometimes energy-momentum) tensor. From now on, we shall call it the stress tensor, meaning the four-dimensional stress-energy-momentum tensor, unless otherwise stated.

From the electromagnetic conservation laws that we have already proved, we can identify all the components of $T_F^{\nu\mu}$. From the energy equation, $P_F^0 = \int d^3x\, T_F^{00}$, we see that T_F^{00} is the energy density and

$$\dot{P}_F^0 = -\int d^3x\, \partial_i T_F^{i0};$$

thus, T_F^{i0} is the Poynting vector:

$$T_F^{i0} = \left(\frac{\mathbf{E} \times \mathbf{B}}{4\pi}\right)_i. \tag{6.6.9}$$

We also know the momentum density

$$p_F^i = \left(\frac{\mathbf{E} \times \mathbf{B}}{4\pi}\right)_i.$$

But, from (6.6.7),

$$p_F^i = T_F^{0i}, \tag{6.6.10}$$

so

$$T_F^{0i} = T_F^{i0}.$$

From the momentum conservation equation, we find

$$\dot{P}_F^i = \int d^3x \, \partial_0 T_F^{0i} = -\int d^3x \, \partial_j T_F^{ji}$$

and $-T_F^{ji}$ is the Maxwell stress tensor:

$$-T_F^{ji} = \frac{1}{4\pi}\left(E_i E_j - \frac{1}{2}E^2\delta_{ij} + B_i B_j - \frac{1}{2}B^2\delta_{ij}\right). \tag{6.6.11}$$

In all, we have a symmetric two-index object $T_F^{\nu\mu}$, that generates a vector on integrating $T_F^{0\mu}$ over $d\mathbf{x}$. It is, in fact, a Lorentz tensor. Its manifestly covariant expression is

$$T_F^{\nu\mu} = \frac{1}{4\pi}\left(\eta^{\mu\alpha}F_{\alpha\lambda}F^{\lambda\nu} + \frac{1}{4}\eta^{\mu\nu}F_{\alpha\beta}F^{\alpha\beta}\right). \tag{6.6.12}$$

Next, we calculate $\partial_\nu T_F^{\nu\mu}$:

$$\partial_\nu T_F^{\nu\mu} = \frac{1}{4\pi}\eta^{\mu\alpha}F_{\alpha\lambda}\partial_\nu F^{\lambda\nu} + Q^\mu \tag{6.6.13}$$

where

$$4\pi Q^\mu = F^{\lambda\nu}\eta^{\mu\alpha}\partial_\nu F_{\alpha\lambda} + \frac{1}{2}\eta^{\mu\nu}F^{\alpha\beta}\partial^\nu F_{\alpha\beta}, \tag{6.6.14}$$

$$= F^{\lambda\nu}\eta^{\mu\alpha}\left(\partial_\nu F_{\alpha\lambda} + \frac{1}{2}\partial_\alpha F_{\lambda\nu}\right)$$

$$= \frac{1}{2}F^{\lambda\nu}\eta^{\mu\alpha}(\partial_\nu F_{\alpha\lambda} + \partial_\lambda F_{\nu\alpha} + \partial_\alpha F_{\lambda\nu}) = 0. \qquad (6.6.15)$$

Thus,

$$\partial_\nu T_F^{\nu\mu} = \eta^{\mu\alpha}F_{\alpha\lambda}j^\lambda \qquad (6.6.16)$$

and $T_F^{\mu\nu}$ is conserved in the absence of sources. We know, however, that global energy momentum conservation holds in the presence of sources, if we add together the energies and momenta of the sources (particles) and fields. We now write the most obvious candidate for a particle stress tensor, $T_P^{\mu\nu}$, that we will add to $T_F^{\mu\nu}$:

$$T_P^{\mu\nu} = mu^\mu u^\nu d \qquad (6.6.17)$$

where d is the scalar density defined in (6.4.11):

$$d = \sqrt{1 - v^2}\,\delta^3\big(\mathbf{x} - \mathbf{x}_p(x_0)\big). \qquad (6.6.18)$$

If more than one particle is present, or there are other interacting fields in play, then the single term in (6.6.17) must be replaced by a sum. We here confine ourselves to this simple system of electromagnetic field plus charged point particles. We shall see in Section 7.4 that for the relativistically invariant theories that we are discussing, one can always find an appropriate conserved stress tensor. However, the electrodynamic plus point particles case is particularly simple, in that there

$$T^{\mu\nu} = T_F^{\mu\nu} + T_P^{\mu\nu} \qquad (6.6.19)$$

is, in fact, the correct conserved stress tensor.

We calculate $\partial_\nu T_P^{\mu\nu}$, remembering that

$$\partial_\mu(u^\mu d) = \partial_\nu J^\nu = 0. \qquad (6.6.20)$$

Therefore, since u^μ depends explicitly only on time, not $\mathbf{x}$,

$$\partial_\nu T_P^{\mu\nu} = m\frac{du^\mu}{dx^0}u^0 d = m\frac{du^\mu}{d\tau}d. \qquad (6.6.21)$$

But, from (6.5.1), with $\kappa = -1$,

$$m\frac{du^{\mu}}{d\tau} = -qF^{\mu\nu}u_{\nu}$$

and

$$\partial_{\nu}T_P^{\mu\nu} = -qF^{\mu\nu}u_{\nu}d = -F^{\mu\nu}j_{\nu} \tag{6.6.22}$$

where j_{ν} is the current density four-vector. Adding $T^{\mu\nu} = T_F^{\mu\nu} + T_P^{\mu\nu}$, from (6.6.16) and (6.6.22), we get

$$\frac{\partial T^{\mu\nu}}{\partial x^{\nu}} = \eta^{\mu\alpha}F_{\alpha\lambda}j^{\lambda} - F^{\mu\nu}j_{\nu} = 0.$$

Thus, we have a symmetric, locally conserved tensor $T^{\mu\nu}$, from which we can construct a globally conserved energy-momentum four-vector:

$$P^{\nu} = \int d^3x\, T^{0\nu}. \tag{6.6.23}$$

In more general field theories, it is easy (see Section 7.3) to construct a stress tensor $\Theta^{\mu\nu}$ that is conserved,

$$\partial_{\mu}\Theta^{\mu\nu} = 0,$$

but is not necessarily symmetrical. It is, however, possible (but not necessarily easy) to find a conserved symmetrical $T^{\mu\nu}$. These techniques will be discussed in Sections 7.4 and 7.8.

The importance of the symmetry is that it makes possible the construction of six more global constants by defining a third-rank tensor

$$M^{\mu\nu\lambda} = (x^{\nu}T^{\mu\lambda} - x^{\lambda}T^{\mu\nu}). \tag{6.6.24}$$

$M^{\mu\nu\lambda}$ is conserved with respect to the μ index:

$$\partial_{\mu}M^{\mu\nu\lambda} = T^{\mu\lambda}\delta_{\mu}^{\nu} - T^{\mu\nu}\delta_{\mu}^{\lambda} = T^{\nu\lambda} - T^{\lambda\nu} = 0$$

provided $T^{\mu\nu} = T^{\nu\mu}$.

Therefore, we have six global constants constructed as usual as

$$L^{\nu\mu} = \int d^3x \, M^{0\nu\mu} = \int d^3x (x^\nu T^{0\mu} - x^\mu T^{0\nu}). \qquad (6.6.25)$$

The space-space components L_{ij} define an angular momentum

$$\mathbf{L} = \int d^3x \, \mathbf{r} \times \mathbf{p}, \qquad (6.6.26)$$

where $p^i = T^{0i}$ is the momentum density. The components L^{0i} are

$$L^{0i} = \int d^3x (x^0 T^{0i} - x^i T^{00}) = P^i x^0 - P^0 x_c^i \qquad (6.6.27)$$

where P^0 is the total energy:

$$P^0 = \int d^3x \, T^{00}, \qquad (6.6.28)$$

x_c^i is the center of energy,

$$x_c^i = \frac{\int d^3x \, T^{00} x^i}{\int d^3x \, T^{00}}, \qquad (6.6.29)$$

and P^i and x^0 are, as usual, total momentum and time. We thus learn that for a relativistic system, $dL_{0i}/dx^0 = 0$, or

$$\frac{dx_c^i}{dx^0} = \frac{P^i}{P^0}. \qquad (6.6.30)$$

This is as close to a center of mass theorem as one can come in a relativistic theory: The center of energy moves with constant velocity, $\mathbf{v}_c = \mathbf{P}/W$. None of the other center-of-mass theorems of nonrelativistic mechanics hold.

CHAPTER 6 PROBLEMS

6.1 A point particle at rest undergoes an acceleration $\mathbf{a} = \hat{\mathbf{i}} a_x + \hat{\mathbf{j}} a_y$, so that its motion is described by the equations $x = \frac{1}{2} a_x t^2$, $y = \frac{1}{2} a_y t^2$.

How does an observer O' moving to the left along the x-axis with velocity v describe the motion near $x = y = t = 0$? That is, how does acceleration transform under a Lorentz transformation?

6.2 Using the result of Problem 6.1, that is, $a'_x = a_x(1 - v^2)^{3/2}$, show that a particle starting from rest at $t = 0$, and undergoing an acceleration $a(\tau)$ in its own rest system as a function of its proper time τ, travels a distance $x = \int_0^\tau \sinh u(\tau)\, d\tau$ in proper time τ. Here, $u(\tau)$ is the "proper velocity":

$$u(\tau) = \int_0^\tau a(\tau)\, d\tau.$$

6.3 Show that the electromagnetic energy radiated per unit time by a point particle is an invariant under a Lorentz transformation; from that, calculate the rate of energy radiation by a point particle moving with velocity $\mathbf{v}$ and acceleration $\mathbf{a}$. Then calculate the instantaneous rate of radiation of momentum for a particle moving with velocity $\mathbf{v}$ and acceleration $\mathbf{a}$. (*Hint:* First show that the momentum radiated in the rest system is zero.)

6.4 From the known electrostatic field (or potential), find by Lorentz transformation the electric and magnetic fields of a point charge moving with constant velocity $\mathbf{v}$ with $|\mathbf{v}| < c$. These fields must, of course, be expressed in terms of the space and time coordinates used by the observer with respect to whom the charge has the velocity $\mathbf{v}$.

6.5 The production of a π^+ meson from a proton by a photon is described as the process

$$\gamma + p \rightarrow \pi^+ + n.$$

At what photon energy E_γ on a proton at rest will the threshold for the process be reached? Call the respective masses m_π, m_p, m_n, and $m_\gamma = 0$.

6.6 Consider a Lorentz transformation in an arbitrary direction $\mathbf{v}$. Write a three-vector equation for the transformation $\mathbf{r}' = \mathbf{f}(\mathbf{r}, \mathbf{v}, t)$ and a three-scalar equation $t' = g(\mathbf{r}, \mathbf{v}, t)$, where $\mathbf{f}$ is a vector function of $\mathbf{r}$, $\mathbf{v}$, and t, and g a scalar function of $\mathbf{r}$, $\mathbf{v}$, and t.

6.7 An observer O' has a very small velocity v_x relative to O; a second observer O'' has a very small velocity v_y relative to O'.

Now consider observers $\bar{O}'$ and $\bar{O}''$ who have interchanged the order, that is, $\bar{O}'$ has v_y, $\bar{O}''$ has v_x. To lowest order in $v_x v_y$, what is the transformation that brings the observers $\bar{O}''$ and O'' into coincidence?

6.8 A charged condenser moves with constant velocity $\mathbf{v}$. Call the charge density $\rho(\mathbf{r})$.

(a) In terms of ρ, write expressions for the electric and magnetic fields generated by ρ.

(b) Write an expression for the force per unit volume on the moving charge density.

(c) Integrate the expression you found in (b) to find the self-force on the condenser. It should be zero.

(d) From your results in (b), write an expression for the torque per unit volume on the moving charge density and integrate it to find the self-torque. The answer should involve the tensor

$$\int \frac{(x_{2i} - x_{1i})(x_{2j} - x_{1j})}{|\mathbf{r}_2 - \mathbf{r}_1|^3} \rho(2)\rho(1) \, d^3x_2 \, d^3x_1.$$

Let the condenser consist of two oppositely charged very small spheres, separated by a vector $\mathbf{d}$. In terms of the total charge Q on each sphere and the vector $\mathbf{d}$, write an expression for the self-torque. Note that it is not zero, and so should have (according to prerelativistic theory) furnished a way of detecting absolute motion. Experiments, of course, failed to do so. Can you say why? After all, the field calculations are correct, as is the Lorentz force.

6.9 Study the relativistic Doppler shifts by noting that the phase of an electromagnetic wave is invariant. Thus,

$$e^{ik_\lambda x^\lambda} = e^{ik'_\lambda x'^\lambda}$$

and k^λ must transform like x^λ. Thus, for a Lorentz transformation in the x direction

$$k'_x = \frac{k_x - v\omega}{\sqrt{1 - v^2}}, \qquad k'_y = k_y, \qquad k'_z = k_z, \qquad \omega' = \frac{\omega - vk_x}{\sqrt{1 - vk^2}}$$

with $\mathbf{k}$ the wave vector and ω the frequency in the rest system of the radiator. Show that the frequency measured by an observer who sees the radiator moving with velocity v is

$$\omega' = \frac{\omega\sqrt{1 - v^2}}{1 + v \cos\theta}$$

where θ is the angle the radiation makes with the motion of the radiator (as seen by the observer).

6.10 (a) Calculate the ratio r of L, the rate of energy loss through radiation, to G, the rate of gain of energy from the accelerating field in an electron linear accelerator. Express $v = L/G$ in terms of the accelerating field E and natural constants. Assume the electron is relativistic so that $v \approx c$.

(b) Calculate the ratio r of the energy loss per cycle of a relativistic electron in a circular accelerator to its energy. The electron has energy E, the average magnetic field is B, and the machine radius is R. Assume $v \sim c$ so that the frequency $\omega \sim c/R$. Eliminate R and so express the ratio in terms of E, B, and natural constants. Finally, express the ratio r in terms of R in kilometers, B in tesla, and E in electron rest energy units. Explain from these results the advantage of linear accelerators over circular acceleration for ultrarelativistic particles.

6.11 The functioning of an accelerator requires that the chosen orbit be stable with respect to small deviations. Consider a particle beam in the z direction and a stabilizing electrostatic potential $\phi = kx^2/2$ in the x direction. Unfortunately, since $\nabla^2\phi = 0$, the y potential must be destabilizing. The overall potential must be a quadrupole: $\phi = k[(x^2/2) - (y^2/2)]$. The alternating gradient principle takes advantage of the fact that a stabilizing passage followed by a destabilizing one can be stabilizing with a proper choice of parameters. Since the potential is harmonic, one can calculate the matrix U_{21} that takes the vector $(x_1, \dot{x}_1/\omega)$ to $(x_2, \dot{x}_2/\omega)$ and combine the U matrices for different lengths of potential, a and b:

$$U_{31}^{a+b} = U_{32}^{b}U_{21}^{a}, \text{ etc.}$$

Let U_{21} describe a converging sector, with $\omega(t_2 - t_1) = \varphi$, and U_{32} a diverging sector with $\omega(t_3 - t_2) = \theta$. Here, $\omega^2 = k/m(1 - v_z^2)^{1/2}$, provided $\dot{x}$ and $\dot{y}$ are much smaller than v_z, the beam velocity.

(a) Calculate $U_{21}(\varphi) = U_{43}(\varphi)$ and $U_{32}(\theta)$.

(b) Combine the sector U's as described above to give $U = U_{43}(\varphi)\, U_{32}(\theta)\, U_{21}(\varphi)$.

(c) Show that $\det U = 1$ and its eigenvalues are $e^{\pm i\mu}$, where $\cos\mu = \cosh\theta\cos 2\varphi$ and $|\cosh\theta\cos 2\phi| < 1$. For $|\cosh\theta\cos 2\phi| > 1$, the eigenvalues are $e^{\pm\lambda}$, where $\cosh\lambda = |\cosh\theta\cos 2\phi|$ and $\cos 2\phi > 0$ or $-e^{\pm\lambda}$ when $\cos 2\phi < 0$.

(d) Note that U is neither unitary nor Hermitian, so the apparent stable behavior coming from the eigenvalues can mask actual growth. To illustrate this phenomenon, find the two components of the vector $U\binom{1}{0}$ and show that one of them can actually be arbitrarily large, even for $|\cosh\theta\cos 2\phi| < 1$. Then show however that for small $2\varphi = \theta$, the sector is stabilizing.

(e) Show that magnetic stabilizing via a quadrupole magnetic field leads to the same equations for x and y.

CHAPTER 7

Lagrangian Field Theory

In classical field theory, we introduce Lagrange equations and Hamilton's principle for several reasons. These include the special simplicity of using generalized coordinates and eliminating constraints, but most important, the ease of building invariance principles and the corresponding conservation laws into our equations. In quantum theory, the canonical formalism plays such a crucial role that the use of Lagrangians is often indispensable.

7.1. REVIEW OF LAGRANGIANS IN MECHANICS

The Lagrangian is a function of generalized coordinates and velocities, q_a and $\dot{q}_a$:

$$L = L(q_a, \dot{q}_a, t), \qquad (7.1.1)$$

where $\dot{q}_a$ is the derivative with respect to t. Here, t can be a variable, usually the time, that is used to parametrize the trajectories of the system.

The second-order equations of motion follow from Hamilton's principle, that is, from the requirement that the action

$$S(q_a) = \int_{t_1}^{t_2} L(q_a(t), \dot{q}_a(t), t) \, dt \qquad (7.1.2)$$

be stationary for infinitesimal variations of $q_a(t) \to q_a(t) + \delta q_a$, or

$$q_a(t) \to q_a(t) + \epsilon \eta_a(t) \qquad (7.1.3)$$

provided $\eta_a(t_1) = \eta_a(t_2) = 0$. Here, ϵ is an infinitesimal and $\eta_a(t)$ an

281

arbitrary function of t. Sometimes, we will use the common shorthand $\epsilon \eta_a(t) = \delta q_a$; however, in case of confusion, it is often best to go back to (7.1.3). For example, (7.1.3) makes it obvious that $(d/dt)\, \delta q_a = \delta(dq_a/dt)$. The notation $S(q_a)$ indicates that S is a *functional* of $q_a(t)$; that is, it is a number that depends on the functions $q_a(t)$.

We carry out the variation

$$\delta S = S(q_a + \delta q_a) - S(q_a)$$

$$= \int_{t_1}^{t_2} \left(\frac{\partial L}{\partial q_a} \delta q_a + \frac{\partial L}{\partial \dot{q}_a} \delta \dot{q}_a \right) dt$$

$$= \int_{t_1}^{t_2} \left(\frac{\partial L}{\partial q_a} - \frac{d}{dt} \frac{\partial L}{\partial \dot{q}_a} \right) \delta q_a + \frac{\partial L}{\partial \dot{q}_a} \delta q_a \bigg|_{t_1}^{t_2}. \tag{7.1.4}$$

Since the integrated term vanishes at the t boundaries, (7.1.4) shows that the condition $\delta S = 0$ for any δq_a requires

$$\frac{\partial L}{\partial q_a} - \frac{d}{dt} \frac{\partial L}{\partial \dot{q}_a} = 0. \tag{7.1.5}$$

For obvious reasons, the left-hand side of (7.1.5) is called the variational derivative of S, written as $\delta S / \delta q_a$:

$$\frac{\partial S}{\partial q_a} \equiv \frac{\partial L}{\partial q_a} - \frac{d}{dt} \frac{\partial L}{\partial \dot{q}_a} + \frac{d^2}{dt^2} \frac{\partial L}{\partial \ddot{q}_a} - \cdots \tag{7.1.6}$$

and so on, if the Lagrangian contains higher derivatives.[1]

The relation between the symmetries of the Lagrangian and conservation laws is given in first instance by Noether's theorem, which follows.

Suppose that when $q_a \rightarrow q_a + \delta q_a$, $\delta L = d\delta A/dt$, where $d\delta A/dt$ is a time derivative of a function of the q's and $\dot{q}$'s. Then, independent of the equation of motion,

$$\sum_a \left[\frac{\partial L}{\partial q_a} \delta q_a + \frac{\partial L}{\partial \dot{q}_a} \delta \dot{q}_a \right] - \frac{d\delta A}{dt} = 0 \tag{7.1.7}$$

or, using the equations of motion, we get

---

[1]If the Lagrangian does have higher derivatives, the action principle most economically formulated states that δS depends only on the variation of q, $\dot{q}$, etc. on the boundary.

$$\frac{d}{dt}\left(\sum_a \frac{\partial L}{\partial \dot{q}_a} \delta q_a - \delta A\right) = 0 \tag{7.1.8}$$

and we have a conserved quantity

$$\delta Q = \sum_a \frac{\partial L}{\partial \dot{q}_a} \delta q_a - \delta A \tag{7.1.9}$$

which is independent of t.

We list the major examples from classical mechanics:

1. $\delta \mathbf{x}_a \equiv \boldsymbol{\delta}$, a fixed displacement, identical for every particle, with L invariant. This would hold for motion in a potential depending only on the relative position of particles. In that case, we would have

$$\delta Q = \sum_a \mathbf{p}_a \cdot \boldsymbol{\delta} = \boldsymbol{\delta} \cdot \sum_a \nabla_{\dot{\mathbf{x}}_a} L \tag{7.1.10}$$

and we have total momentum conserved by virtue of translational invariance. The symbol $\nabla_{\dot{\mathbf{x}}_a}$ stands for the operator

$$\nabla_{\dot{\mathbf{x}}_a} = \hat{\mathbf{e}}_x \frac{\partial}{\partial \dot{x}_a} + \hat{\mathbf{e}}_y \frac{\partial}{\partial \dot{y}_a} + \hat{\mathbf{e}}_z \frac{\partial}{\partial \dot{z}_a}. \tag{7.1.11}$$

2. $\delta \mathbf{x}_a = \delta \boldsymbol{\theta} \times \mathbf{x}_a$, a fixed rotation about the origin, identical for every particle, with L invariant. δQ is given by

$$\delta Q = \sum \nabla_{\dot{\mathbf{x}}_a} L \cdot \delta \boldsymbol{\theta} \times \mathbf{x}_a \tag{7.1.12}$$

or

$$\delta Q = \delta \boldsymbol{\theta} \cdot \sum \mathbf{x}_a \times \nabla_{\dot{\mathbf{x}}_a} L \tag{7.1.13}$$

so that

$$\mathbf{L} = \sum_a \mathbf{x}_a \times \nabla_{\dot{\mathbf{x}}_a} L, \tag{7.1.14}$$

is conserved. We identify (7.1.14) with the angular momentum. (Do not confuse $\mathbf{L}$, the angular momentum, with L, the Lagrangian.)

3. Suppose L is translation-invariant in t—that is, it depends on t only through its dependence on q and $\dot{q}$. Then with $\delta q_a = \dot{q}_a \delta t$ [so that

$q(t) \rightarrow q(t + \delta t), \dot{q}(t) \rightarrow \dot{q}(t + \delta t)]$

$$\delta L = \sum_a \left(\frac{\partial L}{\partial q_a} \dot{q}_a + \frac{\partial L}{\partial \dot{q}_a} \ddot{q}_a \right) \delta t = \frac{dL}{dt} \delta t$$

and

$$W = \sum_a \frac{\partial L}{\partial \dot{q}_a} \dot{q}_a - L \qquad (7.1.15)$$

is conserved. We recognize in (7.1.15) the conservation of energy.

4. Our last example is a Galilean transformation to a moving observer: $\delta x_i = \delta v_i t$, with δv_i the same for all particles. Then, summing over particles a, we obtain

$$\delta L = \sum_a \frac{\partial L}{\partial x_{ai}} \delta v_i t + \sum_a \frac{\partial L}{\partial \dot{x}_{ai}} \delta v_i = \frac{d}{dt} \left(\sum_a m_a x_{ai} \delta v_i \right)$$

for a Lagrangian with translation invariance, whose only velocity dependence is in its kinetic energy. So, our conservation law here is

$$\left(\sum_a \frac{\partial L}{\partial \dot{x}_{ai}} \right) t - \sum_a m_a x_{ai} = \text{constant}$$

or

$$\left(\sum m_a \dot{x}_{ai} \right) t - \sum m_a x_{ai} = \text{constant.} \qquad (7.1.16)$$

Equation (7.1.16) tells us that the center of mass moves with constant velocity.

7.2. RELATIVISTIC LAGRANGIAN FOR PARTICLES IN A FIELD

Relativistic invariance may be achieved by choosing an action that is itself invariant. The obvious choice for a relativistic free particle action is the integrated proper time interval for the particle:

$$S_p = -M \int_{\tau_1}^{\tau_2} d\tau = -M \int \sqrt{dx_\mu\, dx^\mu} = -M \int_{\sigma_1}^{\sigma_2} d\sigma \sqrt{\frac{dx_\mu}{d\sigma}\frac{dx^\mu}{d\sigma}} \qquad (7.2.1)$$

where M is a constant with the dimensions of mass and σ *any* increasing function of τ. It is important to distinguish between dynamical variables (the x^μ here) and Lagrangian integration variables (the σ here). Notice that x^0, the physical time, is here regarded as a dynamical variable. Since the final result is required to give us **x** as a function of x^0, our equations must make this possible. We shall see that this is always the case. The invariance (with respect to the choice of σ) is called reparametrization invariance.

In order to include an electromagnetic interaction, the simplest choice is to add to S the invariant

$$S_{\text{e.m.}} = E \int dx^\mu A_\mu = E \int \frac{dx^\mu}{d\sigma} A_\mu\, d\sigma \qquad (7.2.2)$$

where E is a constant with the dimensions of charge.

Note that (7.2.2) is gauge-invariant: A gauge change,

$$A_\mu \to A_\mu + \partial_\mu \Lambda(x),$$

changes S by

$$\delta S = E \int_{\sigma_1}^{\sigma_2} \frac{\partial \Lambda}{\partial x^\mu} \frac{dx^\mu}{d\sigma}\, d\sigma = \int d\Lambda = \Lambda(2) - \Lambda(1)$$

so that a variation of S keeping the end points $x^\mu(\sigma_1)$ and $x^\mu(\sigma_2)$ fixed is unchanged by the change of gauge.

The Lagrange equations, with σ as the independent variable, are

$$\frac{d}{d\sigma} \frac{\partial L}{\partial \left(\dfrac{dx^\mu}{d\sigma}\right)} = \frac{\partial L}{\partial x^\mu} \qquad (7.2.3)$$

where

$$L = -M \sqrt{\frac{dx^\mu}{d\sigma}\frac{dx_\mu}{d\sigma}} + E \frac{dx^\mu}{d\sigma} A_\mu, \qquad (7.2.4)$$

so that

$$-\frac{d}{d\sigma}\left[\left(\frac{M\dfrac{dx_\mu}{d\sigma}}{\sqrt{\dfrac{dx^\alpha}{d\sigma}\dfrac{dx_\alpha}{d\sigma}}}\right) - EA_\mu\right] = E\frac{dx^\lambda}{d\sigma}\frac{\partial A_\lambda}{\partial x^\mu} \qquad (7.2.5)$$

since $A_\mu(x)$ is evaluated at $x = x(\sigma)$; $\partial A_\mu/\partial x^\lambda$ is the space-time derivative of the field evaluated at $x = x(\sigma)$. Carrying out the σ differentiation of A_μ, we have

$$-M\frac{d}{d\sigma}\left(\frac{\dfrac{dx_\mu}{d\sigma}}{\sqrt{\dfrac{dx^\alpha}{d\sigma}\dfrac{dx_\alpha}{d\sigma}}}\right) = E\left(\frac{dx^\lambda}{d\sigma}\frac{\partial A_\lambda}{\partial x^\mu} - \frac{dx^\lambda}{d\sigma}\frac{\partial A_\mu}{\partial x^\lambda}\right)$$

$$= -E\frac{dx^\lambda}{d\sigma}F_{\lambda\mu}. \qquad (7.2.6)$$

There are two simple choices for $d\sigma$: either $d\sigma^2 = dx^\mu\, dx_\mu = d\tau^2$, or $d\sigma = dx_0$. Either yields the equation

$$M\frac{d}{d\tau}\left(\frac{dx_\mu}{d\tau}\right) = -E\frac{dx^\lambda}{d\tau}F_{\mu\lambda} \qquad (7.2.7)$$

so that (7.2.6) agrees with (6.5.1), provided we set $M = m$, the particle mass, and $E = q$, the particle charge.

There is an alternative action, also Lorentz-invariant, that gives the same final equations for a particle in a given electromagnetic field:

$$S = -\frac{m}{2}\int \frac{dx^\mu}{d\sigma}\frac{dx_\mu}{d\sigma}\, d\sigma + q\int d\sigma\, A_\mu(x)\frac{dx^\mu}{d\sigma},$$

corresponding to a Lagrangian

$$L = -\frac{m}{2}\frac{dx^\mu}{d\sigma}\frac{dx_\mu}{d\sigma} + qA_\mu\frac{dx^\mu}{d\sigma}. \qquad (7.2.8)$$

The Lagrange equations are

$$\frac{d}{d\sigma}\left(m\frac{dx_\mu}{d\sigma} - qA_\mu\right) = -q\frac{dx^\lambda}{d\sigma}\frac{\partial A_\lambda}{\partial x^\mu} \tag{7.2.9}$$

or

$$m\frac{d^2x_\mu}{d\sigma^2} = q\left(\frac{\partial A_\mu}{\partial x^\lambda} - \frac{\partial A_\lambda}{\partial x^\mu}\right)\frac{dx^\lambda}{d\sigma}. \tag{7.2.10}$$

The action S is not reparametrization-invariant; therefore, $d\sigma$ is not a free variable, but is determined by (7.2.10). Multiply (7.2.10) by $dx^\mu/d\sigma$ to obtain an integral of the motion:

$$\frac{d}{d\sigma}\left(\frac{dx^\mu}{d\sigma}\frac{dx_\mu}{d\sigma}\right) = 0, \tag{7.2.11}$$

so that $d\sigma$ must be a constant multiple of $d\tau$. A change in the constant multiple can be compensated for by redefining the coefficient m. Evidently, $\sigma = \tau$ gives us the correct equation, as written in (7.2.10).

Note that the "energy" constant that we would obtain from the new action is

$$\text{"W"} = -\left(m\frac{dx_\mu}{d\sigma} - qA_\mu\right)\frac{dx^\mu}{d\sigma} - L = -\frac{m}{2}\frac{dx_\mu}{d\sigma}\frac{dx^\mu}{d\sigma}, \tag{7.2.12}$$

just the integral we found above in (7.2.11).

There is still, however, Noether's theorem that may be applied to time displacement to obtain a physical energy integral. If the Lagrangian L is invariant under $x^0 \to x^0 + \delta$ (δ a constant), then there will be an integral of the motion. For L to have this invariance, A_μ must be independent of x^0. The constant of the motion is then

$$\delta W = \frac{\partial L}{\partial\left(\dfrac{dx^0}{d\sigma}\right)}\delta. \tag{7.2.13}$$

L depends on $dx^0/d\sigma$ as

$$L = -\frac{m}{2}\left(\frac{dx^0}{d\sigma}\right)^2 + q\frac{dx^0}{d\sigma}A_0 + \text{non}\frac{dx^0}{d\sigma}\text{ terms} \tag{7.2.14}$$

so

$$\delta W = \frac{\partial L}{\partial \left(\dfrac{dx^0}{d\sigma} \right)} \delta = - \left(m \frac{dx^0}{d\sigma} - qA_0 \right) \delta$$

and with $d\sigma = d\tau$

$$W = - \left[\frac{m}{\sqrt{1 - v^2}} + q\phi \right], \tag{7.2.15}$$

in agreement with our earlier formula (6.5.5), as it must be, since the equations of motion are the same. A technical point: (7.2.15) is a constant of the motion, but it is the negative of the usual energy $T + V$. For a particle in an external field, the sign is irrelevant. For a system of interacting fields and particles, the energy must be obtained from the full Lagrangian (or better, full action) for the particles and fields. We will discuss this problem in Section 7.4.

Although the Lagrangians (7.2.4) and (7.2.8) for a particle in an electromagnetic field lead to the same equations of motion, this is a special circumstance and does not necessarily hold for other interactions. In particular, adding a linear gravitational interaction to (7.2.8) leads to a simple—and experimentally correct—linear theory, and, as we shall see, to Einstein's equations for the gravitational field; following a similar procedure with (7.2.4) does not. From here on, we will work only with extensions of (7.2.8).

We wish here to complete the list of Noether currents for the case of a Lorentz-invariant particle Lagrangian. Thus, consider an action

$$S = \int d\sigma L \left(x^\mu(\sigma), \frac{dx^\mu}{d\sigma} \right) \tag{7.2.16}$$

which is invariant under the Lorentz transformation

$$x^\mu \to x'^\mu = \Lambda^\mu{}_\alpha x^\alpha \tag{7.2.17}$$

and

$$\frac{dx^\mu}{d\sigma} \to \frac{dx'^\mu}{d\sigma} = \Lambda^\mu{}_\alpha \frac{dx^\alpha}{d\sigma}; \tag{7.2.18}$$

with $d\sigma$ an invariant, L itself will be invariant. Thus, the Noether theorem will give us the conservation (with respect to σ) of

$$\delta Q = \frac{\partial L}{\partial \left(\dfrac{dx^{\mu}}{d\sigma} \right)} \delta x^{\mu}, \tag{7.2.19}$$

where δx^{μ} is the change $x'^{\mu} - x^{\mu}$ for an infinitesimal Lorentz transformation. We take as an example the "new" Lagrangian

$$-\frac{m}{2} \frac{dx^{\mu}}{d\sigma} \frac{dx_{\mu}}{d\sigma}. \tag{7.2.20}$$

Thus, δQ is

$$\delta Q = -m\eta_{\mu\nu} \frac{dx^{\nu}}{d\sigma} \delta x^{\mu}. \tag{7.2.21}$$

The combination $\eta_{\mu\nu} \delta x^{\mu}$ is particularly simple. Since $\delta x^{\mu} = (\Lambda^{\mu}{}_{\alpha} - \delta^{\mu}{}_{\alpha})x^{\alpha}$, we learn from (6.2.31) that

$$\eta_{\mu\nu} \delta x^{\mu} = \eta_{\mu\nu}(\Lambda^{\mu}{}_{\alpha} - \delta^{\mu}{}_{\alpha})x^{\alpha} \tag{7.2.22}$$

$$= \epsilon_{\nu\alpha} x^{\alpha}. \tag{7.2.23}$$

Returning to (7.2.21), we obtain

$$\delta Q = -m \frac{dx^{\nu}}{d\sigma} \epsilon_{\nu\beta} x^{\beta}, \tag{7.2.24}$$

so we have six tensor constants of the motion:

$$L^{\alpha\nu} = m\left(x^{\alpha} \frac{dx^{\nu}}{d\sigma} - x^{\nu} \frac{dx^{\alpha}}{d\sigma} \right), \tag{7.2.25}$$

where following (7.2.11), $d\sigma$ must be taken to be proportional to $d\tau$, the proper time.

In this trivial free particle case, we recognize in L^{ij} the three components of angular momentum, and in $L^{i0} = $ constant the three equations of motion, $x^i - (dx^i/dx^0)x^0 = $ constant. The interest of the expression (7.2.25) is that in the case of interacting fields and particles, (7.2.25) will again emerge as the particle contribution to a general conservation law. For an example of this, see Problem 3.3.

7.3. LAGRANGIAN FOR FIELDS

Fields are systems with an infinite number of degrees of freedom: the fields and field time derivatives at every point in space at a given time. The Lagrangian is then itself a functional—an integral over space of some function $\mathscr{L}$ of the fields and their derivatives. The function $\mathscr{L}$ is sometimes called the Lagrangian, although it is more appropriately called the Lagrangian density. Thus, we have

$$L = \int d^3x \, \mathscr{L}(\psi_\alpha(x), \partial_\mu \psi_\alpha, \ldots) \tag{7.3.1}$$

where the ψ_α's are all the fields we are considering and the $\partial_\mu \psi_a$'s their derivatives. The index α can refer to a species of field, or to a vector or tensor component. The action S is now

$$S = \int dx_0 L = \int d^4x \, \mathscr{L}. \tag{7.3.2}$$

Since the four-dimensional volume element d^4x is Lorentz-invariant, the action (and hence the equations of motion) will be invariant if $\mathscr{L}$ itself is an invariant function of the fields and their derivatives.

Again, here we must distinguish between dynamical variables (the fields at every point of space) and the Lagrangian integration variable x^μ (the time and space point at which one asks for the value of the field). In order to avoid confusion, we will from now on call the dynamical space-time coordinates of the particles y_ρ^μ, or sometimes y^μ. We will keep the variables x^μ for the arguments of the fields.

By analogy with ordinary Lagrangians, we know that quadratic functions of the fields in $\mathscr{L}$ will lead to linear equations; higher derivatives than the first will lead to higher than second-order differential equations. We confine ourselves here to first derivatives and second-order equations. Higher derivatives are suggested in Problem 7.6. The rule for using the action is the same as for the point Lagrangian: The action must be stationary for arbitrary variations $\delta\psi_\alpha$ that vanish on the integration boundary. So,

$$\delta S = \sum_\alpha \int d^4x \left[\frac{\partial \mathscr{L}}{\partial \psi_\alpha} \delta\psi_\alpha + \frac{\partial \mathscr{L}}{\partial(\partial_\mu \psi_\alpha)} \delta(\partial_\mu \psi_\alpha) \right]. \tag{7.3.3}$$

Integrating by parts and dropping the boundary term, we have

$$\delta S = \sum_{\alpha} \int d^4x \, \delta\psi_\alpha \left[\frac{\partial \mathcal{L}}{\partial \psi_\alpha} - \frac{\partial}{\partial x^\mu} \frac{\partial \mathcal{L}}{\partial(\partial_\mu \psi_\alpha)} \right] = 0 \qquad (7.3.4)$$

or, since $\delta\psi_\alpha$ is arbitrary in the interior,

$$\frac{\partial S}{\partial \psi_\alpha} \equiv \frac{\partial \mathcal{L}}{\partial \psi_\alpha} - \frac{\partial}{\partial x^\mu} \frac{\partial \mathcal{L}}{\partial(\partial_\mu \psi_\alpha)} = 0, \qquad (7.3.5)$$

the Lagrange equations for the fields. The definition of functional derivatives has been extended in (7.3.5) to apply to fields.

The Lagrange equations (7.3.5) permit the construction of a conserved stress-energy tensor, called canonical:

$$\Theta^{\mu\nu} = \sum_{\alpha} \frac{\partial \mathcal{L}}{\partial(\partial_\mu \psi_\alpha)} \partial^\nu \psi_\alpha - \eta^{\mu\nu} \mathcal{L} \qquad (7.3.6)$$

for which

$$\partial_\mu \Theta^{\mu\nu} = \sum_{\alpha} \left[\partial_\mu \frac{\partial \mathcal{L}}{\partial(\partial_\mu \psi_\alpha)} \partial^\nu \psi_\alpha + \frac{\partial \mathcal{L}}{\partial(\partial_\mu \psi_\alpha)} \partial_\mu \partial^\mu \psi_\alpha \right] - \partial^\nu \mathcal{L}.$$

Using (7.3.5) yields

$$\partial_\mu \Theta^{\mu\nu} = \sum_{\alpha} \left\{ \frac{\partial \mathcal{L}}{\partial \psi_\alpha} \partial^\nu \psi_\alpha + \frac{\partial \mathcal{L}}{\partial(\partial_\mu \psi_\alpha)} \partial_\mu \partial^\nu \psi_\alpha - \frac{\partial \mathcal{L}}{\partial \psi_\alpha} \partial^\nu \psi_\alpha \right.$$
$$\left. - \frac{\partial \mathcal{L}}{\partial(\partial_\mu \psi_\alpha)} \partial^\nu (\partial_\mu \psi_\alpha) \right\} - [\partial^\nu] \mathcal{L}$$
$$= -[\partial^\nu] \mathcal{L} \qquad (7.3.7)$$

where the notation $[\partial^\nu] \mathcal{L}$ means that the derivative is only on explicit space time dependence of $\mathcal{L}$. If $[\partial^\nu] \mathcal{L} = 0$, $\partial_\mu \Theta^{\mu\nu} = 0$, and we have a conserved tensor.

We note here that the conservation of (7.3.6) is a special case of the Noether theorem applied to field Lagrangians. Thus, assume that an infinitesimal transformation

$$\psi_\sigma \rightarrow \psi_\sigma + \delta\psi_\sigma \qquad (7.3.8)$$

leads to a change in $\mathscr{L}$ of

$$\delta\mathscr{L} = \partial_\mu(\delta A^\mu) \tag{7.3.9}$$

for some δA^μ. It follows that

$$\partial_\mu\left(\frac{\partial\mathscr{L}}{\partial(\partial_\mu\psi_\sigma)}\delta\psi_\sigma - \delta A^\mu\right) = 0, \tag{7.3.10}$$

since

$$\delta\mathscr{L} = \frac{\partial\mathscr{L}}{\partial\psi_\sigma}\delta\psi_\sigma + \frac{\partial\mathscr{L}}{\partial(\partial_\mu\psi_\sigma)}\partial_\mu\delta\psi_\sigma$$

$$= \partial_\mu\left(\frac{\partial\mathscr{L}}{\partial(\partial_\mu\psi_\sigma)}\delta\psi_\sigma\right); \tag{7.3.11}$$

Combining (7.3.11) and (7.3.9) yields the conservation law (7.3.10). The vector in (7.3.10) that satisfies the conservation law is called the Noether current.

Equation (7.3.6) for $\Theta^{\mu\nu}$ follows by setting $\delta\psi_\sigma = \partial_\lambda\psi_\sigma\delta x^\lambda$, where δx^λ is a constant increment in x^λ and δA^μ is $\mathscr{L}\delta x^\mu$. This will be discussed more fully in Section 7.4.

The conservation of $\Theta^{\mu\nu}$ makes possible the definition of a conserved four-vector that we identify with the field energy–momentum:

$$P^\mu = \int d^3x\,\Theta^{0\mu}, \tag{7.3.12}$$

such that

$$\frac{dP^\mu}{dx^0} = 0. \tag{7.3.13}$$

The tensor $\Theta^{\mu\nu}$ is not symmetric, except for scalar fields. If we could find a tensor $T^{\mu\nu}$ that was both conserved *and* symmetric (and we can), we could, as discussed in Section 6.6, define a tensor field

$$M^{\mu\nu\lambda} = x^\nu T^{\mu\lambda} - x^\lambda T^{\mu\nu}, \tag{7.3.14}$$

which would then be locally conserved:

$$\partial_\mu M^{\mu\nu\lambda} = 0, \tag{7.3.15}$$

leading to the global conservation of the angular momentum tensor

$$L^{\nu\lambda} = \int d^3 x \, M^{0\nu\lambda} = \int d^3 x \, (x^\nu T^{0\lambda} - x^\lambda T^{0\nu}); \qquad (7.3.16)$$

that is,

$$\frac{dL^{\nu\lambda}}{dx^0} = 0. \qquad (7.3.17)$$

Equation (7.3.17) shows the general value of finding a symmetric tensor $T^{\mu\nu}$.

We next show how a conserved, *symmetric* stress-energy tensor *can* be constructed for a Lorentz-invariant Lagrangian density. Since the canonical stress tensor $\Theta^{\mu\nu}$ is symmetric only for a scalar theory, this is a useful procedure. We start from

$$\Theta^{\mu\nu} = \frac{\partial \mathcal{L}}{\partial(\partial_\mu \psi_\sigma)} \partial^\nu \psi_\sigma - \eta^{\mu\nu}\mathcal{L} \qquad (7.3.18)$$

and add to $\Theta^{\mu\nu}$ a tensor

$$\delta\Theta^{\mu\nu} = \partial_\lambda \Phi^{\lambda\mu\nu} \qquad (7.3.19)$$

where $\Phi^{\lambda\mu\nu}$ is antisymmetric in λ and μ, so that

$$T^{\mu\nu} = \Theta^{\mu\nu} + \delta\Theta^{\mu\nu} \qquad (7.3.20)$$

is conserved (since $\partial_\mu \partial_\lambda \Phi^{\lambda\mu\nu} = 0$), and $P^\nu = \int d^3x \, T^{0\nu}$ is left unchanged (since $\int d^3x \, \partial_\lambda \Phi^{\lambda 0\nu} = \int d^3x \, \partial_i \Phi^{i0\nu} = 0$).

In addition, it will be shown that $\Phi^{\lambda\mu\nu}$ may be chosen so that $T^{\mu\nu}$ in (7.3.20) is symmetric under the exchange of μ and ν. We proceed by deriving an identity that follows from the Lorentz invariance of the Lagrangian density, $\mathcal{L}(\psi_\sigma, \partial_\mu \psi_\sigma)$.

Under a Lorentz transformation, the fields ψ_a transform so that

$$\psi'_a = S_{ab}\psi_b \qquad (7.3.21)$$

where S_{ab} is a matrix representation of the Lorentz group. If the transformation is infinitesimal, specified by $\epsilon_{\mu\nu}$, as in (7.2.23), then the matrix S will be of the form

$$S = 1 + \epsilon_{\mu\nu}\Sigma^{\mu\nu} \qquad (7.3.22)$$

where each $\Sigma^{\mu\nu}$ is an antisymmetric matrix in the (a, b) space. We give some examples:

1. ψ a scalar field: $\Sigma = 0$.
2. ψ a spin-one-half field:

$$\Sigma^{\mu\nu} = \frac{i\sigma^{\mu\nu}}{2}, \tag{7.3.23}$$

where $\sigma^{\mu\nu}$ is the set of six relativistic Pauli matrices:

$$\sigma^{\mu\nu} = \frac{[\gamma^{\mu}, \gamma^{\nu}]}{2i}. \tag{7.3.24}$$

3. ψ a covariant vector field ψ_a. Then

$$\delta\psi_a = -\eta^{b\lambda}\epsilon_{\lambda a}\psi_b. \tag{7.3.25}$$

On the other hand, from (7.3.22)

$$\delta\psi_a = \epsilon_{\mu\nu}\Sigma^{\mu\nu,\,b}{}_a\psi_b \tag{7.3.26}$$

so that

$$\Sigma^{\mu\nu,\,b}{}_a = \frac{1}{2}(\delta^{\mu}_a\eta^{\nu b} - \delta^{\nu}_a\eta^{\mu b}) \tag{7.3.27}$$

correctly transforms ψ.

Similarly, a second-rank covariant tensor field transforms like

$$\delta\psi_{ab} = -\epsilon^{a'}{}_a\psi_{a'b} - \epsilon^{b'}{}_b\psi_{ab'} \tag{7.3.28}$$

so that

$$\Sigma^{\mu\nu;\,a'b'}{}_{ab} = \delta^{a'}_a\Sigma^{\mu\nu}{}_b{}^{b'} + \delta^{b'}_b\Sigma^{\mu\nu}{}_a{}^{a'} \tag{7.3.29}$$

with $\Sigma^{\mu\nu}{}_b{}^{b'}$ given by (7.3.27).

The Lorentz invariance of the Lagrangian $\mathcal{L}(\psi, \partial_{\mu}\psi)$ requires that

$$\frac{\partial\mathcal{L}}{\partial\psi}\delta\psi + \frac{\partial\mathcal{L}}{\partial(\partial_{\mu}\psi)}\delta(\partial_{\mu}\psi) = 0 \tag{7.3.30}$$

where

$$\delta\psi = \epsilon_{\mu\nu}\Sigma^{\mu\nu}\psi. \tag{7.3.31}$$

Note the abbreviated notation: $\delta\psi$ and $\partial\mathcal{L}/\partial\psi$ are vectors in the (a, b) space in which Σ is a matrix, as are $\delta(\partial_{\mu}\psi)$ and $\partial\mathcal{L}/\partial(\partial_{\mu}\psi)$. Note that if ψ is covariant, $\partial\mathcal{L}/\partial\psi$ is contravariant and vice-versa.

With

$$\frac{\partial \mathcal{L}}{\partial(\partial_\mu \psi)} = p^\mu, \tag{7.3.32}$$

the Lagrange equation applied to (7.3.30) tells us that

$$(\partial_\mu p^\mu)\, \delta\psi + p^\mu \delta(\partial_\mu \psi) = 0. \tag{7.3.33}$$

The transformation law (7.3.21) together with (6.2.31) informs us that

$$\delta(\partial_\mu \psi) = \partial_\mu \delta\psi - \epsilon^\nu{}_\mu \partial_\nu \psi. \tag{7.3.34}$$

This follows from the recognition that $\delta(\partial_\mu \psi)$ is not equal to $\partial_\mu \delta\psi$; the added Lorentz index μ must also be transformed. Thus, (7.3.33) becomes

$$\partial_\lambda(p^\lambda\, \delta\psi) - p^\mu \eta^{\nu\lambda}\epsilon_{\lambda\mu}\partial_\nu \psi = 0 \tag{7.3.35}$$

or

$$\partial_\lambda(p^\lambda \epsilon_{\mu\nu}\Sigma^{\mu\nu}\psi) = p^\mu \epsilon_{\lambda\mu}\partial^\lambda \psi \tag{7.3.36}$$

so that

$$\frac{p^\mu \partial^\nu \psi - p^\nu \partial^\mu \psi}{2} = \partial_\sigma(p^\sigma \Sigma^{\nu\mu}\psi). \tag{7.3.37}$$

We return now to

$$\Theta^{\mu\nu} = p^\mu \partial^\nu \psi - \eta^{\mu\nu}\mathcal{L} \tag{7.3.38}$$

and note that with

$$\Phi^{\sigma\mu\nu} = (p^\sigma \Sigma^{\mu\nu} - p^\mu \Sigma^{\sigma\nu} - p^\nu \Sigma^{\sigma\mu})\, \psi, \tag{7.3.39}$$

$$T^{\mu\nu} = \Theta^{\mu\nu} + \partial_\sigma \Phi^{\sigma\mu\nu} \tag{7.3.40}$$

has all the desired properties. The divergence of the first term of (7.3.39) added to $p^\mu \partial^\nu \psi$ produces a (μ,ν) symmetric sum. The second and third terms of (7.3.39) are already (μ,ν) symmetric, as is the $\eta^{\mu\nu}$ term in $\Theta^{\mu\nu}$. Therefore $T^{\mu\nu}$ is (μ,ν) symmetric. However, $\Phi^{\sigma\mu\nu}$ is (σ,μ) antisymmetric, so that the new symmetric $T^{\mu\nu}$ is conserved and leads to the same conserved energy and momentum as $\Theta^{\mu\nu}$.

We consider an example: a hypothetical scalar field.

The simplest consistent Lagrangian density is ψ^2, where ψ is the scalar field. This choice yields the variational equation $\psi = 0$; this is not a very interesting result. The next complication is to add a derivative. Since $\mathcal{L}$

must be Lorentz-invariant, we can only add

$$\delta\mathcal{L} = \frac{\partial\psi}{\partial x^{\sigma}}\,\eta^{\sigma\nu}\,\frac{\partial\psi}{\partial x^{\nu}} \tag{7.3.41}$$

to obtain

$$\mathcal{L} = \frac{1}{2}\,\partial_{\lambda}\psi\partial^{\lambda}\psi - \frac{1}{2}\,\mu^2\psi^2. \tag{7.3.42}$$

The arbitrary constant $\frac{1}{2}$ determines our units, which here and from now on are rationalized. This saves a lot of 4π writing! The constant $\mu = \sqrt{\mu^2}$ is real and is, as we will see, the minimum frequency at which the field will oscillate.

    To derive the field equations from the Lagrangian density, it is useful (only this time) to go back to the more explicit notation

$$\partial_{\mu}\psi\partial^{\mu}\psi \equiv \partial_{\mu}\psi\eta^{\mu\nu}\partial_{\nu}\psi$$

so that

$$\frac{\partial\mathcal{L}}{\partial(\partial_{\sigma}\psi)} = \eta^{\sigma\nu}\partial_{\nu}\psi \tag{7.3.43}$$

and

$$\partial_{\sigma}\frac{\partial\mathcal{L}}{\partial(\partial_{\sigma}\psi)} - \frac{\partial\mathcal{L}}{\partial\psi} = (\partial^{\nu}\partial_{\nu} + \mu^2)\psi = 0. \tag{7.3.44}$$

We recognize

$$\partial^{\nu}\partial_{\nu} = \frac{\partial^2}{(\partial x^0)^2} - \nabla^2;$$

Equation (7.3.44) is, of course, the familiar wave equation with solutions $e^{ik^{\lambda}x_{\lambda}}$, where

$$\mu^2 - k_{\lambda}k^{\lambda} = \mu^2 + \mathbf{k}^2 - (k^0)^2 = 0. \tag{7.3.45}$$

The canonical stress-energy tensor for this theory is

$$T^{\sigma\nu} = \partial^{\sigma}\psi\partial^{\nu}\psi - \eta^{\sigma\nu}\left(\frac{\partial^{\lambda}\psi\partial_{\lambda}\psi - \mu^2\psi^2}{2}\right) \tag{7.3.46}$$

which is symmetric and, therefore, entitled to be called $T^{\sigma\nu}$. The four-vector P^{ν} is given by

$$P^\nu = \int \left[\partial^0 \psi \partial^\nu \psi - \eta^{0\nu}\left(\frac{\partial^\lambda \psi \partial_\lambda \psi - \mu^2}{2} \right) \right] d^3x \qquad (7.3.47)$$

with

$$P^0 = \int \left[(\partial^0 \psi)^2 - \frac{(\partial^0 \psi)^2}{2} + \frac{(\nabla \psi)^2}{2} + \frac{\mu^2 \psi^2}{2} \right] d^3x$$

$$= \frac{1}{2} \int d^3x [\dot{\psi}^2 + (\nabla \psi)^2 + \mu^2 \psi^2], \qquad (7.3.48)$$

clearly showing the need for positive μ^2, since otherwise P^0 would be unbounded both below and above and the system would be unstable.

The momentum is

$$P^i = \int \partial^0 \psi \partial^i \psi \, d^3x \qquad \text{or} \qquad \mathbf{P} = - \int d^3x \, \dot{\psi} \nabla \psi. \qquad (7.3.49)$$

The angular momentum density $\mathbf{l(r)}$ is $\mathbf{r} \times \mathbf{p(r)}$, where $\mathbf{p}$ is the momentum density. That is, the angular momentum $\mathbf{L}_V$ contained in a volume V is

$$\mathbf{L}_V = - \int_V \dot{\psi} \mathbf{r} \times \nabla \psi \, d^3x, \qquad (7.3.50)$$

and the angular momentum flux is obtained from the equation

$$\frac{d}{dx^0} L_v^{ji} = \int d^3x (\partial_0 T^{0i} x^j - \partial_0 T^{0j} x^i)$$

$$= \int d^3x [-(\partial_k T^{ki})x^j + (\partial_k T^{kj})x^i]$$

$$= \int_S dS_k (T^{kj} x^i - T^{ki} x^j). \qquad (7.3.51)$$

Thus, the loss (radiation) of angular momentum through a surface S is

$$-d\frac{L_v^{ji}}{dt} = - \int dS_k (T^{ki} x^j - T^{kj} x^i). \qquad (7.3.52)$$

7.4. INTERACTING FIELDS AND PARTICLES

Following the work in Section 7.2, we use particle coordinates $y_\mu^p(\sigma_p)$, with σ an invariant parameter to be determined, as, for example, by (7.2.11). In what follows we drop the label p, with the understanding that we must sum over all participating particles.

The particle Lagrangian L_p is a function of $\dot{y}_\mu(\sigma) = dy_\mu(\sigma)/d\sigma$, and the particle action is the functional

$$S_p = \int L_p \, d\sigma = -\frac{m}{2} \int \dot{y}_\mu(\sigma) \, \dot{y}^\mu(\sigma) \, d\sigma, \tag{7.4.1}$$

as given by (7.2.8).

The field coordinates $\psi_\alpha(x)$ are functions of space and time. The field action is the functional

$$S_\psi = \int d^4x \, \mathscr{L}_\psi(\psi_\alpha(x), \partial_\mu \psi_\alpha(x)) \tag{7.4.2}$$

where $\mathscr{L}_\psi$ is the Lagrangian density. For a scalar field, we have seen in (7.3.42) that we may take

$$\mathscr{L}_\psi = \frac{1}{2} \partial_\lambda \psi \partial^\lambda \psi - \frac{\mu^2}{2} \psi^2. \tag{7.4.3}$$

The Lagrangian corresponding to $\mathscr{L}$ is

$$L_\psi(t) = \int d^3x \, \mathscr{L}(x) \tag{7.4.4}$$

that is a function of t, whereas the particle Lagrangian

$$L_p = -m/2 \, dy^\mu/d\sigma \, dy_\mu/d\sigma$$

is a function of σ.

It is therefore necessary to consider the action as the fundamental functional. The action for the scalar field particle system will be

$$S = S_\psi + S_p + S_I \tag{7.4.5}$$

where S_I is the interaction action. We have seen in (7.2.2) how S_I must be chosen to give the correct equation for charged particle motion in an electromagnetic field:

$$S_I = q \int d\sigma \, A_\mu(y) \frac{dy^\mu}{d\sigma}. \tag{7.4.6}$$

We see that S_I is also a functional of the field $A_\mu(x)$ by rewriting S_I:

$$S_I = q \int d^4x \, d\sigma \, \frac{dy^\mu}{d\sigma} \, \delta^4(y - x) \, A_\mu(x), \tag{7.4.7}$$

in which form the field variable A_μ depends on the field point x, as is required by the Lagrangian procedure.

In general, we will have

$$S_\psi = \int d^4x \, \mathcal{L}_\psi \tag{7.4.8}$$

$$S_p = \int d\sigma L_p \tag{7.4.9}$$

and S_I with two equivalent forms:

$$S_I = \int d\sigma L_I \tag{7.4.10}$$

as in (7.4.6), or

$$S_I = \int d^4x \, \mathcal{L}_I \tag{7.4.11}$$

as in (7.4.7), the first form appropriate to the $y^\mu(\sigma)$ equation, the second to the $\psi_\alpha(x)$ equation.

The Lagrange equations for the combined $\psi_\alpha(x)$, $y^\mu(\sigma)$ system are

$$\partial_\mu \frac{\partial(\mathcal{L}_\psi + \mathcal{L}_I)}{\partial(\partial_\mu \psi)} - \frac{\partial(\mathcal{L}_\psi + \mathcal{L}_I)}{\partial \psi} = 0 \tag{7.4.12}$$

and

$$\frac{d}{d\sigma} \frac{\partial(L_p + L_I)}{\partial \dot{y}^\mu} - \frac{\partial(L_p + L_I)}{\partial y^\mu} = 0. \tag{7.4.13}$$

The interaction action S_I in all the cases we will consider is linear in the ψ_α's; in first approximation, the field action S_ψ is bilinear in the ψ_α's and $\partial_\mu \psi_\alpha$'s. Therefore, the overall sign of S_I, which is equivalent to the sign of the coupling constant, is irrelevant, since it can be changed by a redefinition of ψ_α : $\psi_\alpha \to -\psi_\alpha$. By the same token, the signs of S_ψ and S_p alone are irrelevant. The important sign is the relative sign of S_p and S_ψ and, if there are several independent S_ψ's and S_p's, corresponding to several fields and particles, their relative sign. We choose the relative signs so that the energy of the noninteracting system has a definite sign, which we normally take to be positive.

If the equations of motion are translation-invariant in space and time, we can apply Noether's theorem to find the conserved energy–momentum four-vector. The translational invariance is with respect to the transformation

$$y^\mu \to y^\mu + \delta y^\mu \tag{7.4.14}$$

with δy^μ a constant four-vector, and

$$\psi_\alpha(x) \to \psi_\alpha(x - \delta y) \tag{7.4.15}$$

or, for an infinitesimal transformation,

$$\psi_\alpha(x) \to \psi_\alpha(x) - \delta y^\mu \partial_\mu \psi_\alpha(x). \tag{7.4.16}$$

As in (7.3.9) and (7.3.11), we calculate δS in two ways. First, using the Lagrange equations, we get

$$\delta S = \int d^4x \partial_\mu \left[\frac{\partial(\mathcal{L}_\psi + \mathcal{L}_I)}{\partial(\partial_\mu \psi_\alpha)} \delta \psi_\alpha \right] + \int d\sigma \frac{d}{d\sigma}\left[\frac{\partial(L_p + L_I)}{\partial(\dot{y}^\mu)} \delta y^\mu \right]. \tag{7.4.17}$$

In the d^4x integration, we can carry out the time integral from t_1 to t_2; the space components of ∂_μ integrate to zero. Similarly, we carry out the $d\sigma$ integral from σ_1 to σ_2, where σ_1 is such that $y^0(\sigma_1) = t_1$ and $y^0(\sigma_2) = t_2$. δS becomes

$$\delta S = \left[\int d^3x \frac{\partial \mathcal{L}_\psi}{\partial(\partial_0 \psi_\alpha)} \delta \psi_\alpha \Big|_{t_2}^{t_1} + \frac{\partial(L_p + L_I)}{\partial \dot{y}^\mu} \delta y^\mu \Big|_{\sigma_2}^{\sigma_1} \right] \tag{7.4.18}$$

since $\partial \mathcal{L}_I / \partial(\partial_o \psi_\alpha) = 0$.

Second, remembering that δy^μ is constant and, as in (7.4.1) and (7.4.6), that S_I and S_p are invariant under the transformation, we find, by a direct calculation

$$\delta S = -\int \delta y^\mu \partial_\mu \mathcal{L}_\psi \, d^4x = -\delta y^0 \int \mathcal{L}_\psi \, d^3x \Big|_{t_2}^{t_1}. \tag{7.4.19}$$

Equating (7.4.18) to (7.4.19), we find with $\delta \psi_\alpha = -\delta y^\mu \partial_\mu \psi_\alpha$

$$- \delta y^{\mu} \left\{ \int d^3x \left[\frac{\partial \mathcal{L}_{\psi}}{\partial(\partial_0 \psi_{\alpha})} \partial_{\mu} \psi_{\alpha} - \delta^0_{\ \mu} \mathcal{L}_{\psi} \right] 2 \frac{\partial(L_p + L_I)}{\partial \dot{y}^{\mu}} \right\} = - \delta y^{\mu} P_{\mu}$$

(7.4.20)

with P_{μ} a constant:

$$P_{\mu} = \int d^3x \left[\frac{\partial \mathcal{L}_{\psi}}{\partial(\partial_0 \psi_{\alpha})} \partial_{\mu} \psi_{\alpha} - \delta^0_{\ \mu} \mathcal{L}_{\psi} \right] - \frac{\partial}{\partial \dot{y}^{\mu}} (L_p + L_I).$$

(7.4.21)

Equation (7.4.21) gives us a rule for extending the field theoretic canonical formalism of (7.3.6) to include particle degrees of freedom, the $y_{\mu}(\sigma)$.

We return to the example of the scalar field, considered earlier in Section 7.3, but now interacting. We choose the action (see Problems 7.1 and 7.2)

$$S = \int d^4x \left[\frac{\partial_{\lambda} \psi \partial^{\lambda} \psi - \mu^2 \psi^2}{2} \right] - \frac{m}{2} \int d\sigma \frac{dy^{\mu}}{d\sigma} \frac{dy_{\mu}}{d\sigma} (1 + g\psi(y)).$$

(7.4.22)

The equations of motion are

$$- \partial_{\lambda} \partial^{\lambda} \psi - \mu^2 \psi - \frac{m}{2} \int d\sigma \frac{dy^{\lambda}}{d\sigma} \frac{dy_{\lambda}}{d\sigma} g \, \delta^4(x - y) = 0$$

(7.4.23)

and

$$m \frac{d}{d\sigma} \left(\frac{dy_{\mu}}{d\sigma} (1 + g\psi(y)) \right) - \frac{m}{2} \frac{dy^{\lambda}}{d\sigma} \frac{dy_{\lambda}}{d\sigma} g \frac{\partial \psi(y)}{\partial y^{\mu}} = 0.$$

(7.4.24)

The "W" conservation law is for

$$\frac{dy^{\lambda}}{d\sigma} \frac{dy_{\lambda}}{d\sigma} (1 + g\psi(y)) = \frac{(d\tau)^2}{(d\sigma)^2} (1 + g\psi(y))$$

(7.4.25)

which we may set equal to 1. The energy momentum vector P_{α} is

$$P_{\alpha} = \int d^3x \left\{ \frac{\partial \psi}{\partial x^0} \frac{\partial \psi}{\partial x^{\alpha}} - \delta^0_{\alpha} \frac{(\partial_{\lambda} \psi \partial^{\lambda} \psi - \mu^2 \psi^2)}{2} \right\} + m \frac{dy_{\alpha}}{d\sigma} (1 + g\psi(y)).$$

(7.4.26)

Note the positive energy

$$P_0 = \int d^3x \frac{\dot{\psi}^2 + (\nabla\psi)^2 + \mu^2\psi^2}{2} + \frac{m(1 + g\psi)^{1/2}}{\sqrt{1 - v^2}}, \qquad (7.4.27)$$

provided the expression $(d\tau/d\sigma)^2 (1 + g\psi) = 1$ behaves properly.

If we return to (7.4.21), an obvious question arises next: Is there a conserved tensor $\Theta^\mu{}_\nu$, of which P_ν is the space integral of the 0ν component? The answer is yes. It can be written down by inspection. It is

$$\Theta^\mu{}_\nu = \frac{\partial\mathcal{L}_\psi}{\partial(\partial_\mu\psi_\alpha)} \partial_\nu\psi_\alpha - \delta^\mu{}_\nu \mathcal{L}_\psi - \int \dot{y}^\mu \frac{\partial}{\partial\dot{y}^\nu} (L_p + L_I) \delta^4(x - y) \, d\sigma.$$

$$(7.4.28)$$

It is left as Problem 7.4 to show that $\Theta^\mu{}_\nu$ has the desired properties.

The totally contravariant form of $\Theta^\mu{}_\nu$,

$$\Theta^{\mu\nu} = \frac{\partial\mathcal{L}_\psi}{\partial(\partial_\mu\psi_\alpha)} \partial^\nu\psi_\alpha - \eta^{\mu\nu}\mathcal{L}_\psi - \int \dot{y}^\mu \frac{\partial}{\partial\dot{y}_\nu} (L_p + L_I) \delta^4(x - y) \, d\sigma,$$

$$(7.4.29)$$

is, in general, not symmetric.

The next question: Can we always find a tensor $\delta\Theta^{\mu\nu}$ such that

$$T^{\mu\nu} = \Theta^{\mu\nu} + \delta\Theta^{\mu\nu} \qquad (7.4.30)$$

has the desired properties listed in Section 7.3? That is, is it symmetric, conserved, and having the same P^μ integral as $\Theta^{\mu\nu}$? Again, the answer is yes. The construction technique parallels the treatment leading to (7.3.40).

We make use of the fact that $\mathcal{L}_\psi$ and $L = L_p + L_I$ are separately invariant under the Lorentz transformation $\psi \to \psi + \epsilon_{\mu\nu}\Sigma^{\mu\nu}\psi$, as in (7.3.22), and

$$\dot{y}^\mu \to \dot{y}^\mu + \epsilon^\mu{}_\lambda y^\lambda. \qquad (7.4.31)$$

From the invariance of $\mathcal{L}_\psi$ follows

$$\frac{\partial\mathcal{L}_\psi}{\partial\psi} \delta\psi + p^\mu \delta(\partial_\mu\psi) = 0 \qquad (7.4.32)$$

where p^μ is still given by (7.3.32). From the Lagrange equation for ψ, we learn that

$$\frac{\partial \mathcal{L}_\psi}{\partial \psi} + \int \frac{\delta L}{\delta \psi(x)} d\sigma - \partial_\mu p^\mu = 0, \tag{7.4.33}$$

so that, with $\delta(\partial_\mu \psi) = \partial_\mu \delta \psi - \epsilon^\lambda{}_\mu \partial_\lambda \psi$, (7.4.32) becomes

$$(\partial_\mu p^\mu) \, \delta \psi - \int \frac{\delta L}{\delta \psi(x)} d\sigma \, \delta \psi + p^\mu \partial_\mu \delta \psi - p^\mu \epsilon^\lambda{}_\mu \partial_\lambda \psi = 0$$

or

$$\partial_\mu (p^\mu \, \delta \psi) = \int d\sigma \, \frac{\delta L}{\delta \psi(x)} \delta \psi + p^\mu \epsilon_{\lambda \mu} \partial^\lambda \psi, \tag{7.4.34}$$

and from (7.3.22)

$$\epsilon_{\mu\nu} \partial_\sigma p^\sigma \Sigma^{\mu\nu} \psi = \epsilon_{\mu\nu} \int d\sigma \, \frac{\delta L}{\delta \psi(x)} \Sigma^{\mu\nu} \psi + \epsilon_{\mu\nu} \frac{(p^\nu \partial^\mu \psi - p^\mu \partial^\nu \psi)}{2}$$

so that

$$\partial_\sigma (p^\sigma \Sigma^{\mu\nu} \psi) = \int d\sigma \, \frac{\delta L}{\delta \psi(x)} \Sigma^{\mu\nu} \psi + \frac{p^\nu \partial^\mu \psi - p^\mu \partial^\nu \psi}{2}. \tag{7.4.35}$$

We define $\Phi^{\sigma\mu\nu}$ as in (7.3.39) and add $\delta\Theta^{\mu\nu}$ to $\Theta^{\mu\nu}$, where

$$\delta\Theta^{\mu\nu} = \partial_\sigma \Phi^{\sigma\mu\nu} \tag{7.4.36}$$

and

$$T^{\mu\nu} = \Theta^{\mu\nu} + \delta\Theta^{\mu\nu}. \tag{7.4.37}$$

The symmetry properties of $\Phi^{\sigma\mu\nu}$ show that $\delta\Theta^{\mu\nu}$ is conserved:

$$\partial_\mu \Theta^{\mu\nu} = 0 \tag{7.4.38}$$

and the energy–momentum vector P^μ is unchanged by $\delta\Theta^{\mu\nu}$:

$$\int d^3x \, \delta\Theta^{0\nu} = 0. \tag{7.4.39}$$

It must next be shown that $T^{\mu\nu}$ is symmetric. We see that the last term on the right of (7.4.35), $(p^\nu \partial^\mu \psi - p^\mu \partial^\nu \psi)/2$, added to the first term of $\Theta^{\mu\nu}$ in (7.4.29), $p^\mu \partial^\nu \psi$, is symmetric.

There remains

$$\delta T^{\mu\nu} = -\int \dot{y}^\mu \frac{\partial L}{\partial \dot{y}_\nu} \delta^4(x-y)\, d\sigma + \int \frac{\delta L}{\delta \psi(x)} \Sigma^{\mu\nu} \psi \, d\sigma. \qquad (7.4.40)$$

From the invariance of L, we find

$$\epsilon_{\mu\nu} \frac{\partial L}{\partial \psi(y)} \Sigma^{\mu\nu} \psi(y) + \frac{\partial L}{\partial \dot{y}^\mu} \epsilon^\mu{}_\lambda \dot{y}^\lambda = 0 \qquad (7.4.41)$$

or

$$\frac{\partial L}{\partial \psi(y)} \Sigma^{\mu\nu} \psi(y) + \frac{1}{2}\left(\frac{\partial L}{\partial \dot{y}_\mu} \dot{y}^\nu - \frac{\partial L}{\partial \dot{y}_\nu} \dot{y}^\mu \right) = 0. \qquad (7.4.42)$$

The variational derivative $\delta L/\delta \psi(x)$ is given by

$$\frac{\partial L}{\partial \psi(x)} = \frac{\partial L}{\partial \psi(y)} \delta^4(x-y) \qquad (7.4.43)$$

so that

$$\delta T^{\mu\nu} = -\int \dot{y}^\mu \frac{\delta L}{\partial \dot{y}_\nu} \delta^4(x-y)\, d\sigma - \frac{1}{2}\int d\sigma\, \delta^4(x-y)\left(\frac{\partial L}{\partial \dot{y}_\mu} \dot{y}^\nu - \frac{\partial L}{\partial \dot{y}_\nu} \dot{y}^\mu \right)$$

which is $\mu\nu$ symmetric. This completes the construction of $T^{\mu\nu}$.

In Section 7.8, we will find a second method of constructing a conserved, symmetric stress tensor directly from an action principle.

7.5. VECTOR FIELDS

We return now to electrodynamics. We first ask for a Lagrangian density for a free massive vector field A_μ, analogous to the vector potential of electrodynamics. We require an invariant bilinear function of A_μ, involving at most first derivatives. A first guess might be $\mathscr{L}_G$ (G for a guess):

$$\mathscr{L}_G = -\partial_\alpha A^\nu \partial^\alpha A_\nu + \mu^2 A^\nu A_\nu. \qquad (7.5.1)$$

This choice has an obvious problem: If we consider each ν value separately, we have in (7.5.1) a sum of four scalarlike $\mathscr{L}$'s, three with the sign to give a positive energy and one with the opposite sign. Therefore, there is no lower *or* upper bound on the energy and the system is unstable.

How do we produce an $\mathscr{L}$ with positive energy? There are two ways. The first, which we leave as an exercise (see Problem 7.9), is to write down all possible invariant bilinear functions of A_ν and $\partial^\mu A_\nu$ with arbitrary coefficients, calculate the energy, and adjust the coefficients to make the energy positive-definite.

A second way is easier[2] and perhaps more enlightening. Notice in (7.5.1) that the space components of A_μ carry positive energy, the time component negative energy. Furthermore, the space components form a rotational vector, and the time component a rotational scalar. Perhaps the scalar is present only to preserve Lorentz invariance, and we should try to eliminate it as an independent field. The clue is in the four-vector wave number k_μ,[3] which for a massive propagating field will be timelike:

$$k_\alpha k^\alpha = \mu^2. \tag{7.5.2}$$

Therefore, there is a Lorentz system for which $k^i = 0$, and the Lorentz-invariant condition

$$k_\alpha A^\alpha = 0 \tag{7.5.3}$$

will set $A_0 = 0$ in that coordinate system.

We expect the field to be radiated by a current density j_μ according to an equation resembling

$$(\partial_\alpha \partial^\alpha + \mu^2)A^\lambda = -j^\lambda \tag{7.5.4}$$

(note rationalized units), but requiring that a propagating solution with wave number k^α will satisfy $k_\alpha A^\alpha = 0$.

We try to accomplish this by projecting (7.5.4). With wave number k^μ, (7.5.4) becomes

$$((ik_\alpha)(ik^\alpha) + \mu^2)A^\lambda = -P^\lambda{}_\eta j^\eta \tag{7.5.5}$$

where P is a projection operator that makes (7.5.3) hold for $k^2 = \mu^2$. The projection operator $P^\lambda{}_\eta$ is clearly $\delta^\lambda_\eta - (k^\lambda k_\eta/\mu^2)$, since

$$k_\lambda\left(\delta^\lambda_\eta - \frac{k^\lambda k_\eta}{\mu^2}\right) = k_\eta\left(1 - \frac{k^3}{\mu^2}\right)$$

---

[2]This is especially true in the case of a second-rank tensor field, like the gravitational field, where, as we shall see in Chapter 8, the Lagrangian has *many* possible terms.
    [3]We go back and forth freely between a coordinate space description of the fields and their Fourier-transform space.

which vanishes at $k^2 = \mu^2$. Note that this procedure, as opposed to the direct construction of a Lagrangian density suggested earlier, requires that one start with $\mu \neq 0$.

Equation (7.5.5) now becomes

$$-(\mu^2 - k^2)A^\lambda = j^\lambda - \frac{k \cdot j k^\lambda}{\mu^2}. \tag{7.5.6}$$

We eliminate $k \cdot j$ by operating on (7.5.6) with k_λ:

$$(\mu^2 - k^2)k \cdot A = -k \cdot j\left(1 - \frac{k^2}{\mu^2}\right) \quad \text{or} \quad k \cdot j = -\mu^2 k \cdot A \tag{7.5.7}$$

and (7.5.6) becomes

$$(\mu^2 - k^2)A^\lambda + k^\lambda k \cdot A = -j^\lambda. \tag{7.5.8}$$

In coordinate space, $k_\lambda = (1/i)\,\partial_\lambda$ so

$$\mu^2 A^\lambda + \partial_\alpha \partial^\alpha A^\lambda - \partial^\lambda \partial_\alpha A^\alpha = -j^\lambda$$

or

$$\partial_\alpha(\partial^\alpha A^\lambda - \partial^\lambda A^\alpha) + \mu^2 A^\lambda = -j^\lambda$$

or

$$\partial_\alpha F^{\alpha\lambda} + \mu^2 A^\lambda = -j^\lambda \tag{7.5.9}$$

the natural extension of Maxwell's equations to the case of finite μ. Equation (7.5.9) is called the Proca equation.

The cancellation of the factor $\mu^2 - k^2$ in (7.5.7) would appear to contradict the assumption that would make $k \cdot A = 0$ for a propagating mode. However, the remaining identity, $\mu^2 k \cdot A = -k \cdot j$ or

$$\partial_\alpha A^\alpha = -\frac{\partial_\alpha j^\alpha}{\mu^2}$$

shows that the cancellation does occur, since it is only at the source j_μ that $\partial_\mu A^\mu$ fails to be zero. An equivalent statement is that $k \cdot A$ has no pole at $k^2 = \mu^2$, although A^λ does have such a pole. The residue of the pole represents the propagating radiation. Since $k \cdot A$ has no pole, the propagating solution has $k \cdot A = 0$.

We see in (7.5.9) that there are only three degrees of freedom for the vector potential A^μ, since the second-order time derivatives only act on the space components A^i. Therefore, A^0 is a constrained variable, and there are only three propagating modes. Of course, this results directly from the way in which we derived the equation, that is, from demanding that the propagating mode with $k_i = 0$ have no A^0 component. The three degrees of freedom correspond in quantum theory to the three directions of spin of a massive spin one particle.

The presence of the mass term has destroyed the gauge invariance of the massless theory. We also note that the limit $\mu^2 \to 0$ requires a conserved current j^λ, since in that limit the left-hand side of (7.5.9), $\partial_\mu F^{\mu\lambda}$, is identically conserved by the antisymmetry of $F^{\mu\lambda}$. Of course, in that limit gauge invariance is restored, and any one spatial component of the vector potential can be eliminated by a gauge transformation, leaving two independent modes.

We next write a free field Lagrangian density that will yield (7.5.9) when the A^μ are taken as the independent coordinates.[4] Since it is the spatial vector coordinates that correspond to the actual degrees of freedom of the field, we choose the sign of $\mathcal{L}$ to make their energy positive. Also, for $\mu^2 = 0$, $\mathcal{L}$ should be gauge- and Lorentz-invariant, and bilinear in A^μ. The only choice is

$$\mathcal{L} = -c_1 \frac{1}{4} F_{\lambda\nu} F^{\lambda\nu} + \frac{1}{2} c_2 \mu^2 A_\lambda A^\lambda \tag{7.5.10}$$

where c_1 determines our units and $c_2/c_1 = 1$ satisfies the field equation (7.5.9).

From now on, we work in rationalized units with the free field Lagrangian density

$$\mathcal{L} = -\frac{1}{4} F_{\lambda\nu} F^{\lambda\nu} + \frac{\mu^2}{2} A_\lambda A^\lambda. \tag{7.5.11}$$

We already expect, from our work with particle Lagrangians in external fields, that the scalar potential enters as $-q\phi(\mathbf{y}_p, t)$, where $\mathbf{y}_p$ is the particle coordinate and q its charge. Since

$$q\phi(\mathbf{y}_p, t) = \int d\mathbf{x} \rho(\mathbf{x}) \, \phi(\mathbf{x}, t),$$

where ρ is the particle charge density, we guess that the correct interaction Lagrangian density should be $\mathcal{L}_I = j^\mu A_\mu$. The Lagrangian density (to whose action we will eventually add the particle degrees of freedom) is

---

[4]And j^μ is set equal to zero.

then

$$\mathcal{L} = -\frac{1}{4} F_{\alpha\nu} F^{\alpha\nu} + \frac{\mu^2}{2} A_\alpha A^\alpha + j^\alpha A_\alpha. \tag{7.5.12}$$

With

$$F_{\lambda\nu} = (\partial_\lambda A_\nu - \partial_\nu A_\lambda)$$

we have

$$\frac{\partial\mathcal{L}}{\partial(\partial_\lambda A_\nu)} = -F^{\lambda\nu} \tag{7.5.13}$$

and

$$\frac{\partial\mathcal{L}}{\partial(A_\nu)} = \mu^2 A^\nu + j^\nu \tag{7.5.14}$$

leading as expected to (7.5.9):

$$-\partial_\alpha F^{\alpha\nu} = \mu^2 A^\nu + j^\nu. \tag{7.5.15}$$

We may repeat the argument following (7.5.9) from the Lagrangian point of view, by noting that $\dot{A}_0$ does not appear in $\mathcal{L}$. Therefore, the A_0 equation

$$\frac{\partial\mathcal{L}}{\partial A_0} = 0,$$

is a constraint equation, and only the three A_i equations are dynamical. As before, when we consider the $\mu = 0$ limit, we eliminate one more degree of freedom with the introduction of the arbitrary gauge function.

We wish to verify that the energy P^0 of this theory, with $j_\mu = 0$, is positive. It is

$$P^0 = \int d^3x \left[\frac{\partial\mathcal{L}}{\partial(\partial_0 A_\lambda)} \partial_0 A_\lambda - \mathcal{L} \right], \tag{7.5.16}$$

since the canonical tensor $\Theta^{\mu\nu}$ and the symmetric tensor $T^{\mu\nu}$ give the same conserved energy–momentum vector P^μ. In terms of $(\phi, \mathbf{A}) = -(A^0, A^i)$ with

$$\mathcal{L} = \frac{1}{2}(\nabla\phi + \dot{\mathbf{A}})^2 - \frac{1}{2}(\nabla \times \mathbf{A})^2 + \frac{\mu^2}{2}(\phi^2 - \mathbf{A}^2) \tag{7.5.17}$$

we have

$$
\Theta^{00} = (\nabla\phi + \dot{\mathbf{A}}) \cdot \dot{\mathbf{A}} - \frac{1}{2}(\nabla\phi + \dot{\mathbf{A}})^2 + \frac{1}{2}(\nabla \times \mathbf{A})^2 + \frac{\mu^2}{2}\mathbf{A}^2 - \frac{\mu^2}{2}\phi^2
$$

$$
= \frac{\dot{\mathbf{A}}^2}{2} + \frac{1}{2}(\nabla \times \mathbf{A})^2 + \frac{\mu^2}{2}\mathbf{A}^2 - \frac{1}{2}(\nabla\phi)^2 - \frac{\mu^2}{2}\phi^2
$$

or, with $\boldsymbol{\pi} = \nabla\phi + \dot{\mathbf{A}}$, and using the equation [from (7.5.15), with $j^\lambda = 0$]

$$
\nabla \cdot \boldsymbol{\pi} = \mu^2\phi
$$

we find, after an integration by parts,

$$
P^0 = \int d^3 x\, \Theta^{00} = \int d^3 x\left[\frac{\boldsymbol{\pi}^2}{2} + \frac{\mu^2\mathbf{A}^2}{2} + \frac{1}{2}(\nabla \times \mathbf{A})^2 + \frac{(\nabla \cdot \boldsymbol{\pi})^2}{2\mu^2}\right] \qquad (7.5.19)
$$

which is positive-definite.

It is also of interest to express the energy P^0 in terms of the Fourier transform of the vector field:[5]

$$
A_\mu(x) = \int d^3k[a_\mu(k)\, e^{i(\mathbf{k}\cdot\mathbf{x} - k_0x_0)} + \text{c.c.}].
$$

Since P^0 is time-independent, only cross terms between $e^{\pm i\omega t}$ will be different from zero. In addition, the d^3x integral will only connect $+\mathbf{k}$ to $-\mathbf{k}$ and, hence, a_μ to a_μ^*. From the divergence condition on A_μ, $k_0a_0 = \mathbf{k} \cdot \mathbf{a}$, and from (7.5.18),

$$
P^0 = (2\pi)^3 \int d^3k\left\{|\mathbf{a(k)}|^2 - \frac{|\mathbf{k} \cdot \mathbf{a}|^2}{k_0{}^2}\right\}(k_0{}^2 + \mathbf{k}^2 + \mu^2)
$$

or, with $\mathbf{k}$ in the three direction, and $k/k_0 = \epsilon < 1$

$$
P^0 = (2\pi)^3 \int d^3k\left\{\left(k_0{}^2 + \mathbf{k}^2 + \mu^2\right)(|a_1|^2 + |a_2|^2)\right.
$$

$$
\left. + (1 - \epsilon^2)(k_0{}^2 + \mathbf{k}^2 + \mu^2)|a_3|^2\right\}. \qquad (7.5.20)
$$

[5]Note here, and remember in the future that $e^{i(\mathbf{k}\cdot\mathbf{x} - k^0x^0)} = e^{-ik\cdot x}$.

We see that as $\mu \to 0$, $k \to k_0$ and $\epsilon \to 1$, so

$$P^0 \to 2(2\pi)^3 \int d^3 k \mathbf{k}^2 [|a_1|^2 + |a_2|^2] \qquad (7.5.21)$$

and the longitudinal mode a_3 carries no energy, unless a_3 becomes singular as $\mu \to 0$. This does not happen in the vector theory we are now considering. We shall see later that the situation with the gravitational field is not so simple.

From the general result (7.4.37) and the preceding discussion we can calculate the symmetric stress tensor $T^{\mu\nu}$ for the electromagnetic field and a charged particle in interaction with each other. The action is

$$S = -\frac{1}{4} \int d^4 x \, F_{\mu\nu} F^{\mu\nu} - \int d\sigma \left\{ \frac{m}{2} \frac{dy^\mu}{d\sigma} \frac{dy_\mu}{d\sigma} - q \frac{dy^\mu}{d\sigma} A_\mu \right\}. \qquad (7.5.22)$$

The stress tensor is

$$T^{\mu\nu} = \left(\eta^{\mu\alpha} F_{\alpha\lambda} F^{\lambda\nu} + \frac{1}{4} \eta^{\mu\nu} F_{\alpha\beta} F^{\alpha\beta} \right) + m \int d\sigma \frac{dy^\mu}{d\sigma} \frac{dy^\nu}{d\sigma} \delta^4(x - y(\sigma))$$
$$(7.5.23)$$

in agreement with (6.6.19). (See Problem 7.5.)

Evidently, we have chosen the signs of the free field and free particle terms in the action correctly, since the energy density

$$T^{00} = \frac{1}{2} (\mathbf{E}^2 + \mathbf{B}^2) + \frac{m\delta^3(\mathbf{x} - \mathbf{y}(\sigma))}{\left(1 - \left(\dfrac{d\mathbf{y}}{dy_0} \right)^2 \right)^{1/2}} \qquad (7.5.24)$$

is positive.

As we have remarked earlier, it is an exceptional circumstance that the electromagnetic and particle stress tensor $T^{\mu\nu}$ is the sum of the free particle and free field stress tensors, even though the equations for $\mathbf{E}$, $\mathbf{B}$, and $\mathbf{y}$ involve the interaction between them. It is clear from the discussion leading to (7.4.37) that $T^{\mu\nu}$ will normally have an explicit interaction term in other field theories.

One must realize that the fact that the energy density (7.5.24) is positive (strictly, nonnegative) is a necessary, but not sufficient condition for the equations of the theory to have sensible (i.e., finite) solutions. In fact, the equations of the electrodynamics of point particles do *not* appear to have sensible solutions. In classical theory, this is manifested by the electromagnetic contribution to the inertia of the point particle becoming infinite (a consequence of the singularity of the electric and magnetic fields at the position of the particle). This problem was discovered by Lorentz, who tried to remove it by renormalization. We explain.

Lorentz calculated the self-force (i.e., the force of the retarded fields generated by the electron on itself) for an electron, instantaneously at rest, but undergoing arbitrary motion as a function of time. To make the calculation finite, he assumed a charge density for the electron

$$\rho(\mathbf{x}, t) = ef(\mathbf{x} - \mathbf{y}(t)) \tag{7.5.25}$$

where

$$\int f(\mathbf{x}) \, d^3x = 1. \tag{7.5.26}$$

Lorentz found for the force

$$\mathbf{F} = -\delta m \frac{d^2\mathbf{y}}{dt^2} + \frac{2}{3} \frac{e^2}{c^3} \frac{d^3\overline{\mathbf{y}}}{dt^3}, \tag{7.5.27}$$

where δm is the electromagnetic self-mass

$$\delta m = \frac{2}{3} e^2 \int d^3x \, d^3y \, \frac{f(\mathbf{x})f(\mathbf{y})}{|x - y|}, \tag{7.5.28}$$

and $\overline{\mathbf{y}}(t)$ approaches $\mathbf{y}(t)$) as the characteristic radius R of the cut off function f approaches zero. In the limit of a point electron, the integral (7.5.28) diverges like $1/R$. The second term in (7.5.27) (which we have already seen in Section 5.9 and Problem 4.8) is independent of R as $R \to 0$. In this limit, there are no other terms.

Lorentz observed that the equation of motion for the electron would follow from (7.5.27):

$$m_0 \frac{d^2\mathbf{y}}{dt^2} = -\delta m \frac{d^2\mathbf{y}}{dt^2} + \frac{2}{3} \frac{e^2}{c^3} \frac{d^3\mathbf{y}}{dt^3} \tag{7.5.29}$$

(where m_0 is the nonelectromagnetic mass of the electron) or

$$m \frac{d^2\mathbf{y}}{dt^2} = \frac{2}{3} \frac{e^2}{c^3} \frac{d^3\mathbf{y}}{dt^3} \tag{7.5.30}$$

where

$$m = m_0 + \delta m. \tag{7.5.31}$$

Since the equation of motion for the electron only involves m, and not m_0 and δm separately, one might optimistically hope that a finite m, achieved either by a negatively infinite m_0, or by a cut-off at small r, would leave us with a sensible theory.

These two alternatives must be considered separately. The first, a negatively infinite m_0, does not work, because (7.5.30) has exponentially growing solutions. We have already seen a hint of this problem (without the infinite δm) in Section 5.9, with (5.9.11) and (5.9.12).

One can, in fact, show that the run-away solutions and the divergent self-mass are related problems: A charge density that makes the integral (7.5.28) finite does not have run-away solutions. Unfortunately, such a cut-off function would violate special relativity, since the action, as in Problem 7.8, would not be Lorentz-invariant. The application of relativistic quantum theory improves the situation but not enough: It turns the linear divergence of (7.5.28), that is, the linear dependence on $1/R$, into a logarithmic divergence. The run-away solutions disappear, but new problems arise.

How should one deal with this situation? There is probably a modest consensus favoring the following view.

We know that the electron mass is finite; therefore, our theory is wrong. At some small distance, R_0, the equations must become less singular so that the equivalent integral (7.5.28) converges. One can try to guess a value for R_0. A popular guess is the Planck radius R_p. R_p is the Compton wavelength and radius of a body whose gravitational self-energy is equal to its rest energy.[6] This radius is

$$R_p = \left(\frac{G\hbar}{c^3}\right)^{1/2} \approx 10^{-32} \text{ cm.} \tag{7.5.32}$$

Here, G is Newton's gravitational constant and $\hbar$ Planck's constant divided by 2π.

Why pick R_p? For one thing, the quantum theory of gravity must come into play at that radius. We have no satisfactory quantum theory of gravitation; perhaps a correct quantum theory of gravity would provide the necessary cut-off. Since the divergence is only logarithmic, the self-mass is relatively insensitive to the cut-off. The self-mass value given by quantum electrodynamics, cut-off at the Planck radius, is

$$\frac{\delta m}{m} = \frac{4}{3}\frac{\alpha}{\pi}\log\frac{\dfrac{\hbar}{m_e c}}{R_p} \tag{7.5.33}$$

[6]Incidentally, a particle with this radius would be a black hole.

where α is the fine structure constant:

$$\frac{1}{\alpha} \approx 137 \tag{7.5.34}$$

and $\hbar/m_e c$ is the Compton wavelength of the electron:

$$\frac{\hbar}{m_e c} \simeq \frac{1}{3} \times 10^{-10} \text{ cm.} \tag{7.5.35}$$

The resulting logarithm, $\log 10^{22}$, is small enough so that $\delta m/m$ is still smaller than 1.

Of course the cut-off, the distance where our present theories fail, could be much greater than 10^{-32} cm. We know experimentally from electron-positron collisions that electrodynamics holds at least as far down as 10^{-15} cm, so the cut-off could be anywhere from 10^{-15} to 10^{-32} cm!

To conclude this brief discussion of the boundaries to our understanding, we summarize.

Our present theory of electromagnetic fields interacting with electrons does not lead to finite results. However, modifications of the theory at interaction distances that might be as small as 10^{-32} cm might provide a consistent, finite theory, with no perceptible effect on present-day physics, including atomic and nuclear scale phenomena. We proceed with this assumption, even though we do not know how to construct such a theory.

7.6. GENERAL COVARIANCE

We next take up the subject of general covariance, that is, the study of objects that transform like tensors under general coordinate transformations. We need this knowledge in order to formulate a consistent theory of gravitational fields (Sections 8.6 and 8.7). We consider it now because it permits us to construct a symmetric stress tensor $T^{\mu\nu}$ directly from a generally covariant action (Section 7.8). In addition, it permits us to write known dynamical equations in arbitrary coordinate systems.

We start from the notion of physical tensors under Lorentz transformations and define an extension to general coordinate transformations. Whatever the coordinate transformation $x'(x)$ be, *define* a contravariant vector to transform like dx^μ, that is, like

$$dx'^\mu = \frac{\partial x'^\mu}{\partial x^\nu} dx^\nu. \tag{7.6.1}$$

Higher-rank contravariant tensors transform like products of vectors, so that

$$T'^{\mu\nu} = \frac{\partial x'^\mu}{\partial x^\lambda} \frac{\partial x'^\nu}{\partial x^\sigma} T^{\lambda\sigma} \tag{7.6.2}$$

etc.

Tensor equations hold in all coordinate systems. Thus, if $V^\mu = 0$, so will

$$V'^\mu = \frac{\partial x'^\mu}{\partial x^\nu} V^\nu \tag{7.6.3}$$

and conversely. To see the converse, multiply (7.6.3) by $\partial x^\lambda / \partial x'^\mu$. There results

$$\frac{\partial x^\lambda}{\partial x'^\mu} V'^\mu = \frac{\partial x^\lambda}{\partial x'^\mu} \frac{\partial x'^\mu}{\partial x^\nu} V^\nu = V^\lambda. \tag{7.6.4}$$

The tensor transformation property defined by (7.6.1) is consistent. That is, if

$$V'^\mu = \frac{\partial x'^\mu}{\partial x^\nu} V^\nu$$

and

$$V''^\lambda = \frac{\partial x''^\lambda}{\partial x'^\eta} V'^\eta,$$

then

$$V''^\lambda = \frac{\partial x''^\lambda}{\partial x'^\eta} \frac{\partial x'^\eta}{\partial x^\nu} V^\nu$$

or, by the chain rule,

$$V''^\lambda = \frac{\partial x''^\lambda}{\partial x^\nu} V^\nu.$$

There are physical invariants under Lorentz transformations. We define them to be invariants under general coordinate transformations. Thus, if ϕ is an invariant field,

$$\phi'(x'(x)) = \phi(x) \qquad \text{and} \qquad \phi(x + \delta x) - \phi(x) = \frac{\partial \phi}{\partial x^\lambda} \delta x^\lambda \qquad (7.6.5)$$

is also an invariant. Evidently,

$$\frac{\partial \phi'}{\partial x'^\lambda} = \frac{\partial \phi}{\partial x^\mu} \frac{\partial x^\mu}{\partial x'^\lambda}; \qquad (7.6.6)$$

this transformation rule is called covariant. Just as in the case of Lorentz tensors, the contraction of a covariant with a contravariant index produces an invariant. We see a special case in (7.6.5); the general rule follows from the defined transformation properties.

Contravariant, covariant, and mixed tensors can be found by multiplication. As usual, symmetry properties are preserved under tensor transformations.

An interesting object is δ^μ_ν, which is a mixed tensor since

$$\delta'^\mu_\nu = \frac{\partial x'^\mu}{\partial x^\lambda} \frac{\partial x^\sigma}{\partial x'^\nu} \delta^\lambda_\sigma = \delta^\mu_\nu. \qquad (7.6.7)$$

The space-time interval

$$d\tau^2 = d\xi^\mu \, \eta_{\mu\nu} \, d\xi^\nu \qquad (7.6.8)$$

where the ξ^μ's are the normal rectangular coordinates in some Lorentz system is an invariant. In a general coordinate system, it will have the form

$$d\tau^2 = \frac{\partial \xi^\mu}{\partial x^\sigma} \frac{\partial \xi^\nu}{\partial x^\lambda} \, \eta_{\mu\nu} \, dx^\sigma \, dx^\lambda$$

$$\equiv g_{\sigma\lambda} \, dx^\sigma \, dx^\lambda. \qquad (7.6.9)$$

Since $d\tau^2$ is an invariant, $g_{\sigma\lambda}$ is a tensor and symmetric. It is called the metric tensor.[7] The tensor $g_{\mu\nu}$ can be used to lower indices. Thus, if V^μ is a contravariant vector, $V_\mu \equiv g_{\mu\nu}V^\nu$ is its covariant representation. This can be seen as follows. Let U^μ and $U_\mu = \eta_{\mu\nu}U^\nu$ be the contravariant and covariant representations of a vector in a Minkowskian coordinate system.

---

[7]Conversely, if $g_{\mu\nu}$ is a tensor, $d\tau^2$ as defined in (7.6.9) is an invariant.

Remember: If $U^\mu = (U^0, \mathbf{U})$, $U_\mu = (U^0, -\mathbf{U})$. Then, by the definition of contravariance and covariance,

$$V^\mu = \frac{\partial x^\mu}{\partial \xi^\lambda} U^\lambda \tag{7.6.10}$$

and

$$V_\mu = \frac{\partial \xi^\lambda}{\partial x^\mu} U_\lambda = \frac{\partial \xi^\lambda}{\partial x^\mu} \eta_{\lambda\sigma} U^\sigma. \tag{7.6.11}$$

Since

$$U^\sigma = \frac{\partial \xi^\sigma}{\partial x^\nu} V^\nu,$$

Equation (7.6.11) yields

$$V_\mu = \frac{\partial \xi^\lambda}{\partial x^\mu} \eta_{\lambda\sigma} \frac{\partial \xi^\sigma}{\partial x^\nu} V^\nu = g_{\mu\nu} V^\nu.$$

The metric tensor has its contravariant counterpart, defined here by

$$g^{\mu\lambda} g_{\lambda\nu} = \delta^\mu_\nu; \tag{7.6.12}$$

from (7.6.12) we see that lowering both indices of $g^{\mu\nu}$ produces $g_{\mu\nu}$, since

$$g_{\mu\sigma} g_{\nu\lambda} g^{\sigma\lambda} = g_{\mu\sigma} \delta^\sigma_\nu = g_{\mu\nu}. \tag{7.6.13}$$

The determinant of $g_{\mu\nu}$ is another interesting object. Of course, the determinant of $\eta_{\mu\nu}$ is -1 and remains unchanged under Lorentz transformations. The same is not true of det $g_{\mu\nu}$. We define

$$g = -\det g_{\mu\nu} \tag{7.6.14}$$

and calculate g':

$$g' = -\det g'_{\mu\nu} = -\det\left(\frac{\partial x^\sigma}{\partial x'^\mu} \frac{\partial x^\lambda}{\partial x'^\nu} g_{\sigma\lambda}\right)$$

$$= \left[\det\left(\frac{\partial x^\alpha}{\partial x'^\beta}\right)\right]^2 g$$

and

$$\sqrt{g'} = \left(\det \frac{\partial x^\alpha}{\partial x'^\beta}\right) \sqrt{g}. \tag{7.6.15}$$

An object that transforms like $\sqrt{g}$ is called a scalar density of weight -1. The number of powers of $\det(\partial x'/\partial x)$ that multiply a normal tensor transformation law is called the weight of the tensor density. Thus, for example, $(1/\sqrt{g})T^{\mu\nu}$ is a second-rank tensor density of weight 1.

The transformation property of $\sqrt{g}$ provides us with an invariant volume element. Since

$$d^4x = \det\left(\frac{\partial x^\alpha}{\partial x'^\beta}\right) d^4x'$$

$$\sqrt{g}\, d^4x = \det\left(\frac{\partial x^\alpha}{\partial x'^\beta}\right) \sqrt{g}\, d^4x'$$

$$= \sqrt{g'}\, d^4x', \tag{7.6.16}$$

$\sqrt{g}\, d^4x$ is an invariant volume element.

We have learned how to rewrite some Lorentz covariant formulas so that they are generally covariant. For example, suppose ϕ is a scalar field and C_μ a covariant vector. Then

$$\frac{\partial \phi}{\partial x^\mu} = C_\mu \tag{7.6.17}$$

is a generally covariant equation, as is

$$A_\mu\, g^{\mu\nu}\, B_\nu = \text{constant} \tag{7.6.18}$$

if A_μ and B_μ are covariant vectors. The equations of electrodynamics, however, involve space-time derivatives of vector and tensor fields. In a Minkowskian coordinate system, these form Lorentz covariant tensors of one higher rank; they are not tensors under general coordinate transformations. To take a specific example, we recall that the space-time derivatives of a contravariant vector U^μ in a Minkowskian coordinate system form the components of a mixed Lorentz tensor:

$$\frac{\partial U^\mu}{\partial \xi^\lambda} = Q^\mu{}_\lambda. \tag{7.6.19}$$

We can therefore *define* a tensor $T^\mu{}_\lambda$ in a new coordinate systems by its transformation from $Q^{\mu}{}_{\lambda}$:

$$T^{\mu}{}_{\lambda} = \frac{\partial x^{\mu}}{\partial \xi^{\sigma}} \frac{\partial \xi^{\tau}}{\partial x^{\lambda}} Q^{\sigma}{}_{\tau}$$

$$= \frac{\partial x^{\mu}}{\partial \xi^{\sigma}} \frac{\partial \xi^{\tau}}{\partial x^{\lambda}} \frac{\partial U^{\sigma}}{\partial \xi^{\tau}}. \tag{7.6.20}$$

On the other hand, we have already agreed to *define* the vector V^{μ} in a general coordinate system by the equation

$$V^{\mu} = \frac{\partial x^{\mu}}{\partial \xi^{\sigma}} U^{\sigma}. \tag{7.6.21}$$

If we now calculate $\partial V^{\mu}/\partial x^{\lambda}$, we find

$$\frac{\partial V^{\mu}}{\partial x^{\lambda}} = \frac{\partial x^{\mu}}{\partial \xi^{\sigma}} \frac{\partial U^{\sigma}}{\partial \xi^{\tau}} \frac{\partial \xi^{\tau}}{\partial x^{\lambda}} + \frac{\partial}{2x^{\lambda}} \left(\frac{\partial x^{\mu}}{\partial \xi^{\sigma}} \right) U^{\sigma} \tag{7.6.22}$$

or, since

$$U^{\sigma} = \frac{\partial \xi^{\sigma}}{\partial x^{\mu}} V^{\mu}, \tag{7.6.23}$$

$$\frac{\partial V^{\mu}}{\partial x^{\lambda}} = \frac{\partial x^{\mu}}{\partial \xi^{\sigma}} \frac{\partial \xi^{\tau}}{\partial x^{\lambda}} \frac{\partial U^{\sigma}}{\partial \xi^{\tau}} + \frac{\partial}{\partial x^{\lambda}} \left(\frac{\partial x^{\mu}}{\partial \xi^{\sigma}} \right) \frac{\partial \xi^{\sigma}}{\partial x^{\nu}} V^{\nu} \tag{7.6.24}$$

$$= T^{\mu}{}_{\lambda} - \Gamma^{\mu}{}_{\lambda \nu} V^{\nu} \tag{7.6.25}$$

or

$$T^{\mu}{}_{\lambda} = \frac{\partial V^{\mu}}{\partial x^{\lambda}} + \Gamma^{\mu}{}_{\lambda \nu} V^{\nu}, \tag{7.6.26}$$

where

$$\Gamma^{\mu}{}_{\lambda \nu} = - \frac{\partial \xi^{\sigma}}{\partial x^{\nu}} \frac{\partial^{2} x^{\mu}}{\partial \xi^{\sigma} \partial \xi^{\tau}} \frac{\partial \xi^{\tau}}{\partial x^{\lambda}}. \tag{7.6.27}$$

$T^{\mu}{}_{\lambda}$ is called the covariant derivative of V^{μ}. It is the tensor that in a Minkowskian coordinate system is the tensor $\partial U^{\mu}/\partial \xi^{\lambda}$.

The extra term in (7.6.26) arises from the correct formulation of parallel displacement. If we displace the vector U^{μ} from ξ to $\xi + d\xi$, the new vector is still U^{μ}. However, in a general coordinate system, the first vector is

$$V^\mu = \frac{\partial x^\mu}{\partial \xi^\sigma} U^\sigma \tag{7.6.28}$$

and the displaced vector is

$$V^\mu + \delta V^\mu = \frac{\partial x^\mu}{\partial \xi^\sigma} (x + \delta x) U^\sigma$$

$$= V^\mu + \frac{\partial}{\partial x^\lambda} \left(\frac{\partial x^\mu}{\partial \xi^\sigma} \right) \delta x^\lambda U^\sigma$$

$$= V^\mu + \frac{\partial}{\partial x^\lambda} \left(\frac{\partial x^\mu}{\partial \xi^\sigma} \right) \delta x^\lambda \frac{\partial \xi^\sigma}{\partial x^\nu} V^\nu \tag{7.6.29}$$

or

$$\delta V^\mu = -\Gamma^\mu_{\lambda\nu} \delta x^\lambda V^\nu. \tag{7.6.30}$$

The covariant derivative subtracts the parallel displaced vector from the vector at the new point; this results in a covariant derivative, as given by (7.6.26).

The three-index quantity $\Gamma^\mu_{\lambda\nu}$ is called the affine connection. Although $\Gamma^\mu_{\lambda\nu}$ has tensor indices, it is not a tensor. For example, it vanishes in a Minkowskian coordinate system ($x = \xi$). We note here that $\Gamma^\mu_{\lambda\nu}$ is symmetric in λ and ν.

We introduce some convenient notation: The ordinary derivative of V^μ is written as

$$V^\mu{}_{,\lambda} \equiv \frac{\partial V^\mu}{\partial x^\lambda}. \tag{7.6.31}$$

The covariant derivative is written as

$$T^\mu{}_\lambda = V^\mu{}_{;\lambda}. \tag{7.6.32}$$

The rule for covariantly differentiating a higher-rank tensor follows trivially from the above procedure: There is one extra "kinematic" derivative for every $\partial x / \partial \xi$ in the transformation replacing (7.6.21). Therefore, one must replace the single extra term in (7.6.26) with a sum:

$$V^{\mu\nu}{}_{;\lambda} = \frac{\partial V^{\mu\nu}}{\partial x^\lambda} + \Gamma^\mu_{\lambda\sigma} V^{\sigma\nu} + \Gamma^\nu_{\lambda\sigma} V^{\mu\sigma} \tag{7.6.33}$$

etc. for higher tensors.

Finally, we derive the equivalent rule for a covariant index by noting that if A_μ and B^μ are vectors, then $A_\mu B^\mu$ is a scalar, $\partial_\lambda(A_\mu B^\mu)$ is a vector, and

$$\partial_\lambda(A_\mu B^\mu) = (\partial_\lambda A_\mu) B^\mu + A_\mu(\partial_\lambda B^\mu). \tag{7.6.34}$$

On the other hand, we can consider (7.6.34) in a Minkowskian coordinate system and then transform each term separately. The second will become

$$A_\mu B^\mu{}_{;\lambda} \tag{7.6.35}$$

and the first

$$A_{\mu;\lambda} B^\mu. \tag{7.6.36}$$

Therefore,

$$A_\mu(B^\mu{}_{;\lambda} - B^\mu{}_{,\lambda}) + B^\mu(A_{\mu;\lambda} - A_{\mu,\lambda}) = 0$$

or

$$A_\mu \Gamma^\mu_{\lambda\sigma} B^\sigma + B^\sigma(A_{\sigma;\lambda} - A_{\sigma,\lambda}) = 0$$

so that, since A and B are independent,

$$A_{\sigma;\lambda} = A_{\sigma,\lambda} - \Gamma^\mu_{\sigma\lambda} A_\mu. \tag{7.6.37}$$

We can express the Γ's in terms of the metric tensor by noting that the covariant derivative of $g_{\mu\nu}$ must be zero, since $g_{\mu\nu}$ is a tensor, and $g_{\mu\nu} = \eta_{\mu\nu}$, a constant, in a Minkowskian system. Therefore,

$$g_{\mu\nu;\lambda} = \frac{\partial g_{\mu\nu}}{\partial x^\lambda} - \Gamma^\tau_{\mu\lambda} g_{\tau\nu} - \Gamma^\tau_{\nu\lambda} g_{\tau\mu} = 0. \tag{7.6.38}$$

We can solve (7.6.38) for Γ. Interchange ν and λ in (7.6.38) and subtract. Then interchange μ and λ in (7.6.38) and subtract this from the first subtraction. There results

$$g_{\mu\nu,\lambda} - g_{\nu\lambda,\mu} - g_{\mu\lambda,\nu} + 2\Gamma^\tau_{\mu\nu} g_{\tau\lambda} = 0 \tag{7.6.39}$$

or

$$\Gamma^\sigma_{\mu\nu} = \frac{g^{\sigma\lambda}}{2}(g_{\nu\lambda,\mu} + g_{\mu\lambda,\nu} - g_{\mu\nu,\lambda}). \tag{7.6.40}$$

Thus, given the components of g, Γ can be calculated. We will see in the next section that we can find a coordinate transformation that takes any metric tensor to Minkowskian form at a point, with vanishing derivatives, and thus vanishing Γ. This *a posteriori* will justify the procedure of transforming from a Minkowskian system to derive the formulas for co-variant differentiation, whether or not the underlying space is Minkowskian.[8]

It is possible, but nontrivial, to derive (7.6.40) directly from (7.6.9).

We now know how to make any Lorentz covariant expression generally covariant. We simply replace all derivatives by covariant derivatives, and all d^4x integrals by $d^4x\sqrt{g}$ integrals.

There are some special cases worth noting:

1. $A_{\mu;\nu} - A_{\nu;\mu} = (A_{\mu,\nu} - \Gamma^{\tau}_{\mu\nu}A_{\tau}) - (A_{\nu,\mu} - \Gamma^{\tau}_{\nu\mu}A_{\tau}) = A_{\mu,\nu} - A_{\nu,\mu}.$

$$(7.6.41)$$

This is a very helpful formula in electrodynamics.

2. $F_{\mu\nu;\lambda} + F_{\lambda\mu;\nu} + F_{\nu\lambda;\mu}$ (for $F_{\mu\nu}$ antisymmetric)

$$= F_{\mu\nu,\lambda} - \Gamma^{\tau}_{\mu\lambda} F_{\tau\nu} - \Gamma^{\tau}_{\nu\lambda} F_{\mu\tau}$$

$$+ F_{\lambda\mu,\nu} - \Gamma^{\tau}_{\lambda\nu} F_{\tau\mu} - \Gamma^{\tau}_{\mu\nu} F_{\lambda\tau}$$

$$+ F_{\nu\lambda,\mu} - \Gamma^{\tau}_{\nu\mu} F_{\tau\lambda} - \Gamma^{\tau}_{\lambda\mu} F_{\nu\tau}; \qquad (7.6.42)$$

by virtue of the antisymmetry of F and the $\mu\nu$ symmetry of $\Gamma^{\tau}_{\mu\nu}$, all the Γ terms cancel, leaving

$$F_{\mu\nu;\lambda} + F_{\lambda\mu;\nu} + F_{\nu\lambda;\mu} = F_{\mu\nu,\lambda} + F_{\lambda\mu,\nu} + F_{\nu\lambda,\mu}. \qquad (7.6.43)$$

This is also a very helpful formula in electrodynamics.

3. The divergence of a vector (or tensor) $V^{\mu}{}_{;\mu}$. From (7.6.26),

$$V^{\mu}{}_{;\lambda} = \partial_{\lambda} V^{\mu} + \Gamma^{\mu}_{\lambda\nu} V^{\nu} \qquad \text{and} \qquad V^{\mu}{}_{;\mu} = \partial_{\mu} V^{\mu} + \Gamma^{\mu}_{\mu\nu} V^{\nu}.$$

$$(7.6.44)$$

From (7.6.40),

$$\Gamma^{\mu}_{\mu\nu} = \frac{g^{\mu\lambda}}{2} g_{\mu\lambda,\nu} \qquad (7.6.45)$$

which can in turn be calculated from the determinant $g = -\det(g_{\mu\nu})$:

$$g + \delta g = -[\det(g_{\mu\nu} + \delta g_{\mu\nu})] \qquad (7.6.46)$$

---

[8]That is, whether or not there is a coordinate system in which $g_{\mu\nu} = \eta_{\mu\nu}$ everywhere.

and

$$\frac{1}{g}(g + \delta g) = \det(g^{\sigma\lambda}) \det(g_{\mu\nu} + \delta g_{\mu\nu}) \tag{7.6.47}$$

$$= \det g^{\sigma\lambda}(g_{\lambda\nu} + \delta g_{\lambda\nu})$$

$$= \det(\delta^{\sigma}_{\nu} + g^{\sigma\lambda}\delta g_{\lambda\nu}) \tag{7.6.48}$$

so

$$\frac{\delta g}{g} = g^{\sigma\lambda}\delta g_{\lambda\sigma} \quad \text{and} \quad \Gamma^{\mu}_{\mu\nu} = \frac{1}{2g}\frac{\partial g}{\partial x^{\nu}}. \tag{7.6.49}$$

For $V^{\mu}_{\ :\mu}$ we then have

$$V^{\mu}_{\ :\mu} = \partial_{\mu}V^{\mu} + \frac{1}{2g}\frac{\partial g}{\partial x^{\nu}}V^{\nu} = \frac{1}{\sqrt{g}}\partial_{\mu}(\sqrt{g}\,V^{\mu}). \tag{7.6.50}$$

The divergence of a tensor follows in a similar way:

$$F^{\mu\nu}_{\ \ :\mu} = \frac{\partial F^{\mu\nu}}{\partial x^{\mu}} + \Gamma^{\mu}_{\mu\sigma}F^{\sigma\nu} + \Gamma^{\nu}_{\mu\sigma}F^{\mu\sigma}; \tag{7.6.51}$$

for the special case $F^{\mu\sigma} = -F^{\sigma\mu}$, the last term in (7.6.51) is zero, and we have, for the electromagnetic field tensor,

$$F^{\mu\nu}_{\ \ :\mu} = \frac{1}{\sqrt{g}}\partial_{\mu}(\sqrt{g}\,F^{\mu\nu}). \tag{7.6.52}$$

4. We can use (7.6.50) to derive the general expression for the n-dimensional Laplacian (or pseudo-Laplacian, if the space is Minkowskian rather than Euclidean):

$$\frac{\partial}{\partial\xi^{\lambda}}\eta^{\lambda\tau}\frac{\partial}{\partial\xi^{\tau}}\psi = \frac{1}{\sqrt{g}}\partial_{\mu}(\sqrt{g}\,g^{\mu\sigma}\partial_{\sigma}\psi) \tag{7.6.53}$$

where ψ is a scalar.

We can illustrate (7.6.53) with the familiar case of $\nabla^2\psi$ in orthogonal coordinates in three dimensions: With

$$(ds)^2 = h_1^2\,dq_1^2 + h_2^2\,dq_2^2 + h_3^2\,dq_3^2 = g_{\mu\nu}\,dq^{\mu}\,dq^{\nu} \quad \text{and} \quad \sqrt{g} = h_1h_2h_3, \tag{7.6.54}$$

we have

$$\nabla^2 \psi = \frac{1}{h_1 h_2 h_3} \left(\frac{\partial}{\partial q_1} h_1 h_2 h_3 \frac{1}{h_1^2} \frac{\partial}{\partial q_1} + \cdots \right). \tag{7.6.55}$$

7.7. LOCAL TRANSFORMATION TO A PSEUDO-EUCLIDEAN SYSTEM

We will now show that any symmetric tensor that is analytic in the neighborhood of a point can be transformed to pseudo-Euclidean form at that point. Let $g_{\mu\nu}$ be that tensor and $x_0 = 0$ that point. The theorem states that we can find a coordinate system in which $g_{\mu\nu}(x_0) = \eta_{\mu\nu}$ and $\partial g_{\mu\nu}/\partial x_\lambda|_{x=x_0} = 0$. Here, $\eta_{\mu\nu}$ is the appropriate pseudo-Euclidean tensor, in that it must have the same number of positive and negative eigenvalues (all ± 1 in this case) as $g_{\mu\nu}$. As was noted earlier, this shows that the existence of a coordinate system in which the tensor $g_{\mu\nu} = \eta_{\mu\nu}$ *everywhere* is not necessary for the arguments of Section 7.6. We start in a coordinate system with metric $g_{\mu\nu}(x)$. In the neighborhood of $x = 0$,

$$g_{\mu\nu} = g_{\mu\nu}(0) + \partial_\lambda g_{\mu\nu}(0) \, x^\lambda. \tag{7.7.1}$$

We transform to a new coordinate system

$$x'^\mu = \alpha^\mu_\nu x^\nu + \frac{\beta^\mu_{\nu\lambda} x^\nu x^\lambda}{2} + \cdots \quad \text{and} \quad x^\mu = a^\mu_\nu x'^\nu + \frac{b^\mu_{\nu\lambda} x'^\nu x'^\lambda}{2} + \cdots \tag{7.7.2}$$

for x and x' close to zero. The equation for $g'_{\mu\nu}$ is

$$g'_{\mu\nu} = \frac{\partial x^\lambda}{\partial x'^\mu} \frac{\partial x^\sigma}{\partial x'^\nu} g_{\lambda\sigma}$$

or

$$g'_{\mu\nu} = (a^\lambda_\mu + b^\lambda_{\mu\alpha} x'^\alpha)(a^\sigma_\nu + b^\sigma_{\nu\delta} x'^\delta)[g_{\lambda\sigma}(0) + \partial_\alpha g_{\lambda\sigma}(0) \, a^\alpha_\beta x'^\beta]. \tag{7.7.3}$$

We must now solve for a's and b's that make $g'_{\mu\nu}(0) = \eta_{\mu\nu}$ and $\partial g'_{\mu\nu}/\partial x'^\lambda = 0$. The first condition specifies 10 values of $g'_{\mu\nu}(0)$, and there are 16 a^λ_μ's. Thus, six parameters are left over, corresponding to the six parameters of a Lorentz transformation, which are undetermined, since $\eta_{\mu\nu}$ is left invariant by a Lorentz transformation. There are 40 $\partial_\lambda g_{\mu\nu}$'s and 40 $b^\mu_{\lambda\nu}$'s, so that there are just enough conditions to determine the local transformation $\partial x'/\partial x$ up to a Lorentz transformation. We must still show that the equations have a solution. We divide the proof into three parts. First, we diagonalize $g'_{\mu\nu}$ at $x = 0$. Since $g_{\mu\nu}$ is a real, symmetric tensor, we can diagonalize it by an orthogonal transformation. An orthogonal transformation can always be written as e^A, where A is real and antisymmetric; hence, there are six independent parameters that are determined by this process. The metric will now locally be of the form $g_{\mu\nu} = \delta_{\mu\nu}\lambda^{(\mu)}$, where the $\lambda^{(\mu)}$ are the eigenvalues of the original matrix. We can now reduce the λ's to ± 1 by a scale change, determining four more parameters. It must be the case that three of the eigenvalues of g are negative and one positive; otherwise, we cannot transform to $\eta_{\mu\nu}$ without a singular transformation.

The remaining 40 equations are expanded to first order in x', with $a^\lambda_\mu = \delta^\lambda_\mu$ and $g_{\lambda\sigma}(0) = \eta_{\lambda\sigma}$:

$$g'_{\mu\nu} = \eta_{\mu\nu} + x'^\rho[b^\lambda_{\mu\rho}\,\eta_{\lambda\nu} + b^\lambda_{\nu\rho}\,\eta_{\lambda\mu} + \partial_\rho g_{\mu\nu}(0)], \qquad (7.7.4)$$

and therefore,

$$b^\lambda_{\mu\rho}\,\eta_{\lambda\nu} + b^\lambda_{\nu\rho}\,\eta_{\lambda\mu} + \partial_\rho g_{\mu\nu} = 0. \qquad (7.7.5)$$

The solution to Eq. (7.7.5) is easily seen to be [repeating the work leading to (7.6.40)]

$$b^\lambda_{\mu\rho} = -\frac{\eta^{\lambda\tau}}{2}(\partial_\rho g_{\mu\tau} + \partial_\mu g_{\rho\tau} - \partial_\tau g_{\mu\rho}). \qquad (7.7.6)$$

If we try to go to one higher power of x, that is, make all second derivatives vanish, we will, in general, fail. The number of conditions is now the vanishing of $\partial_\lambda\partial_\sigma g_{\mu\nu}$, or 10 second derivatives of a 10-component tensor; thus, 100 conditions. The coordinate transformation is $x^\mu = a^\mu_{\nu\lambda\eta}x'^\nu x'^\lambda x'^\eta$; the available number of transformation parameters is four (for $\mu = 0, 1, 2, 3$) times the number of components of a symmetric tensor of rank ℓ (with $\ell = 3$) in four dimensions. This number is calculated in Appendix B. It is given by (B.2.7):

$$S(\ell, 4) = \sum_{n_1=0}^{\ell} \sum_{n_2=0}^{\ell-n_1} \sum_{n_3=0}^{\ell-n_1-n_2} 1 \qquad (7.7.7)$$

$$= \frac{(\ell + 1)(\ell + 2)(\ell + 3)}{6} \qquad (7.7.8)$$

and for $\ell = 3$, $S(3, 4) = 20$. Thus, there are $4 \times 20 = 80$ adjustable parameters in the transformation, and $10 \times 10 = 100$ equations to satisfy. This leaves 20 combinations of second derivatives that cannot, in general, be set equal to zero. As we shall see in the next chapter, this is just the number of components of the curvature tensor in four dimensions. (See also Problem 8.6.)

In three dimensions the number of nonzero second derivatives is six; in two it is one, and in one dimension it is zero. In each case, this number is the number of components of the curvature tensor.

We see that we cannot, in general, carry out a local coordinate transformation to a pseudo-Euclidean coordinate system, up to vanishing second derivatives of $g_{\mu\nu}$, although we can do so, up to vanishing of all first derivatives of $g_{\mu\nu}$.

The metric tensor can therefore inform us of intrinsic properties of the space: for example, as just seen, the impossibility of finding a coordinate system for which the metric is pseudo-Euclidean. We recall here two examples with which the reader is surely familiar: an invariant interval

$$(ds)^2 = (dr)^2 + r^2(d\theta)^2 + r^2 \sin^2 \theta (d\varphi)^2, \qquad (7.7.9)$$

corresponding to a diagonal metric tensor

$$g_{rr} = 1, \qquad g_{\theta\theta} = r^2, \qquad g_{\varphi\varphi} = r^2 \sin^2 \theta, \qquad (7.7.10)$$

will permit a transformation

$$z = r \cos \theta, \qquad x = r \sin \theta \cos \varphi, \qquad y = r \sin \theta \sin \varphi \qquad (7.7.11)$$

which expresses $(ds)^2$ as

$$(ds)^2 = (dx)^2 + (dy)^2 + (dz)^2 \qquad (7.7.12)$$

for *all* r, θ, φ.

A space of two variables, θ and φ, with $0 \le \theta \le \pi$ and $0 \le \varphi \le 2\pi$, and invariant interval

$$ds^2 = R^2[(d\theta)^2 + \sin^2 \theta (d\varphi)^2] \qquad (7.7.13)$$

will permit no such transformation. Of course, (7.7.13) describes a spherical surface embedded in three-dimensional Euclidean space. We will see later, when we discuss curvature, precisely how the metric tensor determines intrinsic geometry.

7.8. ALTERNATIVE CONSTRUCTION OF A COVARIANTLY CONSERVED, SYMMETRIC STRESS-ENERGY TENSOR

The method takes advantage of the possibility of writing a Lorentz-invariant action in a generally invariant form. Thus, we introduce a symmetric tensor $g_{\mu\nu}$ and rewrite the special relativistic action with the substitutions

$$d^4x \rightarrow \sqrt{g}\, d^4x \tag{7.8.1}$$

$$A_{\mu,\nu} \rightarrow A_{\mu;\nu}, \tag{7.8.2}$$

etc.

The action S is now a general invariant, so a general coordinate transformation does not change it. The main point is then the following: A general, but infinitesimal coordinate transformation changes all the dynamical variables. However, provided y_p^μ, A_μ, etc. satisfy the Lagrange equations, the infinitesimal variations δy_p^μ, δA_μ, etc. will leave the action invariant. Therefore, the only interesting consequence of the transformation is the change of $g_{\mu\nu}$, which is *not* a Lagrangian variable; this change alone must therefore leave S invariant. We shall see that the statement $\delta S = 0$ is equivalent to the conservation of a specific symmetric tensor $T^{\mu\nu}$.

The algorithm is the following. Write S in generally invariant form; then make an infinitesimal variation in $g_{\mu\nu}$, $g_{\mu\nu} \rightarrow g_{\mu\nu} + \delta g_{\mu\nu}$. $T^{\mu\nu}$ is given by the equation

$$\delta S = -\frac{1}{2} \int d^4x \sqrt{g}\, T^{\mu\nu} \delta g_{\mu\nu}, \tag{7.8.3}$$

which obviously defines a symmetric tensor. We will show that $T^{\mu\nu}$ is conserved:

$$T^{\mu\nu}{}_{;\mu} = 0 \tag{7.8.4}$$

is a correct equation with $_{;\mu}$ representing the covariant divergence. All we need for the special relativistic case is this equation with $g_{\mu\nu}$ set equal to $\eta_{\mu\nu}$, where (7.8.4) would become

$$T^{\mu\nu}{}_{,\nu} = \partial_\nu T^{\mu\nu} = 0. \tag{7.8.5}$$

Note that (7.8.3) would change sign were we to choose the metric $\eta_{\mu\nu}$ with positive space components instead of a positive time component.

Consider as an example the action for charged particles with charge

q_p interacting with the electromagnetic field. We know the action

$$S = -\sum_p \frac{m_p}{2} \int \frac{dx_p^\mu}{d\sigma_p} \frac{dx_p^\nu}{d\sigma_p} \eta_{\mu\nu} \, d\sigma_p - \frac{1}{4} \int d^4x F_{\mu\nu} F^{\mu\nu}$$

$$+ \sum_p q_p \int \frac{dy_p^\mu}{d\sigma_p} A_\mu(y_p) \, d\sigma_p \qquad (7.8.6)$$

where σ_p is an invariant parameter associated with the pth charged particle. We remember that this choice of S is *not* reparametrization-invariant,
so that $d\sigma_p$ will be determined by the equation of motion. It is proportional
to $d\tau_p$; choosing $d\sigma = d\tau$ makes m_p the observed particle mass.

We can easily write a generally covariant form for (7.8.6):

$$S_g = -\sum_p \frac{m_p}{2} \int d\sigma_p \, g_{\mu\nu} \frac{dx_p^\mu}{d\sigma_p} \frac{dx_p^\nu}{d\sigma_p} - \frac{1}{4} \int d^4x \sqrt{g} F_{\mu\nu} g^{\mu\sigma} g^{\nu\lambda} F_{\sigma\lambda}$$

$$+ \sum_p q_p \int d\sigma_p \frac{dy_p^\mu}{d\sigma_p} A_\mu(x_p). \qquad (7.8.7)$$

We make a small change in $g_{\mu\nu}$:

$$g_{\mu\nu} \to g_{\mu\nu} + \delta g_{\mu\nu} \qquad (7.8.8)$$

and from

$$\delta(g^{\mu\lambda} g_{\lambda\nu}) = \delta(\delta_\nu^\mu) = 0,$$

observe

$$\delta g^{\mu\sigma} = -g^{\mu\lambda} \delta g_{\lambda\eta} g^{\eta\sigma} \qquad (7.8.9)$$

and recall from (7.6.49)

$$\frac{\delta g}{g} = g^{\mu\nu} \delta g_{\mu\nu}. \qquad (7.8.10)$$

The result is

$$\delta S = -\sum_p \frac{m_p}{2} \int d\sigma_p \int d^4x \delta^4(x - y_p(\sigma_p)) \frac{dy_p^\mu}{d\sigma_p} \frac{dy_p^\nu}{d\sigma_p} \delta g_{\mu\nu}(x)$$

$$- \frac{1}{8} \int d^4x \sqrt{g} g^{\mu\nu} \delta g_{\mu\nu} F_{\lambda\sigma} F^{\lambda\sigma}$$

$$+ \frac{1}{4} \int d^4x \sqrt{g} \delta g_{\mu\nu} (F^\mu{}_\lambda F^{\nu\lambda} + F^{\lambda\mu} F_\lambda{}^\nu), \qquad (7.8.11)$$

so that, setting $g^{\mu\nu} = \eta^{\mu\nu}$, we find

$$T^{\mu\nu} = \sum_p m_p \int d\sigma_p \frac{dx_p^\mu}{d\sigma_p} \frac{dx_p^\nu}{d\sigma_p} \delta^4(x - y_p(\sigma_p))$$

$$+ \frac{1}{4}[\eta^{\mu\nu}F_{\lambda\sigma}F^{\lambda\sigma} - 2(F^\mu{}_\lambda F^{\nu\lambda} + F^{\lambda\mu}F_\lambda{}^\nu)] \qquad (7.8.12)$$

Note that the last two terms are equal and symmetric in μ and ν. With $d\sigma_p = d\tau_p$,

$$\int d\sigma_p \delta^4(x - y_p) = \int d\tau_p \delta^4(x - y_p(\tau)) = \frac{d\tau_p}{dx_p^0} \delta^3(x - y_p)$$

$$= \sqrt{1 - v_p^2}\,\delta^3(x - y_p) \qquad (7.8.13)$$

so that

$$T^{\mu\nu} = \sum_p m_p \sqrt{1 - v_p^2} \frac{dy_p^\mu}{d\tau_p} \frac{dy_p^\nu}{d\tau_p} \delta^3(x - y_p) + \frac{1}{4}\eta^{\mu\nu}F_{\lambda\sigma}F^{\lambda\sigma} - F^\mu{}_\lambda F^{\nu\lambda},$$

$$(7.8.14)$$

in agreement with our earlier result (7.5.23).

We call attention to an important property of (7.8.14), the vanishing trace of the electromagnetic stress tensor:

$$T_\mu{}^\mu = \sum_p m_p \sqrt{1 - v_p^2}\,\delta^3(x - x_p) + F_{\lambda\sigma}F^{\lambda\sigma} - F_{\mu\lambda}F^{\mu\lambda}$$

$$= \sum_p m_p \sqrt{1 - v_p^2}\,\delta^3(x - y_p). \qquad (7.8.15)$$

Note also that the particle contribution to (7.8.15) can be written as

$$\sum_p \frac{m_p^2}{E_p} \delta^3(x - x_p)$$

and therefore vanishes like m_p^2 as $m_p \to 0$, E_p remaining finite. In the case of a massive vector field, with a term

$$\delta\mathcal{L} = \frac{\mu^2}{2} A_\alpha A^\alpha$$

added to the Lagrangian density, the trace of the vector field stress tensor no longer vanishes.

We show that (7.8.3) defines a conserved $T^{\mu\nu}$. Assume

$$S = \int d^4x \sqrt{g(x)} \, \mathcal{L}(\psi_\sigma(x), x, y_p) \qquad (7.8.16)$$

where $\mathcal{L}$ is an invariant density and ψ_σ are the dynamical variables in the Lagrangian, including the metric $g_{\mu\nu}$. The meaning of the ψ_σ is clear enough for A_μ, $g_{\mu\nu}$, etc.; for the particle variables, we would write

$$S_p = -\frac{m}{2} \int d\sigma \, \frac{dy_p^\mu}{d\sigma_p} g_{\mu\nu}(y_p) \frac{dy_p^\nu}{d\sigma_p}$$

$$= -\frac{m}{2} \int d^4x \sqrt{g} \, d\sigma_p \frac{dy_p^\mu}{d\sigma_p} g_{\mu\nu}(x) \frac{dy_p^\nu}{d\sigma_p} \frac{\delta^4(x - y_p)}{\sqrt{g}}. \qquad (7.8.17)$$

The action S is invariant under a general coordinate transformation $x^\sigma = x^\sigma(x')$, under which

$$d^4x \to d^4x', \qquad g(x) \to g'(x'), \qquad A_\mu \to A'_\mu(x') = \frac{\partial x^\sigma}{\partial x'^\mu} A_\sigma(x),$$

etc., so

$$S = \int d^4x' \sqrt{g'(x')} \, \mathcal{L}(A'_\mu(x'), g'_{\mu\nu}(x'), x', y'_p, \ldots). \qquad (7.8.18)$$

The integration variable x' in (7.8.18) can be changed to x without changing the value of S. Thus, in one dimension,

$$\int dx \, f(x) = \int dx' f(x') \qquad (7.8.19)$$

with x' and x single-valued functions of each other with the same end points, and f an arbitrary function. Similarly,

$$S = \int d^4x \sqrt{g'(x)} \, \mathcal{L}^T(x). \qquad (7.8.20)$$

The superscript $\mathcal{L}^T$ instructs us to calculate $\mathcal{L}$ as a function of the transformed field, but with x' set equal to x. For example, if ψ is a scalar field and $x = x' + \delta x$, then

$$\psi'(x') = \psi(x)$$

and

$$\psi'(x) = \psi'(x') + \psi'(x) - \psi'(x')$$

$$= \psi(x) + \frac{\partial \psi}{\partial x^\lambda} \delta x^\lambda \qquad (7.8.21)$$

to first order in δx; similar definitions hold for the other variables.

We now return to (7.8.20). As shown above, we replace x' by x, which leaves S unchanged. Next, we expand S in the first-order changes in A_μ, y_p^μ, etc. Since A_μ, y_p^μ, etc. obey the Lagrangian equations, their first-order changes leave S invariant. The only change that could affect S is in $g_{\mu\nu}$ with

$$g'_{\mu\nu}(x) = g'_{\mu\nu}(x') + g'_{\mu\nu}(x) - g'_{\mu\nu}(x') \qquad (7.8.22)$$

$$= \frac{\partial x^\sigma}{\partial x'^\mu} \frac{\partial x^\lambda}{\partial x'^\nu} g_{\sigma\lambda} + \left(\frac{\partial}{\partial x^\sigma} g_{\mu\nu} \right) \delta x^\sigma \qquad (7.8.23)$$

$$= \left(\delta_\mu^\sigma + \frac{\partial \delta x^\sigma}{\partial x^\mu} \right) \left(\delta_\nu^\lambda + \frac{\partial \delta x^\lambda}{\partial x^\nu} \right) g_{\sigma\lambda} + \delta x^\sigma \partial_\sigma g_{\mu\nu}$$

$$= g_{\mu\nu} + \frac{\partial \delta x^\sigma}{\partial x^\mu} g_{\sigma\nu} + \frac{\partial \delta x^\lambda}{\partial x^\nu} g_{\mu\lambda} + \delta x^\sigma \partial_\sigma g_{\mu\nu} \qquad (7.8.24)$$

still accurate to first order in δx. However, this transformation does not affect S, since S is invariant. Thus, we have from the definition (7.8.3)

$$\delta S \equiv -\frac{1}{2} \int d^4x \sqrt{g}\, T^{\mu\nu} \delta g_{\mu\nu} \qquad (7.8.25)$$

$$= -\frac{1}{2} \int d^4x \sqrt{g}\, T^{\mu\nu} \left(\frac{\partial \delta x^\sigma}{\partial x^\mu} g_{\sigma\nu} + \frac{\partial \delta x^\sigma}{\partial x^\nu} g_{\mu\sigma} + \delta x^\sigma \frac{\partial}{\partial x^\sigma} g_{\mu\nu} \right)$$

$$= \frac{1}{2} \int d^4x\, \delta x^\sigma \{ \partial_\mu(g_{\sigma\nu} \sqrt{g}\, T^{\mu\nu}) + \partial_\nu(g_{\sigma\mu} \sqrt{g}\, T^{\mu\nu}) - (\partial_\sigma g_{\mu\nu}) \sqrt{g}\, T^{\mu\nu} \}$$

$$= 0. \qquad (7.8.26)$$

Since the integral in (7.8.26) is generally invariant, (7.8.26) can be written as

$$0 = \int d^4x \sqrt{g}\, \delta x^\sigma Q_\sigma \qquad (7.8.27)$$

where Q_σ is a vector under general coordinate transformations. Since locally, in the ξ system, $Q_\sigma = (\partial/\partial\xi^\mu)T^\mu{}_\sigma$, Q_σ must be the vector:

$$Q_\sigma = T^\mu{}_{\sigma\,:\mu}. \qquad (7.8.28)$$

Equation (7.8.27) then shows that $T^\mu{}_\sigma$, and hence $T^{\mu\sigma}$, is conserved.

CHAPTER 7 PROBLEMS

7.1 Consider a particle of mass m moving in an external scalar potential ψ. Consider the action

$$S = \int L\,dt$$

with $L = -(m + g\psi)\sqrt{1 - \mathbf{v}^2}$ and g a dimensionless coupling constant.

(a) Derive the equations of motion for $\mathbf{x}(t)$.

(b) Show that they form the space components of a consistent four-vector equation.

7.2 Consider next the action

$$S = \int L'\,d\sigma$$

with $L' = -(m + g\psi)\,dy^\mu/d\sigma\,dy_\mu/d\sigma$ and σ an invariant parameter replacing time.

(a) Derive the equations of motion for $y^\mu(\sigma)$.

(b) Show that $(m + g\psi)\,dy^\mu/d\sigma\,dy_\mu/d\sigma$ is constant and that the constant may be freely chosen without affecting the equations of motion.

(c) Compare the equations derived from L' with those derived from L in Problem 7.1 and show that there are no scale changes of g and m that bring them into agreement. Thus, if we assume that we complete the scalar theory with the usual field Lagrangian for ψ, the two Lagrangians L and L' lead to different theories.

(d) Show, however, that there is a nonlinear transformation of the field ψ in the Lagrangian L' that will bring it into agreement with the Lagrangian L. This shows that with a *given* external field ψ, one can use either Lagrangian for determining the particle motion, providing one makes the appropriate transformation of ψ.

*7.3 Complete the construction of a scalar field theory outlined in the text: (7.4.22–7.4.27).

(a) Derive the Lagrangian equations for the field $\psi(x)$ and the particle coordinates y^μ.

(b) In the non-relativistic limit ($g\psi/m \ll 1$, $\partial\psi/\partial t \ll |\nabla\psi|$, $d\sigma \approx dy^0$), show that the interparticle force is attractive between like particles. (Do not try to calculate the force of the particle on itself.)

7.4 Verify that $\Theta^\mu{}_\nu$ as given by (7.4.28) is conserved, $\partial_\mu\Theta^\mu{}_\nu = 0$, and $P_\nu = \int \Theta^0{}_\nu \, d^3x$ is correctly given by (7.4.21).

*7.5 From the action for an electromagnetic field interacting with a charged particle,

$$S = \int d^4x \left\{ -\frac{1}{4}F_{\mu\nu}F^{\mu\nu} \right\} + q \int d\sigma \frac{dy^\mu}{d\sigma} A_\mu(y) - \frac{m}{2} \int d\sigma \frac{dy^\mu}{d\sigma}\frac{dy_\mu}{d\sigma},$$

where $A_\mu(x)$ and y^μ are the dynamical variables:

(a) Construct the canonical stress tensor $\Theta^{\mu\nu}$ given by (7.4.29).

(b) Construct from it the symmetric stress tensor $T^{\mu\nu}$ as defined by (7.4.37).

(c) Show that in the $\mu^2 = 0$ case the energy $P^0 = \int T^{00} \, d^3x$ is given by (7.5.24).

*7.6 Consider a Lagrangian density $\mathscr{L}$ that is a function of a set of fields $\psi_\alpha(x)$ and their first and second space-time derivatives, $\partial_\mu\psi_\alpha(x)$ and $\partial_\mu\partial_\nu\psi_\alpha(x)$.

(a) Derive the Lagrange equations for this case.

(b) Assuming that $\mathscr{L}$ has no explicit x dependence, apply the Noether procedure [as in (7.4.9–7.4.17)] to find an expression for the energy-momentum four-vector P^α.

(c) P^α has the form

$$P^\alpha = \int d^3x \, \Theta^{0\alpha}$$

where $\Theta^{0\alpha}$ is the 0–α component of a tensor $\Theta^{\beta\alpha}$. Construct the tensor $\Theta^{\beta\alpha}$ and show that it is locally conserved: $\partial_\beta\Theta^{\beta\alpha} = 0$.

(d) Construct a single invariant Lagrangian of the form $\partial_\mu A^\mu$, with A^μ a function of a single scalar field ψ and its first derivatives. Imagine that this function is added to an existing Lagrangian. Verify that the Lagrange equations are unchanged by the addition. Show that the tensor $\Theta^{\mu\nu}$ is changed, symmetric, and conserved, but does not contribute a change to the energy-momentum vector P^μ. (The fact that $\Theta^{\mu\nu}$ is symmetric will not, in general, be true for more complex theories.)

*7.7 A Lagrangian density $\mathcal{L}$ is changed by adding a $\Delta\mathcal{L} = \partial_\mu A^\mu(\psi_\sigma, \partial_\lambda \psi_\sigma)$, where the ψ_σ are the set of fields described by the Lagrangian.

   In order to simplify the problem, choose A^μ so that the $\Delta\mathcal{L}$ (as well as $\mathcal{L}$) depends only on the fields and their first derivatives.

   (a) Find the condition on A^μ such that $\Delta\mathcal{L}$ not depend on second derivatives.

   (b) A^μ (and $\mathcal{L}$) are taken to be independent of x^μ except through the x^μ dependence of the fields. Show that the Lagrange equations for the fields are unchanged by the addition of $\Delta\mathcal{L}$.

   (c) Construct the change in the canonical stress-energy tensor $\Delta\Theta^{\alpha\beta}$ and show that it is conserved (but *not* zero).

   (d) Show, however, that the change in the energy-momentum four-vector

$$\Delta P^\beta = \int d^3x\, \Delta\Theta^{0\beta}$$

   is zero.

*7.8 The action that yields the model of Problem 3.2 is

$$S = \int d^3x\, dt\, \mathcal{L} + \int dt\, L_1(y, \dot{y})$$

where

$$\mathcal{L} = \frac{1}{2}(\nabla\varphi + \dot{\mathbf{A}})^2 - \left(\frac{\nabla \times \mathbf{A}}{2}\right)^2 - q[\phi(\mathbf{x}, t) - \dot{\mathbf{y}} \cdot \mathbf{A}(\mathbf{x}, t)]f(\mathbf{x} - \mathbf{y})$$

and $L_1 = \frac{1}{2}m\dot{y}^2$.
   Define the field Lagrangian $L = \int \mathcal{L}\, d^3x$.

   (a) Show that the above action yields the equations of Problem 3.2.

   (b) Show that, provided $f(\mathbf{x} - \mathbf{y}) = f(|\mathbf{x} - \mathbf{y}|)$, the Lagrangian $L + L_1$ is invariant under a rotation; that is, a transformation

$$\phi'(\mathbf{x}') \equiv \phi(\mathbf{x}(\mathbf{x}'))$$

   where to first order in $\boldsymbol{\epsilon}$

$$\mathbf{x}' = \mathbf{x} + \boldsymbol{\epsilon} \times \mathbf{x}$$

so

$$\phi' = \phi - \boldsymbol{\epsilon} \times \mathbf{x} \cdot \nabla\phi$$

$$\mathbf{A}' = \mathbf{A} + \boldsymbol{\epsilon} \times \mathbf{A} - \boldsymbol{\epsilon} \times \mathbf{x} \cdot \nabla\mathbf{A}$$

and

$$\mathbf{y}' = \mathbf{y} + \boldsymbol{\epsilon} \times \mathbf{y}.$$

(c) Use Noether's theorem to construct the conservation law emerging from this symmetry. The conserved quantity $\delta Q = \boldsymbol{\epsilon} \cdot \mathbf{L}$; the axial vector $\mathbf{L}$ is a conserved angular momentum.

(d) Show that $\mathbf{L} = \mathbf{L}_{\text{e.m.}} + \mathbf{L}_{\text{mech}} + \mathbf{L}_{\text{int}}$, where $\mathbf{L}_{\text{e.m.}}$ is the electromagnetic angular momentum:

$$\mathbf{L}_{\text{e.m.}} = \frac{1}{4\pi} \int d^3x\, \mathbf{r} \times (\mathbf{E} \times \mathbf{B}),$$

$\mathbf{L}_{\text{mech}}$ the particle angular momentum:

$$\mathbf{L}_{\text{mech}} = \sum \mathbf{y}_i \times m_i \dot{\mathbf{y}}_i,$$

and $\mathbf{L}_{\text{int}}$ an "interaction" angular momentum:

$$\mathbf{L}_{\text{int}} = q \int (\mathbf{y} - \mathbf{r}) \times \mathbf{A} f(\mathbf{r} - \mathbf{y}).$$

(e) Show that $\mathbf{L}_{\text{int}}$ is gauge-invariant [only, of course, when $f(\mathbf{r} - \mathbf{y})$ is invariant under rotations: $f = f(|\mathbf{r} - \mathbf{y}|)$.

*7.9 Show that the Lagrange density

$$\mathcal{L} = -\frac{1}{2}(\partial_\mu A_\nu \partial^\mu A^\nu - \alpha \partial_\mu A^\mu \partial_\nu A^\mu)$$

does not lead to a positive energy except for $\alpha = 1$.

(a) Do this in two steps. First, with $\alpha = 1$, show that the A^0 equation is an equation of constraint: $\partial_i(\partial^i A^0 - \partial^0 A^i) = 0$.

(b) From this, using the technique that led to (7.5.19), show that the energy integral $P^0 = \int \Theta^{00}\, d^3x$ is positive-definite.

(c) Now take $\alpha \neq 1$. The equation for A^0 is no longer a constraint equation, since it gives the $\ddot{A}^0$ in terms of the fields (potentials) and their first derivatives. Show that the energy integral is now unbounded in both positive and negative directions.

7.10 Consider a freely propagating electromagnetic field. Let the vector potential have a well-behaved Fourier transform $a_\mu(k)$. $a_\mu(k)$ satisfies the Maxwell equations

$$k^2 a_\mu - k_\mu k^\lambda a_\lambda = 0.$$

(a) Show from this that if $k^2 = k_0^2 - \mathbf{k}^2 \neq 0$, the field Fourier transform $k_\mu a_\nu - k_\nu a_\mu$ is zero.

(b) If $k^2 = 0$, evidently $k^\lambda a_\lambda = 0$, and we are in a Lorentz gauge. Clearly, the requirement that the Fourier transform for A be well behaved is a strong condition.

7.11 (a) Consider a flat two-dimensional space with polar coordinates $x^\rho = \rho$ and $x^\varphi = \varphi$ which we define to be contravariant. The invariant distance squared is

$$(ds)^2 = (d\rho)^2 + \rho^2 (d\varphi)^2.$$

Find $g_{\mu\nu}$ ($\mu, \nu = 1, 2$), $g^{\mu\nu}$, and express the invariant $\nabla^2 \psi = (g^{\mu\nu} \partial_\nu \psi)_{;\mu}$ in terms of ρ, and derivatives of the scalar ψ with respect to ρ and φ.

(b) What are the covariant coordinates x_ρ and x_φ?

(c) Repeat the exercise for spherical coordinates in three dimensions. There

$$(ds)^2 = (dr)^2 + r^2 (d\theta)^2 + r^2 \sin^2 \theta (d\varphi)^2.$$

7.12 Consider a vector $V = \hat{\mathbf{i}} V_x + \hat{\mathbf{j}} V_y + \hat{\mathbf{k}} V_z$, whose spherical components are $\overline{V}_r$. $\overline{V}_\theta$, and $\overline{V}_\varphi$. That is,

$$V = \hat{\mathbf{r}} \overline{V}_r + \hat{\theta} \overline{V}_\theta + \hat{\varphi} \overline{V}_\varphi$$

where $\hat{\mathbf{r}}$, $\hat{\theta}$, and $\hat{\varphi}$ are unit vectors in the corresponding directions.

(a) Give $\overline{V}_r$, $\overline{V}_\theta$, and $\overline{V}_\varphi$ in terms of the rectangular components of **V** and **r**.

(b) Give the covariant components of **V**: V_r, V_θ, and V_φ in terms of $\overline{V}_r$, $\overline{V}_\theta$, $\overline{V}_\varphi$.

(c) The same as (b), but give contravariant components.

*7.13 We consider here a pair of scalar fields, ψ_1 and ψ_2, with a Lagrangian

$$\mathcal{L}_\psi = \frac{1}{2} \partial_\mu \psi_1 \partial^\mu \psi_1 + \frac{1}{2} \partial_\mu \psi_2 \partial^\mu \psi_2 - V(\psi_1^2 + \psi_2^2).$$

We note the symmetry of $\mathcal{L}_\psi$ under a rotation like mixing of ψ_1 and ψ_2, that is,

$$\psi'_1 = \psi_1 \cos \alpha + \psi_2 \sin \alpha, \qquad \psi'_2 = -\psi_1 \sin \alpha + \psi_2 \cos \alpha.$$

The infinitesimal symmetry is

$$\delta \psi_1 = \delta \alpha \psi_2, \qquad \delta \psi_2 = -\delta \alpha \psi_1.$$

The Noether charge is

$$\delta Q = \int \left\{ \frac{\partial \mathscr{L}}{\partial(\partial_0 \psi_1)} \delta\psi_1 + \frac{\partial \mathscr{L}}{\partial(\partial_0 \psi_2)} \delta\psi_2 \right\} d^3x$$

which suggests a locally conserved current

$$j^\mu = (\psi_2 \partial^\mu \psi_1 - \psi_1 \partial^\mu \psi_2).$$

A more convenient representation is obtained by introducing a complex field

$$\psi = \frac{\psi_1 + i\psi_2}{\sqrt{2}}$$

and its complex conjugate

$$\psi^* = \frac{\psi_1 - i\psi_2}{\sqrt{2}},$$

in terms of which

$$\mathscr{L}_\psi = \partial_\mu \psi^* \partial^\mu \psi - V(\psi^* \psi),$$

and the invariance described above becomes the invariance of $\mathscr{L}$ under the phase transformation

$$\psi' = e^{i\alpha}\psi$$

with real, constant α.

(a) Show that the independent variation $\delta\psi_1$ and $\delta\psi_2$ can be replaced by independent variation $\delta\psi$ and $\delta\psi^*$; then verify that the Lagrange equations are the same.

(b) Find the Noether current arising from the phase invariance $\psi' = e^{i\alpha}\psi$.

The phase transformation is "gauged" by permitting α to be space-time-dependent and introducing a vector field A_μ into

$$\mathscr{L}_\psi = (\partial^\mu + ieA^\mu)\,\psi^*(\partial_\mu - ieA_\mu)\psi - V(\psi^*\psi). \tag{1}$$

Evidently, $\mathscr{L}_\psi$ is invariant under the combined transformations

$$\psi' = e^{ie\alpha}\psi, \qquad \psi'^* = e^{-ie\alpha}\psi^*, \qquad \text{and} \qquad A'_\mu = A_\mu + \partial_\mu \alpha. \tag{2}$$

One must add the Lagrangian $\mathscr{L}_A$ of the vector field to have a complete theory of the interaction of the A and ψ fields. Of course, $\mathscr{L}_A$ must also be invariant under the gauge transformation of A for the symmetry to hold. However, independent of $\mathscr{L}_A$,

(c) There is a conserved current for the new $\mathscr{L}_\psi$. It is

$$j^\mu = \frac{\partial \mathcal{L}_\psi}{\partial A_\mu}. \tag{3}$$

Show by explicit calculation that this current is conserved for ψ and ψ^* satisfying their equations of motion.

(d) Show that the gauge invariance of the Lagrangian (1) implies the conservation of the current given by (3). The general theorem is the following: Given a Lagrangian $\mathcal{L}_\psi(\psi_\alpha, A_\mu)$ that is invariant under the transformation $\psi_\alpha \rightarrow \psi_\alpha + \delta\psi_\alpha$ and $A_\mu \rightarrow A_\mu + \delta_\mu\Lambda$, then $j^\mu = \partial\mathcal{L}_\psi/\partial A_\mu$ is a conserved current, that is, $\partial_\mu j^\mu = 0$, provided the Lagrange equations are satisfied by the ψ fields. *Hint:* If the Lagrange equations for the ψ_α are satisfied, the action $S = \int d^4x \, \mathcal{L}_\psi$ is also invariant under the transformation $\psi_\alpha \rightarrow \psi_\alpha + \delta\psi_\alpha$ with no change in A_μ.

CHAPTER 8

Gravity

8.1. THE NATURE OF THE GRAVITATIONAL FIELD

The essential phenomenology that leads to a theory of gravity was given by Newton:

> The force of gravity between two bodies is always attractive and proportional to $m_1 m_2/r_{12}^2$, where m_1 and m_2 are the inertial masses of the two bodies and r_{12} the distance between them.

From our discussions of massive field theories, we recognize that the gravitational field must be massless and presumably satisfies some equation like

$$\nabla^2 \phi - \frac{\partial^2 \phi}{\partial t^2} = 4\pi\rho \tag{8.1.1}$$

where ρ is the source density and ϕ some component of the gravitational potential (for simplicity, we will call it *the* gravitational potential). The asymptotic potential arising from a body with source density ρ_1 will be, just as for a scalar field,

$$\phi_1(\mathbf{r}) = -\frac{1}{r} \int d\mathbf{r}\, \rho_1(\mathbf{r}), \tag{8.1.2}$$

and the asymptotic interaction energy with body two will be

$$-\frac{1}{r_{12}} \int d\mathbf{r}_2\, \rho_2(\mathbf{r}_2) \int d\mathbf{r}_1\, \rho_1(\mathbf{r}_1), \tag{8.1.3}$$

with the minus sign for attraction. Therefore, we expect to have

338

$$\left.\begin{array}{l} \int d\mathbf{r}\, \rho_1(\mathbf{r}) = \lambda m_1 \\[2mm] \int d\mathbf{r}\, \rho_2(\mathbf{r}) = \lambda m_2 \end{array}\right\} \tag{8.1.4}$$

with λ a universal constant.

The remarkable feature of (8.1.4)—the equality (to within a choice of units) of gravitational and inertial mass—is known to hold to very high accuracy. All bodies at the same point in space fall with the same acceleration to about one part in 10^{11}. We will try to find a relativistic field theory that accounts for the simple phenomenology outlined above. We will be led to a theory that will turn out to be a linear approximation to Einstein's theory; when we try to make the theory internally consistent, we will be led to the complete Einstein theory. This approach was initiated by Feynman,[1] Gupta,[2] and Thirring.[3]

What kind of field can carry gravity? Clearly, a vector field is out of the question, since it generates a repulsive force between like particles.

We have seen (Problem 7.3) that the force produced by a scalar field is attractive, and therefore a scalar field is a candidate for the carrier of gravity. However, it does not work. Note that the density $\rho(\mathbf{r})$ in (8.1.1) must be proportional to energy density, since $\int \rho(\mathbf{r})\, d\mathbf{r} = \lambda m$. However, the density that couples to a scalar field is a scalar source, which for a point particle, we have seen, is

$$\rho_s(\mathbf{r}) = \sqrt{1 - \mathbf{v}^2}\, \delta^3(\mathbf{r} - \mathbf{y}_p), \tag{8.1.5}$$

and $\int \rho_s(\mathbf{r})\, d\mathbf{r} = \sqrt{1 - \mathbf{v}^2}$, which is not the energy. In contrast, the T^{00} component of the free particle stress tensor precisely integrates to the energy. From

$$T_p^{\mu\nu} = m \frac{dx^\mu}{d\tau} \frac{dx^\nu}{d\tau} \sqrt{1 - v^2}\, \delta^3(\mathbf{x} - \mathbf{y}_p) \tag{8.1.6}$$

we find

$$\int d\mathbf{r}\, T_p^{00} = \frac{m}{\sqrt{1 - \mathbf{v}^2}}. \tag{8.1.7}$$

---

[1] R. P. Feynman, 1962 CalTech lecture notes.
[2] S. N. Gupta, *Phys. Rev.* **96**, 1683 (1954).
[3] W. E. Thirring, *Ann. Phys.* **16**, 96 (1961).

For a collection of elementary particles forming a nucleus, atom, molecule, planet, etc., the stress tensor $T^{\mu\nu}$ is guaranteed to give the energy via

$$W = \int T^{00} \, d\mathbf{r} \tag{8.1.8}$$

no matter how complex the system.[4]

How decisive is the failure of the scalar coupling to yield the inertial mass? Suppose we consider a nucleus and take a scalar density $\rho_s = m_p \sqrt{1 - v^2} \, \delta^3(\mathbf{r} - \mathbf{y}_p)$ for each particle. The integrated coupling of the scalar field to the nucleus would be

$$M_s = \sum_p m_p \sqrt{1 - v_p^2} + (?) \tag{8.1.9}$$

where (?) stands for interactive effects, perhaps arising from other scalar sources in the nucleus. There is no reason to expect that (?) would correct the error in the factor $\sqrt{1 - v_p^2}$, which is $\sim mv_p^2/m$, or roughly the binding energy of the nucleus over its rest energy, Mev over Gev, or $\sim 10^{-3}$, maybe 10^{-4} or 10^{-5} with a little conspiratorial help, but huge compared to the 10^{-11} equality of gravitational and inertial masses.

Another general way of seeing that there is difficulty is to suppose that the hypothetical scalar gravity is coupled to the natural scalar source density in a nucleus or atom, the trace of the "matter" stress tensor:

$$T = T_\mu{}^\mu, \tag{8.1.10}$$

where 'matter' includes the electrons and quarks and the gluon and electromagnetic fields. Here, we recall that the trace of the electromagnetic stress tensor vanishes; therefore, its contribution to $T_\mu{}^\mu$ cannot match its contribution to T^{00}, so we would have a correction to the equality of gravitational and inertial mass of order $T^{00}_{\text{e.m.}}/T^{00}_{\text{total}}$, a few tenths of a percent or more. In addition, gravity would in first approximation not deflect light.

It would clearly take a remarkable conspiracy to cancel out all these problems and restore the known equality of the two masses, gravitational and inertial. We must consider the pure scalar theory of gravity decisively ruled out. We turn, therefore, to the next simplest possibility, a symmetric tensor field $\phi_{\mu\nu}$.

---

[4]We here exclude internal gravitational energy.

8.2. THE TENSOR FIELD

We expect to find equations resembling

$$\partial_\alpha \partial^\alpha \phi_{\rho\nu} + \cdots = -\lambda T_{\rho\nu} \tag{8.2.1}$$

where $T_{\rho\nu}$ is the matter stress tensor, and $+ \cdots$ allows the addition of terms like $\partial_\rho \partial^\alpha \phi_{\alpha\nu}$, etc., as required to give a positive energy. λ is a coupling constant; the minus sign is a convention.

As in our discussion of vector fields, we have two ways to proceed. We can write the most general Lagrangian density that will give equations like (8.2.1)—that means bilinear in $\phi_{\rho\nu}$ and $\partial_\alpha \phi_{\mu\nu}$—and adjust the constants to give positive energy. This is possible, but difficult. It is much simpler to repeat the process we used earlier, that is, to require that the source $T_{\rho\nu}$ radiate only fields possessing in their characteristic coordinate system (rest system for particles, wave number zero for fields) only the five components associated with a three-dimensional symmetric traceless second-rank tensor. As in Section 7.5, we work with the four-dimensional Fourier transform of the field variables. In order to carry out this program, we must start with a massive field. The equation will be

$$\left(\partial_\alpha \partial^\alpha + \mu^2\right) \phi^{\rho\nu} = -\lambda P^{\rho\nu}{}_{\lambda\tau} T^{\lambda\tau} \tag{8.2.2}$$

where P is a projection operator that eliminates the unwanted components.

How do we eliminate the unwanted components of $\phi_{\rho\nu}$? In the $k_i = 0$ coordinate system, we require that the three-dimensional scalars and vectors that we can form from the tensor $\phi_{\rho\nu}$ all vanish:

$$\phi_{00} = 0, \qquad \phi_{i0} = \phi_{0i} = 0, \qquad \text{and} \qquad \phi_{ii} = \sum_{i=1}^{3} \phi_{ii} = 0 \tag{8.2.3}$$

for a propagating mode, that is, for $k^2 = k_\rho k^\rho = \mu^2$. This elimination can be expressed covariantly:

$$k^\rho \phi_{\rho\nu} = 0 \qquad \text{and} \qquad \phi_\rho{}^\rho = 0 \tag{8.2.4}$$

reduce to (8.2.3) when $k^i = 0$.

We have thus reduced the ten component symmetric tensor (by dropping the three-vector, ϕ_{0i}, two three-scalars, ϕ_{00} and ϕ_{ii}) to a five-component traceless, symmetric three-tensor.

We write the most general covariant operator $P^{\rho\nu}{}_{\lambda\tau}$, symmetric in

$\rho\nu$ when operating via $P^{\rho\nu}{}_{\lambda\tau}T^{\lambda\tau}$ on a symmetric tensor, and satisfying $k_\rho P^{\rho\nu}{}_{\lambda\sigma} = 0$ and $\eta_{\rho\nu}P^{\rho\nu}{}_{\lambda\sigma} = 0$ when $k^2 = \mu^2$.

The most general structure for P is

$$P^{\rho\nu}{}_{\lambda\tau} = \delta^\rho_\lambda \delta^\nu_\tau + \frac{\alpha}{\mu^4} k^\rho k^\nu k_\lambda k_\tau + \frac{\beta}{\mu^2} \eta^{\rho\nu} k_\lambda k_\tau$$

$$+ \gamma \frac{k^\rho k^\nu}{\mu^2} \eta_{\lambda\tau} + \xi \eta^{\rho\nu} \eta_{\lambda\tau} + \frac{\epsilon}{\mu^2} (k^\rho k_\lambda \delta^\nu_\tau + k^\nu k_\lambda \delta^\rho_\tau) \quad (8.2.5)$$

where α, β, γ, ξ, and ϵ are adjustable constants. We first trace $\rho\nu$, setting $k^2 = \mu^2$:

$$\eta_{\rho\nu}P^{\rho\nu}{}_{\lambda\tau} = \eta_{\lambda\tau} + \alpha \frac{k_\lambda k_\tau}{\mu^2} + \frac{4\beta}{\mu^2} k_\lambda k_\tau$$

$$+ \gamma\eta_{\lambda\tau} + 4\xi\eta_{\lambda\tau} + \frac{2\epsilon}{\mu^2} k_\tau k_\lambda = 0 \quad (8.2.6)$$

or

$$\left(1 + \gamma + 4\xi\right) = 0 \quad \text{and} \quad \left(\alpha + 4\beta + 2\epsilon\right) = 0. \quad (8.2.7)$$

Next, we multiply with k_ρ:

$$k_\rho P^{\rho\nu}{}_{\lambda\sigma} = k_\lambda \delta^\nu_\tau + \frac{\alpha}{\mu^2} k^\nu k_\lambda k_\tau + \frac{\beta}{\mu^2} k^\nu k_\lambda k_\tau$$

$$+ \gamma k^\nu \eta_{\lambda\tau} + \xi k^\nu \eta_{\lambda\tau} + \epsilon k_\lambda \delta^\nu_\tau + \frac{\epsilon}{\mu^2} k^\nu k_\lambda k_\tau = 0 \quad (8.2.8)$$

so that

$$1 + \epsilon = 0, \quad \alpha + \beta + \epsilon = 0, \quad \text{and} \quad \gamma + \xi = 0. \quad (8.2.9)$$

The solution of (8.2.7) and (8.2.9) is

$$\alpha = \frac{2}{3}, \quad \beta = \frac{1}{3}, \quad \epsilon = -1, \quad \gamma = \frac{1}{3}, \quad \xi = -\frac{1}{3}. \quad (8.2.10)$$

Our equation for $\phi^{\rho\nu}$ follows from (8.2.2):

$$\left(k^2 - \mu^2\right)\phi^\rho\nu = \lambda\left\{T^{\rho\nu} + \frac{2}{3}\frac{k^\rho k_\nu}{\mu^4}k_\lambda k_\tau T^{\lambda\tau} + \frac{1}{3}\eta^{\rho\nu}\frac{k_\lambda k_\tau}{\mu^2}T^{\lambda\tau}\right.$$

$$+ \frac{1}{3}\frac{k^\rho k^\nu}{\mu^2}\eta_{\lambda\tau}T^{\lambda\tau} - \frac{1}{3}\eta^{\rho\nu}\eta_{\lambda\tau}T^{\lambda\tau}$$

$$\left. - \frac{1}{\mu^2}\left(k^\rho k_\lambda T^{\lambda\nu} + k^\nu k_\lambda T^{\lambda\rho}\right)\right\} \tag{8.2.11}$$

where the coupling constant λ will be determined later, and k^2 is no longer subject to the constraint $k^2 = \mu^2$.[5]

Equation (8.2.11) has the property that the modes $k^\rho\phi_{\rho\nu}$ and $\eta_{\rho\nu}\phi^{\rho\nu}$ will not be radiated—that is, the pole $1/(k^2 - \mu^2)$ in the solution of (8.2.11) will be canceled for those modes.

We now turn (8.2.11) into an equation for $\phi^{\rho\nu}$; in its present form, it is not simple to take the limit $\mu \to 0$. We follow the same procedure as in the vector case. We express the objects that appear divided by μ^2 in terms of $\phi^{\rho\nu}$. These objects are $T = \eta_{\rho\nu}T^{\rho\nu}$, $T^\rho = k_\lambda T^{\rho\lambda}$, and $\overline{T} = k_\rho T^\rho$. In terms of these, (8.2.11) is

$$\left(k^2 - \mu^2\right)\phi^\rho\nu = \lambda\left\{T^{\rho\nu} + \frac{2}{3}\frac{k^\rho k^\nu}{\mu^4}\overline{T} + \frac{1}{3}\eta^{\rho\nu}\frac{\overline{T}}{\mu^2}\right.$$

$$\left. + \frac{1}{3}\frac{k^\rho k^\nu}{\mu^2}T - \frac{1}{3}\eta^{\rho\nu}T - \frac{1}{\mu^2}\left(k^\rho T^\nu + k^\nu T^\rho\right)\right\}. \tag{8.2.12}$$

Call $\eta_{\rho\nu}\phi^{\rho\nu} = \phi$, $k_\lambda\phi^{\lambda\nu} = \phi^\nu$, and $k_\rho k_\lambda\phi^{\lambda\rho} = \overline{\phi}$. Now trace (8.1.2):

$$\left(k^2 - \mu^2\right)\phi = \lambda\left\{T + \frac{2}{3}\frac{k^2}{\mu^4}\overline{T} + \frac{4}{3}\frac{\overline{T}}{\mu^2} + \frac{1}{3}\frac{k^2}{\mu^2}T - \frac{4}{3}T - \frac{2}{\mu^2}\overline{T}\right\}. \tag{8.2.13}$$

Next, multiply (8.2.12) by k_ρ:

$$\left(k^2 - \mu^2\right)\phi = \lambda\left\{T^\nu + \frac{2}{3}\frac{k^2}{\mu^4}k^\nu\overline{T} + \frac{1}{3}\frac{k^\nu}{\mu^2}\overline{T} + \frac{1}{3}\frac{k^2}{\mu^2}k^\nu T\right.$$

$$\left. - \frac{1}{3}k^\nu T - \frac{1}{\mu^2}\left(k^2 T^\nu + k^\nu\overline{T}\right)\right\} \tag{8.2.14}$$

[5]We apologize for the use of λ both as a coupling constant and a Lorentz tensorial index. Unfortunately, we have run out of suitable Greek letters.

and now multiply (8.2.14) by k_ν:

$$\left(k^2 - \mu^2\right)\overline{\phi} = \lambda\left\{\overline{T} + \frac{2}{3}\frac{k^4}{\mu^4}\overline{T} + \frac{1}{3}\frac{k^2}{\mu^2}\overline{T} + \frac{1}{3}\frac{k^4}{\mu^2}T - \frac{1}{3}k^2 T - \frac{2k^2}{\mu^2}\overline{T}\right\}.$$

(8.2.15)

From (8.2.13) and (8.2.15), we solve for T and $\overline{T}$. We rewrite

$$\frac{\left(k^2 - \mu^2\right)}{\lambda}\phi = T\left(1 + \frac{1}{3}\frac{k^2}{\mu^2} - \frac{4}{3}\right) + \overline{T}\left(\frac{2}{3}\frac{k^2}{\mu^4} + \frac{4}{3\mu^2} - \frac{2}{\mu^2}\right)$$

(8.2.16)

and

$$\frac{\left(k^2 - \mu^2\right)}{\lambda}\overline{\phi} = \frac{T}{3}k^2\left(\frac{k^2}{\mu^2} - 1\right) + \overline{T}\left(1 + \frac{2}{3}\frac{k^4}{\mu^4} - \frac{5}{3}\frac{k^2}{\mu^2}\right)$$

(8.2.17)

so that

$$\lambda T = 3\phi\left(\mu^2 - \frac{2}{3}k^2\right) + 2\overline{\phi} \qquad \text{and} \qquad \lambda\overline{T} = \mu^2\left(k^2\phi - \overline{\phi}\right).$$

(8.2.18)

Substituting back into (8.2.14), we learn that

$$\frac{\lambda T^\nu}{\mu^2} = k^\nu\phi - \phi^\nu$$

(8.2.19)

and finally putting T, $\overline{T}$ and T^ν back into (8.2.12),

$$\left(k^2 - \mu^2\right)\phi^{\rho\nu} - k^\rho\phi^\nu - k^\nu\phi^\rho + k^\rho k^\nu\phi$$
$$- \eta^{\rho\nu}\left[\left(k^2 - \mu^2\right)\phi - \overline{\phi}\right] = \lambda T^{\rho\nu}$$

(8.2.20)

or in coordinate space, with $k_\rho = 1/i\,\partial_\rho$,

$$\left(\partial_\lambda\partial^\lambda + \mu^2\right)\phi^{\rho\nu} - \partial^\rho\partial_\lambda\phi^{\lambda\nu} - \partial^\nu\partial_\lambda\phi^{\lambda\rho} + \partial^\rho\partial^\nu\phi_\lambda{}^\lambda$$
$$- \eta^{\rho\nu}\left[\left(\partial_\lambda\partial^\lambda + \mu^2\right)\phi_\sigma{}^\sigma - \partial_\lambda\partial_\sigma\phi^{\lambda\sigma}\right] = -\lambda T^{\rho\nu}.$$

(8.2.21)

Equation (8.2.21) tells us the number of propagating modes of $\phi^{\rho\nu}$. Of course, for $\mu \neq 0$ we know the number will be five by construction, just as the number for a massive vector field was three. We first note that whether or not μ is zero, (8.2.21) involves no second time derivatives of the four quantities $\phi^{0\rho}$; hence, the equations for them are equations of constraint. For $\mu \neq 0$, however, there is one more condition, obtained by operating on (8.2.21) with ∂_ρ:

$$\mu^2\left(\partial_\rho \phi^{\rho\nu} - \partial^\nu \phi_\lambda^\lambda\right) = -\lambda \partial_\rho T^{\rho\nu}. \tag{8.2.22}$$

Thus, for $\nu = j$, we have constraint equations on $\partial_0 \phi^{0j}$. However, for $\nu = 0$, we find

$$\partial^0\left(\phi_\lambda^\lambda - \phi^{00}\right) = \partial^0 \phi_i^i = \frac{\lambda \partial_\rho T^{\rho 0}}{\mu^2} - \partial_i \phi^{i0}, \tag{8.2.23}$$

eliminating one more propagating mode. We will see later that for $\mu = 0$ a different mechanism takes over.

In order to have a theory that treats consistently the generation of gravity by matter and the action of gravity on matter, we need a Lagrangian that includes gravity, matter, and their interaction. We turn now to a Lagrangian formulation.

8.3. LAGRANGIAN FOR THE GRAVITATIONAL FIELD

We write the Lagrangian density as

$$\mathcal{L}_g = \phi_{\rho\nu,\lambda}\phi_{\rho'\nu',\lambda'}\left\{a\eta^{\rho\rho'}\eta^{\nu\nu'}\eta^{\lambda\lambda'} + \cdots\right\} + \mu^2 \phi_{\rho\nu}\phi_{\rho'\nu'}\left\{\alpha\eta^{\rho\rho'}\eta^{\nu\nu'} + \cdots\right\} \tag{8.3.1}$$

where all possible pairings of the six and four indices must be included.

The number of independent pairings of the four indices associated with the μ^2 terms in (8.3.1) is two: $\phi_{\mu\nu}\phi^{\mu\nu}$ and $\phi_\rho^\rho\phi_\nu^\nu$. The total number of pairings of the six indices in (8.3.1) is $5 \times 3 = 15$. Of these, only four are independent: λ paired with λ' gives two (as in the μ^2 terms above). λ and λ' paired with $\rho\nu$ gives one; λ paired with $\rho\nu$ and λ' with $\rho'\nu'$ gives the last one. All other pairings are reducible to these four by symmetry, relabeling, or integration by parts.

Our Lagrange density therefore must take the form

$$\mathcal{L}_g = \frac{1}{2}\partial_\lambda \phi_{\rho\nu}\partial^\lambda \phi^{\rho\nu} + a\partial_\lambda \phi_{\rho\nu}\partial^\nu \phi^{\rho\lambda} + b\partial_\lambda \phi_\rho^\rho\partial^\lambda \phi_\nu^\nu$$

$$+ c\partial_\rho \phi^{\rho\nu}\partial_\nu \phi_\lambda^\lambda$$

$$+ \frac{\mu^2}{2}\left(\alpha\phi_{\rho\nu}\phi^{\rho\nu} + \beta\phi_\rho^\rho\phi_\nu^{\ \nu}\right), \tag{8.3.2}$$

where we have chosen the factor $\frac{1}{2}$ to give the conventional positive coefficient for the "natural" term, $(\dot{\phi}_{ij})^2/2$. Note that because $\phi^{\rho\nu} = \phi^{\nu\rho}$, $\phi_\rho{}^\nu = \phi^\nu{}_\rho$ and can be written ϕ_ρ^ν with no ambiguity. In order to work out the Lagrange equations, it is convenient to keep track of indices by writing (8.3.2) in the form suggested by (8.3.1). That is,

$$\mathcal{L}_g = \partial_\lambda \phi_{\rho\nu} \partial_{\lambda'} \phi_{\rho'\nu'} \times \left\{ \frac{1}{2} \eta^{\rho\rho'} \eta^{\nu\nu'} \eta^{\lambda\lambda'} + a\, \eta^{\rho\rho'} \eta^{\lambda\nu'} \eta^{\lambda'\nu} + b\eta^{\rho\nu} \eta^{\rho'\nu'} \eta^{\lambda\lambda'} \right.$$

$$\left. + c\eta^{\lambda\rho} \eta^{\rho'\nu'} \eta^{\lambda'\nu} \right\} + \frac{\mu^2}{2} \phi_{\rho\nu} \phi_{\rho'\nu'} \left[a\eta^{\rho\rho'} \eta^{\nu\nu'} + \beta\eta^{\rho\nu} \eta^{\nu'\rho'} \right] \qquad (8.3.3)$$

and

$$\frac{\partial \mathcal{L}_g}{\partial(\partial_\lambda \phi_{\rho\nu})} = \partial_{\lambda'} \phi_{\rho'\nu'} \left[\{ \ \} + \{ \ \}' \right] \qquad (8.3.4)$$

where $\{ \ \}$ is the first bracket in (8.3.3) and $\{ \ \}'$ the same bracket with primed and unprimed indices exchanged.

Similarly,

$$\frac{\partial \mathcal{L}_g}{\partial \phi_{\rho\nu}} = \frac{\mu^2}{2} \phi_{\rho'\nu'} \left\{ [\ ] + [\ ]' \right\}. \qquad (8.3.5)$$

Thus, the Lagrange equations are

$$\partial_\lambda \partial_{\lambda'} \phi_{\rho'\nu'} \left[\eta^{\rho\rho'} \eta^{\nu\nu'} \eta^{\lambda\lambda'} + 2a\eta^{\rho\rho'} \eta^{\lambda\nu'} \eta^{\lambda'\nu} \right.$$

$$\left. + 2b\eta^{\rho\nu} \eta^{\rho'\nu'} \eta^{\lambda\lambda'} + c\eta^{\lambda\rho} \eta^{\rho'\nu'} \eta^{\lambda'\nu} + c\eta^{\lambda'\rho'} \eta^{\rho\nu} \eta^{\lambda\nu'} \right]$$

$$= \frac{\mu^2}{2} \phi_{\rho'\nu'} \left[2\alpha\, \eta^{\rho\rho'} \eta^{\nu\nu'} + 2\beta\, \eta^{\rho\nu} \eta^{\nu'\rho'} \right] \qquad (8.3.6)$$

or

$$\partial_\lambda \left[\partial^\lambda \phi^{\rho\nu} + 2a\partial^\nu \phi^{\rho\lambda} + 2b\eta^{\rho\nu}\partial^\lambda \phi_\sigma{}^\sigma + c \left[\eta^{\lambda\rho}\partial^\nu \phi_\sigma{}^\sigma + \partial_\sigma \phi^{\sigma\lambda} \eta^{\rho\nu} \right] \right]$$

$$= \mu^2 \left[\alpha \phi^{\rho\nu} + \beta \eta^{\rho\nu} \phi_\sigma{}^\sigma \right]. \qquad (8.3.7)$$

We must still symmetrize in $\rho\nu$, since the only permitted variation of $\phi^{\rho\nu}$, $\delta\phi^{\rho\nu}$, must be symmetric in ρ and ν. The equation is then (with $\phi \equiv \phi_\sigma^\sigma$)

$$\partial_\lambda \partial^\lambda \phi^{\rho\nu} + a\left(\partial_\lambda \partial^\nu \phi^{\rho\lambda} + \partial_\lambda \partial^\rho \phi^{\nu\lambda}\right)$$
$$+ c\partial^\rho \partial^\nu \phi + \eta^{\rho\nu}\left(2b\partial_\lambda \partial^\lambda \phi + c\partial_\lambda \partial_\sigma \phi^{\sigma\lambda}\right)$$
$$= \mu^2\left(\alpha \phi^{\rho\nu} + \beta\eta^{\rho\nu}\phi\right). \tag{8.3.8}$$

Comparing with (8.2.21), we find $a = -1$, $c = 1$, $b = -1/2$, $\alpha = -1$ and $\beta = 1$, so $\mathscr{L}$ is as follows:

$$\mathscr{L}_g = \frac{1}{2}\Big[\partial_\lambda \phi_{\rho\nu}\partial^\lambda \phi^{\rho\nu} - 2\partial_\lambda \phi_{\rho\nu}\partial^\nu \phi^{\rho\lambda}$$
$$+ 2\partial_\rho \phi^{\rho\nu}\partial_\nu \phi - \partial_\lambda \phi\partial^\lambda \phi - \mu^2 \phi_{\rho\nu}\phi^{\rho\nu} + \mu^2\phi^2\Big]. \tag{8.3.9}$$

Before continuing, we wish to verify that the energy of the free field described by the Lagrangian density (8.3.9) is positive.

Note first that the second term in (8.3.9) may be rewritten as

$$\partial_\lambda \phi_{\rho\nu}\partial^\nu \phi^{\rho\lambda} = \partial^\nu \phi_{\rho\nu}\partial_\lambda \phi^{\rho\lambda} + \text{total derivatives.}$$

We may therefore calculate the energy from the canonical stress tensor, dropping every term except

$$\mathscr{L}'_g = \frac{1}{2}\Big[\partial_\lambda \phi_{\rho\nu}\partial^\lambda \phi^{\rho\nu} - \mu^2 \phi_{\rho\nu}\phi^{\rho\nu}\Big] \tag{8.3.10}$$

since all the other terms are bilinear products of expressions, each of which vanishes by the equations of motion. The energy is therefore

$$P_0 = \frac{1}{2}\int d^3x\Big[\dot{\phi}_{\rho\nu}\dot{\phi}^{\rho\nu} + \partial_i \phi_{\rho\nu}\partial_i \phi^{\rho\nu} + \mu^2 \phi_{\rho\nu}\phi^{\rho\nu}\Big]. \tag{8.3.11}$$

We proceed here as we did in (7.5.20) by Fourier expansion:

$$\phi_{\rho\nu} = \int d^3k\, \psi_{\rho\nu}\, e^{i(\mathbf{k}\cdot\mathbf{x}-k_0x_0)} + \text{c.c.} \tag{8.3.12}$$

so that P_0 becomes

$$P_0 = (2\pi)^3 \int d^3k\,(\mathbf{k}^2 + \mu^2 + k_0^2)\, \psi_{\rho\nu}\psi^{*\rho\nu}. \tag{8.3.13}$$

$\psi_{\rho\nu}$ is restricted by the conditions

$$k^\rho \psi_{\rho\nu} = 0 \qquad \text{and} \qquad \psi_{\rho\nu} \eta^{\rho\nu} = 0. \tag{8.3.14}$$

Thus,

$$k^0 \psi_{0\nu} + k^i \psi_{i\nu} = 0, \tag{8.3.15}$$

$$k^0 \psi_{00} + k^i \psi_{i0} = 0, \tag{8.3.16}$$

and

$$k^0 \psi_{0j} + k^i \psi_{ij} = 0, \tag{8.3.17}$$

so that

$$\psi_{00} = \frac{k^i}{k^0} \frac{k^j}{k^0} \psi_{ji} \tag{8.3.18}$$

and

$$\psi_{0i} = \psi_{i0} = -\frac{k^j \psi_{ij}}{k_0}, \tag{8.3.19}$$

We see that ψ_{00} and ψ_{0i} are not independent degrees of freedom, nor is $\psi_{ii} = \psi_{00}$. With **k** in the three-direction, there remain as independent components ψ_{33}, ψ_{32}, ψ_{31}, ψ_{12}, and $\psi_{11} - \psi_{22}$. The term $\psi_{11} + \psi_{22} = \psi_{00} - \psi_{33}$ is already determined by ψ_{33}. Substituting (8.3.18) and (8.3.19) in (8.3.13), we find for $\psi_{\rho\nu}\psi^{*\rho\nu}$

$$\psi_{\rho\nu}\psi^{*\rho\nu} = \left| \frac{k^i k^j \psi_{ij}}{k_0^2} \right|^2 - 2\left| \frac{k^i \psi_{ij}}{k_0} \right|^2 + |\psi_{ij}|^2 \tag{8.3.20}$$

$$= |\psi_{33}|^2 \epsilon^4 - 2\epsilon^2 |\psi_{3j}|^2 + |\psi_{ij}|^2$$

$$= |\psi_{33}|^2 (\epsilon^4 - 2\epsilon^2 + 1) + 2(1 - \epsilon^2)\left(|\psi_{31}|^2 + |\psi_{32}|^2 \right)$$

$$+ 2|\psi_{12}|^2 + |\psi_{11}|^2 + |\psi_{22}|^2 \tag{8.3.21}$$

which is positive-definite, since $\epsilon = k/k_0 < 1$.

It is interesting to consider the limit of this theory as $\mu^2 \to 0$ and $(1 - \epsilon^2) = \mu^2/\omega^2 \to 0$. In order to make a nonzero contribution to the energy, $|\psi_{31}|^2$ and $|\psi_{32}|^2$ must go like $1/\mu^2$, and $|\psi_{33}|^2$ like $1/\mu^4$ as $\mu^2 \to 0$. We can find their actual behavior from (8.2.12). For the limit $\mu^2 \to 0$ to exist at all requires a conserved source, so we take

$$k_\rho T^{\rho\nu} = 0. \tag{8.3.22}$$

The equation for $\psi^{\rho\nu}$ becomes

$$(k^2 - \mu^2)\psi^{\rho\nu} = \lambda\left[T^{\rho\nu} + \frac{1}{3}\left(\frac{k^\rho k^\nu}{\mu^2} - \eta^{\rho\nu}\right)T\right], \qquad (8.3.23)$$

where $T = \eta_{\rho\nu} = T^{\rho\nu}$, as before. We note the singularity of ψ^{ij} as $\mu^2 \to 0$: It is uniquely in the amplitude ψ_{33}, *since* **k** is in the three-direction, and makes $|\psi_{33}|^2$ go like $1/\mu^4$ as $\mu^2 \to 0$. The mode

$$\psi^{11} + \psi^{22} = \psi^{00} - \psi^{33}$$

$$= (\epsilon^2 - 1)\psi^{33} \to \text{constant} \qquad (8.3.24)$$

as $\mu^2 \to 0$. The modes ψ^{31} and ψ^{32} are finite in the limit and therefore carry no energy. The independent modes that survive and carry energy are therefore $\psi^{11} - \psi^{22}$, ψ^{12}, and ψ^{33}. As we shall see later, the $\mu^2 = 0$ theory, based on (8.2.21) with $\mu^2 = 0$, has only two independent modes — essentially, $\psi^{11} - \psi^{22}$ and ψ^{12}. The limit of the nonzero μ^2 theory as $\mu^2 \to 0$ exists and is different from the $\mu^2 = 0$ theory.

We will come back later to a discussion of the free Lagrangian — propagating modes, zero mass limit, stress tensor, etc. We wish first to study the system gravitational field plus point particles.

8.4. PARTICLES IN A GRAVITATIONAL FIELD

We write the action as the sum of three terms:

$$S = S_g + S_p + S_I \qquad (8.4.1)$$

where

$$S_g = \int d^4x \, \mathscr{L}_g, \qquad (8.4.2)$$

and

$$S_p = -\sum_p \frac{m_p}{2}\int \frac{dy_p^\mu}{d\sigma_p}\eta_{\mu\nu}\frac{dy_p^\nu}{d\sigma_p}\,d\sigma_p \qquad (8.4.3)$$

as discussed in Section 7.2; for the interaction we take the simplest invariant, nonderivative coupling:

$$S_I = -\lambda\sum_p m_p\int \frac{dy_p^\mu}{d\sigma_p}\frac{dy_p^\nu}{d\sigma_p}\phi_{\mu\nu}(y_p)\,d\sigma_p. \qquad (8.4.4)$$

m_p is inserted by hand to guarantee the equivalence principle, that is, acceleration at a point independent of the particle. We could also add a coupling to the trace $\phi = \eta^{\mu\nu}\phi_{\mu\nu}$, but that would be in some ways equivalent to adding a scalar field and subject to the same experimental problems.

It is obviously simplifying to define

$$g_{\mu\nu}(y_p) = \eta_{\mu\nu} + 2\lambda\,\phi_{\mu\nu}(y_p). \tag{8.4.5}$$

The particle Lagrangian can then include the interaction by writing

$$S_p' = -\sum_p \frac{m_p}{2} \int \frac{dy_p^\mu}{d\sigma_p}\frac{dy_p^\nu}{d\sigma_p} g_{\mu\nu}(y_p)\,d\sigma_p. \tag{8.4.6}$$

The equations of motion are

$$-\frac{d}{d\sigma_p}\,g_{\mu\nu}\frac{dy_p^\nu}{d\sigma_p} = -\frac{1}{2}\frac{dy_p^\lambda}{d\sigma_p}\frac{dy_p^\eta}{d\sigma_p}\partial_\mu g_{\lambda\eta} \tag{8.4.7}$$

or

$$g_{\mu\nu}\frac{d^2y_p^\nu}{d\sigma_p^2} = \frac{1}{2}\frac{dy_p^\lambda}{d\sigma_p}\frac{dy_p^\eta}{d\sigma_p}\partial_\mu g_{\lambda\eta} - \frac{dy_p^\nu}{d\sigma_p}\frac{dy_p^\lambda}{d\sigma_p}\frac{\partial g_{\mu\nu}}{\partial y^\lambda}$$

$$= -\frac{1}{2}\frac{dy_p^\nu}{d\sigma_p}\frac{dy_p^\lambda}{d\sigma_p}\left(\frac{\partial g_{\mu\nu}}{\partial y^\lambda} + \frac{\partial g_{\mu\lambda}}{\partial y^\nu} - \frac{\partial g_{\lambda\nu}}{\partial y^\mu}\right)$$

or

$$\frac{d^2y_p^\mu}{d\sigma_p^2} + \Gamma^\mu_{\nu\lambda}\frac{dy_p^\lambda}{d\sigma_p}\frac{dy_p^\nu}{d\sigma_p} = 0, \tag{8.4.8}$$

where Γ is defined in (7.6.40).

We determine $d\sigma_p$ by calculating the σ "energy" "W" from

$$L_p' = -\frac{m_p}{2}\frac{dy_p^\mu}{d\sigma_p}\frac{dy_p^\nu}{d\sigma_p}g_{\mu\nu}$$

as in (7.2.12). It is

$$\text{``W''} = \frac{dy_p^\mu}{d\sigma_p}\frac{\partial L_p'}{\partial\left(\dfrac{dy_p^\mu}{d\sigma_p}\right)} - L_p'$$

$$= -\frac{m_p}{2}\frac{dy_p^\mu}{d\sigma_p}\frac{dy_p^\nu}{d\sigma_p}g_{\mu\nu} \tag{8.4.9}$$

so that $(d\sigma_p)^2 = \text{const.} \times dy_p^\mu\, dy_p^\nu\, g_{\mu\nu}(y_p)$. As in (7.2.11), we may choose const. $= 1$ by adjusting m_p. Thus,

$$d\sigma_p = \left(dy_p^\mu\, dy_p^\nu\, g_{\mu\nu}(y_p)\right)^{1/2} \tag{8.4.10}$$

and in the equation of motion (8.4.8) $d\sigma_p$ is to be understood as given by (8.4.10).

The field equations now follow. Once we have chosen the Lagrangian to give the effect of the field on the particle, the particle's effect as a source of the field is determined. Then, the field equations follow from $\delta S/\delta \phi_{\mu\nu}$ (with μ^2, the gravitational field mass,[6] set equal to zero). Thus, with

$$S = \int \frac{d^4x}{2} \left[\partial_\lambda \phi_{\mu\nu} \partial^\lambda \phi^{\mu\nu} - \partial_\lambda \phi_{\mu\nu} \partial^\nu \phi^{\mu\lambda} - \partial_\lambda \phi_{\mu\nu} \partial^\mu \phi^{\nu\lambda} \right.$$

$$\left. + 2\partial_\mu \phi^{\mu\nu} \partial_\nu \phi - \partial_\lambda \phi \partial^\lambda \phi \right] - \sum_p \frac{m_p}{2} \int \frac{dy_p^\mu\, dy_p^\nu}{d\sigma_p\, d\sigma_p} g_{\mu\nu}(y_p)\, d\sigma_p,$$

we have

$$-\frac{\partial S}{\delta \phi_{\mu\nu}} = \partial_\lambda \partial^\lambda \phi^{\mu\nu} - \partial^\mu \partial_\lambda \phi^{\lambda\nu} - \partial^\nu \partial_\lambda \phi^\lambda$$

$$+ \partial^\mu \partial^\nu \phi - \eta^{\mu\nu} \left[\partial_\lambda \partial^\lambda \phi - \partial_\lambda \partial_\sigma \phi^{\lambda\sigma} \right]$$

$$+ \frac{\delta}{\delta \phi_{\mu\nu}} \sum_p \int d\sigma_p \left(\lambda m_p \frac{dy_p^\mu\, dy_p^\nu}{d\sigma_p\, d\sigma_p} \phi_{\mu\nu}(y_p) \right) = 0. \tag{8.4.11}$$

The last term in (8.4.11) is transformed via

$$\int d\sigma \phi_{\mu\nu}(y_p) = \int d^4x \delta^4(x - y_p)\, \phi_{\mu\nu}(x)\, d\sigma$$

to

$$\lambda \sum_p m_p \frac{dy_p^\mu\, dy_p^\nu}{d\sigma_p\, d\sigma_p} \int d\sigma_p \delta^4(x - y_p(\sigma_p)) \tag{8.4.12}$$

[6]More precisely, in classical field theory, $1/\mu$ is the Compton wavelength of the field.

which we will call $\lambda \Sigma_p T_p^{\mu\nu}$ where $T_p^{\mu\nu}$ is the particle stress tensor:

$$T_p^{\mu\nu} = m_p \frac{dy_p^\mu}{d\sigma_p} \frac{dy_p^\nu}{d\sigma_p} \int d\sigma_p \, \delta^4(x - y_p(\sigma_p)) \qquad (8.4.13)$$

with

$$d\sigma_p = \left(dx_p^\mu dx_p^\nu g_{\mu\nu}(y_p) \right)^{1/2}.$$

Equation (8.4.11) becomes

$$\partial_\lambda \partial^\lambda \phi^{\mu\nu} - \partial^\mu \partial_\lambda \phi^{\lambda\nu} - \partial^\nu \partial_\lambda \phi^{\lambda\mu} + \partial^\mu \partial^\nu \phi$$
$$- \eta^{\mu\nu}(\partial_\lambda \partial^\lambda \phi - \partial_\sigma \partial_\lambda \phi^{\sigma\lambda}) = -\lambda T_p^{\mu\nu}. \qquad (8.4.14)$$

Equation (8.4.14) has two remarkable properties. First, if we operate with ∂_μ on the left-hand side, we get zero identically. Therefore, the equation is inconsistent, since $\partial_\mu T^{\mu\nu}$ is not zero. It is not zero because there is exchange of energy and momentum between matter and field, so that the matter stress tensor alone cannot be conserved. However, if the field is weak, then to lowest order, $T_p^{\mu\nu}$ is conserved. In fact, from (8.4.13),

$$\partial_\mu T_p^{\mu\nu} = -m_p \delta^3(\mathbf{x} - \mathbf{y}_p) \frac{dy_p^\lambda}{dx^0} \frac{dy_p^\mu}{d\sigma_p} \Gamma^\nu_{\mu\lambda}(y_p)$$

$$= -\Gamma^\nu_{\mu\lambda} T_p^{\mu\lambda}.$$

It is clear what program might be followed. Instead of $T_p^{\mu\nu}$ on the right-hand side of (8.4.14), we should have $T^{\mu\nu}$, the total stress tensor of our linear theory. This way of proceeding poses two difficult problems. First, it is not clear how we should choose the $T^{\mu\nu}$ that is carried by the gravitational field. We saw in Section 7.7 how to construct a conserved tensor from a Lagrangian density. However, since the Lagrangian density can be modified by adding a derivative (without changing the Lagrangian equations and therefore without changing the theory), there is an infinite set of possible $T^{\mu\nu}$'s, all conserved and symmetric. Second, if we modify the right-hand side of (8.4.14), the new equation will conserve a different stress tensor. To find the new stress tensor, we will need to find a Lagrangian for the new equation, and so on, *ad infinitum*. This process can be carried out, but we will not do so here.[7] Remarkably, both problems are solved exactly by Einstein's general theory of relativity. We shall see how this is done in Sections 8.7 and 8.8. For the moment, we will consider

---

[7]It is discussed in some detail in Feynman's notes, previously cited in footnote 1.

(8.4.14) as an approximate equation and deduce its immediate consequences. Therefore, imagine an extra term $\delta T^{\mu\nu}$ on the right-hand side
of (8.4.14); this extra term makes the equations consistent, but is small.
We shall need to find it later in order to calculate the precession of
planetary orbits.

The second remarkable property of (8.4.14) is that it no longer (be-
cause $\mu = 0$) determines $\phi_{\mu\nu}$. It is, in fact, invariant under the transformation

$$\phi^{\mu\nu} \rightarrow \phi^{\mu\nu} + \partial^\mu \xi^\nu + \partial^\nu \xi^\mu \tag{8.4.15}$$

where the four ξ^μ are arbitrary functions of x.

The transformation (8.4.15) is analogous to the gauge transformation
of the potential A_μ in electrodynamics. There are two important differ-
ences, however. The first is that the gravitational field $\Gamma^\mu_{\lambda\sigma}$ which, according to (8.4.8), determines the motion of a test particle is *not* invariant
under the transformation, unlike the electromagnetic case. The second is
that a coordinate transformation can eliminate the change in the gravitational field. To see this, we observe that the action S'_p in (8.4.6) is
invariant under the gauge change (8.4.15) together with a transformation
to new coordinates[8]

$$x'_\rho = x_\rho - 2\lambda \xi_\rho(x),$$

all to first order in $\lambda \phi_{\mu\nu}$ and $\lambda \xi_\rho$.

The existence of this invariance provides us with a rigorous but difficult
way of dealing with the gauge problem: Formulate all experiments in
terms of gauge invariants. In practice, one does less: One usually assumes
that experiments are carried out in gravity-free regions, where the only
coordinate ambiguity is that associated with Lorentz transformations. That
is a not very subtle and not very satisfactory way of instructing the reader
not to worry too much about the choice of gauge.

We wish before proceeding to make sure that the weak field, low-
velocity limits of (8.4.8) and (8.4.14) yield Newtonian mechanics. We try
to solve (8.4.14) for a heavy, stationary source (like the sun), with mass
$M_\odot$.

A convenient gauge that decouples the tensor components of (8.4.1)
is defined by the condition

$$\partial_\mu \phi^{\mu\nu} = \frac{1}{2} \partial^\nu \phi. \tag{8.4.16}$$

---

[8]The alert reader will note here the first appearance of general covariance, albeit in
approximate form.

This gauge is called harmonic. It is analogous to the Lorentz gauge in electrodynamics.

To see that we can always go to this gauge, suppose $\partial_\mu \phi^{\mu\nu} - \frac{1}{2}\partial^\nu \phi = \psi^\nu \neq 0$. Then we let $\phi^{\mu\nu} \to \phi^{\mu\nu} + (\partial^\mu \xi^\nu + \partial^\nu \xi^\mu)$ and demand

$$\partial_\mu(\partial^\mu \xi^\nu + \partial^\nu \xi^\mu) - \partial^\nu \partial^\lambda \xi_\lambda = -\psi^\nu \tag{8.4.17}$$

or

$$\partial_\mu \partial^\mu \xi^\nu = -\psi^\nu \tag{8.4.18}$$

which is an equation we can solve for ξ^ν. Note that like the Lorentz gauge, the harmonic gauge is really a class of gauges, for transformation within which the gauge parameter ξ^μ satisfies the four-dimensional harmonic equation $\partial_\lambda \partial^\lambda \xi^\mu = 0$. Hence, the name harmonic gauge.

If we insert (8.4.16) into (8.4.14), we find

$$\partial_\lambda \partial^\lambda \phi^{\mu\nu} - \frac{\eta^{\mu\nu}}{2}\partial_\lambda \partial^\lambda \phi = -\lambda T_\odot^{\mu\nu} \tag{8.4.19}$$

where $T_\odot^{\mu\nu}$ is the stress tensor of the sun.

A further simplification results from the trace of (8.4.19):

$$\partial_\lambda \partial^\lambda \phi - \frac{4}{2}\partial_\lambda \partial^\lambda \phi = -\lambda T_\odot \tag{8.4.20}$$

with $T_\odot = T_{\odot\mu}^\mu$.

Therefore,

$$\partial_\lambda \partial^\lambda \phi = \lambda T_\odot \tag{8.4.21}$$

and

$$\partial_\lambda \partial^\lambda \phi^{\mu\nu} = -\lambda T_\odot^{\mu\nu} + \frac{\eta^{\mu\nu}}{2}\partial_\lambda \partial^\lambda \phi$$

$$= -\lambda\left(T_\odot^{\mu\nu} - \frac{\eta^{\mu\nu}}{2}T_\odot\right). \tag{8.4.22}$$

For our source, $dy_\odot^i/d\sigma_\odot \approx 0$, and for weak fields and low velocities, $d\sigma_\odot \approx dy_\odot^0$, yielding

$$T_\odot^{00} \approx M_\odot \delta^3(\mathbf{x} - \mathbf{y}_p(y_p^0)); \tag{8.4.23}$$

all other components of $T_\odot^{\mu\nu}$ are approximately zero. Let us define $t^{\mu\nu}$ by the equation

$$T_\odot^{\mu\nu} - \frac{\eta^{\mu\nu}}{2} T_\odot = \frac{M_\odot}{2} \delta^3(\mathbf{x} - \mathbf{y}_\odot) t^{\mu\nu}. \tag{8.4.24}$$

Then from (8.4.24), $t^{00} = t^{11} = t^{22} = t^{33} = 1$; all other components are zero. The static solution of (8.4.24) is then the solution of

$$-\nabla^2 \phi^{\mu\nu} = -\lambda \frac{M_\odot}{2} \delta^3(\mathbf{x} - \mathbf{y}_\odot) t^{\mu\nu}$$

or

$$\phi^{\mu\nu}(\mathbf{x}) = -\frac{\lambda}{4\pi} \frac{M_\odot}{2} \frac{t^{\mu\nu}}{|\mathbf{x} - \mathbf{y}_\odot|}. \tag{8.4.25}$$

We next substitute (8.4.25) into (8.4.8). We take the static field, low-velocity limit of (8.4.8). That is,

$$\frac{d^2 y_P^\mu}{d\sigma^2} + \Gamma_{00}^\mu \left(\frac{dy_P^0}{d\sigma}\right)^2 = 0. \tag{8.4.26}$$

The equation for y_P^0 is then

$$\frac{d^2 y_P^0}{d\sigma^2} + \Gamma_{00}^0 \left(\frac{dy_P^0}{d\sigma}\right)^2 = 0. \tag{8.4.27}$$

But for a static, diagonal $\phi^{\mu\nu}$, as ours is, $\Gamma_{00}^0 = 0$. Thus, $(d^2 y_P^0/d\sigma^2) = 0$, and we may take $\sigma = y_P^0$.

The space components of (8.4.26) are therefore

$$\frac{d^2 y_P^i}{d(y_P^0)^2} + \Gamma_{00}^i = 0. \tag{8.4.28}$$

In the weak field static approximation,

$$\Gamma_{00}^i = \frac{1}{2} \partial_i g_{00} = \lambda \, \partial_i \phi_{00}, \tag{8.4.29}$$

and

$$\frac{d^2 y_P^i}{d(y_P^0)^2} = -\lambda \, \partial_i \phi_{00}$$

with

$$\phi_{00} = -\frac{\lambda}{8\pi} \frac{M_{\odot}}{|\mathbf{x} - \mathbf{y}_{\odot}|} \tag{8.4.30}$$

so

$$\frac{\lambda^2}{8\pi} = G, \tag{8.4.31}$$

Newton's gravitational constant.

8.5. INTERACTION OF THE GRAVITATIONAL FIELD

We consider a given gravitational field, $g_{\mu\nu} = \eta_{\mu\nu} + 2\lambda\phi_{\mu\nu}$. We have seen in Section 8.4 the action of a point particle in the field $g_{\mu\nu}$:

$$S_p = -\frac{m_p}{2} \int d\sigma \frac{dy^\mu}{d\sigma} g_{\mu\nu} \frac{dy^\nu}{d\sigma} \tag{8.5.1}$$

with the resulting equations of motion [as in (8.4.8)]

$$\frac{d^2 y^\mu}{d\sigma^2} + \Gamma^\mu_{\lambda\tau} \frac{dy^\lambda}{d\sigma} \frac{dy^\tau}{d\sigma}. \tag{8.5.2}$$

We have left out the p subscript in (8.5.2) since from now on we will be considering only one particle at a time.

If we let $g_{\mu\nu}$ transform like a tensor under general coordinate transformations (under which $dy^\mu/d\sigma$ transforms like a vector), the action (8.5.1) will be exactly, rather than approximately, invariant, and the equations of motion (8.5.2) covariant under such transformations. Further, the arguments given in Section 7.7 that show that $g_{\mu\nu}$ can be transformed to $g_{\mu\nu} = \eta_{\mu\nu}$ at a point, and $\Gamma^\mu_{\lambda\tau} = 0$ at that point, hold here as well. Therefore, at each point there will be a coordinate system (the elevator system) in which the equations of motion are

$$\frac{d^2}{d\sigma^2} \xi^\mu = 0 \tag{8.5.3}$$

accurate up to and including linear terms in the expansion of $g_{\mu\nu}$ about $\xi = \xi_0$. At the center of this freely falling elevator system, masses move under the effect of all other than gravitational forces: The gravitational force has been eliminated. For this to hold over the entire elevator, the elevator must be small enough so that the effect of the neglected quadratic terms in the expansion of $g_{\mu\nu}$ about $\xi = \xi_0$ is small compared to that of the other forces at work.

The principle that all effects of gravity should vanish in the elevater system was called by Einstein the principle of equivalence of gravitation and inertia.

At this point, we must ask about other interactions. In particular, it does not seem possible that all particles move freely in the elevator frame unless, in that frame, all the laws of nature take their gravitation free form.

This apparent extension of the equivalence principle is certainly necessary to make the original limited principle consistent. For example, if it were not true, the electromagnetic contribution to mass, which is different for different atoms, would show up in the Eötvös experiment (and its children and grandchildren), now accurate to one part in 10^{11}. The action (8.5.1) shows us how to accomplish this more general goal: We use the gravitational tensor $g_{\mu\nu}$ to make the Lorentz-invariant action generally invariant and the Lorentz-covariant equations generally covariant. Thus, we take for the action of the electromagnetic field interacting with charged particles, as in (7.8.7):

$$S = - \sum_p \frac{m_p}{2} \int \frac{dy_p^{\mu}}{d\sigma_p} g_{\mu\nu} \frac{dy_p^{\nu}}{d\sigma_p} d\sigma_p + \sum_p \int q_p \frac{dy_p^{\mu}}{d\sigma_p} A_{\mu} d\sigma_p$$

$$- \frac{1}{4} \int d^4x \sqrt{g} (\partial_{\mu} A_{\nu} - \partial_{\nu} A_{\mu}) g^{\mu\lambda} g^{\nu\tau} (\partial_{\lambda} A_{\tau} - \partial_{\tau} A_{\lambda}). \quad (8.5.4)$$

In (8.5.4) we have taken the fundamental field A_{μ} to be a covariant vector and the particle displacement dy^{μ} to be a contravariant vector. The fields $F_{\mu\nu} = (\partial_{\mu} A_{\nu} - \partial_{\nu} A_{\mu})$ are given by ordinary derivatives, since, as we have seen, the ordinary curl is a tensor.

The equations for the electromagnetic field are then

$$F^{\mu\nu}_{\ ;\nu} = -j^{\nu} \qquad \text{with}$$

$$j^{\mu}(xy) = q_p \int d\sigma_p \frac{\delta^4(x - y_p(\sigma_p))}{\sqrt{g}} \frac{dy_p}{d\sigma_p} \qquad \text{and from (7.6.43),}$$

$$\partial_{\mu} F_{\nu\lambda} + \partial_{\lambda} F_{\mu\nu} + \partial_{\nu} F_{\lambda\mu} = 0. \qquad\qquad\qquad\qquad\qquad (8.5.5)$$

The relativistic quantum wave equation for a charged scalar field ψ

with charge e would be the covariant extension of

$$\left[(\partial_\mu - ieA_\mu)(\partial^\mu - ieA^\mu) + m^2\right]\phi = 0, \qquad (8.5.6)$$

or

$$[g^{\mu\nu}(\partial_\nu - ieA_\nu)\phi]_{;\mu} - ieA_\mu g^{\mu\nu}(\partial_\nu - ieA_\nu)\phi + m^2\phi = 0. \quad (8.5.7)$$

We shall use this equation shortly to study the gravitational red shift.

We take up now three consequences of the tensor theory developed so far. These are the bending of light, the precession of elliptic Keplerian orbits, and the gravitational red shift.

We discuss the bending of light and the orbit precession together. In both cases, what is at issue is the orbit equation for an object in a given static gravitational field. We treat the light ray as a rapidly moving particle ($v \sim 1$).

We start from (8.4.7):

$$\frac{d}{d\sigma}\left(g_{\mu\nu}\frac{dy^\nu}{d\sigma}\right) = \frac{1}{2}\partial_\mu g_{\lambda\tau}\frac{dy^\lambda}{d\sigma}\frac{dy^\tau}{d\sigma}. \qquad (8.5.8)$$

We consider the case where

$$g_{00} = g_0, \qquad g_{ij} = -g_s\,\delta_{ij}, \qquad g_{0i} = 0, \qquad (8.5.9)$$

and g_0 and g_s are functions of $|\mathbf{y}| = r$ alone. We found, from (8.4.25) and (8.4.31),

$$g_0 = 1 - \frac{2M_\odot G}{r} \qquad \text{and} \qquad g_s = 1 + \frac{2M_\odot G}{r}. \qquad (8.5.10)$$

We consider here the more general case [in which (8.5.9) still holds], where we have corrected the right-hand side of (8.4.14), as discussed below in Section 8.8. Under these conditions, (8.5.8) has three integrals. First,

$$\frac{d}{d\sigma}\left(g_0\frac{dy^0}{d\sigma}\right) = 0, \qquad (8.5.11)$$

so

$$g_0\frac{dy^0}{d\sigma} = \text{constant} \equiv W. \qquad (8.5.12)$$

Second, from (8.4.10)

$$\left(\frac{dy^0}{d\sigma}\right)^2 g_0 - \left(\frac{d\mathbf{y}}{d\sigma}\right)^2 g_s = 1 \tag{8.5.13}$$

and third, from the space component vector equation

$$-\frac{d}{d\sigma} g_s \frac{d\mathbf{y}}{d\sigma} = \frac{1}{2} \hat{\mathbf{y}} \left(\frac{d}{dy} g_{\lambda\tau}\right) \frac{dy^\lambda}{d\sigma} \frac{dy^\tau}{d\sigma},$$

where $\hat{\mathbf{y}} = \mathbf{y}/y$ and $y = |\mathbf{y}|$, we derive angular momentum conservation:

$$\mathbf{L} = g_s \mathbf{y} \times \frac{d\mathbf{y}}{d\sigma} = \text{constant}. \tag{8.5.14}$$

We learn from (8.5.14) that the motion stays in a plane; we choose polar coordinates r and θ in that plane. Then (8.5.13) becomes (with $dA/d\sigma \equiv \dot{A}$)

$$g_0 \dot{y}^{o2} - g_s(\dot{r}^2 + r^2\dot{\theta}^2) = 1, \tag{8.5.15}$$

(8.5.14) becomes

$$g_s r^2 \dot{\theta} = L, \tag{8.5.16}$$

and substituting $\dot{y}^0$ from (8.5.12) in (8.5.15) yields

$$\frac{W^2}{g_0} - g_s(\dot{r}^2 + r^2\dot{\theta}^2) = 1 \tag{8.5.17}$$

or

$$(\dot{r}^2 + r^2\dot{\theta}^2) = \frac{1}{g_s}\left(\frac{W^2}{g_0} - 1\right). \tag{8.5.18}$$

We change to θ as an independent variable, using (8.5.16):

$$\left(\left(\frac{dr}{d\theta}\right)^2 + r^2\right)\frac{L^2}{g_s^2 r^4} = \frac{1}{g_s}\left(\frac{W^2}{g_0} - 1\right). \tag{8.5.19}$$

The substitution $u = 1/r$ gives

$$\left[\left(\frac{du}{d\theta}\right)^2 + u^2\right] = \frac{g_s}{L^2}\left(\frac{W^2}{g_0} - 1\right). \tag{8.5.20}$$

The differential equation for the orbit is found by differentiating (8.5.20):

$$\frac{d^2u}{d\theta^2} + u = \frac{\partial}{\partial u}\left[\frac{1}{2}\frac{g_s}{L^2}\left(\frac{W^2}{g_0} - 1\right)\right]. \tag{8.5.21}$$

We see from (8.5.10) that expanding g_s and g_0 in powers of $1/r = u$ is equivalent to expanding in powers of $1/c^2$. Thus, since we are considering a slowly moving body (a planet), we may expand

$$\frac{1}{2}\frac{g_s}{L^2}\left(\frac{W^2}{g_0} - 1\right) = A + Bu + Cu^2 + \cdots. \tag{8.5.22}$$

Equation (8.5.21) becomes

$$\frac{d^2u}{d\theta^2} + u(1 - 2C) = B \tag{8.5.23}$$

or

$$u - \frac{B}{1 - 2C} = u_0 \cos{(1 - 2C)^{1/2}(\theta - \theta_0)}$$

or for small C, with θ_0 set to zero (choice of axis)

$$u = B + u_0 \cos[(1 - C)(\theta)] \tag{8.5.24}$$

for a precession angle $2\pi C$ per orbital year.

We have seen earlier that our linear field equations are not consistent — the right-hand side is not conserved if the particle is accelerating. When we learn how to deal with this problem, we will find corrections to $g_{\mu\nu}$ that contribute to C in (8.5.22). We will return to (8.5.24) when we have that information [see (8.7.23)].

We turn now to the bending of light by the sun. We treat the light as a fast particle with $v_0 \to 1$. In this calculation we need only keep the weak field limit, that is, the linear term in the expansion of g_s and g_0 in powers of the potential $2M_\odot G/r$. The constants W and L that appear in the equation can be calculated from their values far from the source $M_\odot$: $L = bv_0/\sqrt{1 - v_0^2}$ and $W = 1/\sqrt{1 - v_0^2}$. Thus, the orbit equation for $v_0 \to 1$ is [from (8.5.20)]

$$\left(\frac{du}{d\theta}\right)^2 + u^2 = \frac{1}{b^2}\frac{g_s}{g_0}$$

$$\cong \frac{1}{b^2}\left(1 + \frac{4M_\odot}{r}G\right) + \mathcal{O}\left(\left(\frac{M_\odot G}{r}\right)^2\right) \qquad (8.5.25)$$

and

$$u'' + u = \frac{2}{b^2}M_\odot G \qquad (8.5.26)$$

where b is the impact parameter of the light ray passing the sun.
The solution of (8.5.26) is

$$u = \frac{2M_\odot G}{b^2} - \alpha\cos\theta, \qquad (8.5.27)$$

where α is a constant. We have again chosen our axes to make $\theta_0 = 0$.
We determine the coefficient α by differentiating (8.5.27) with respect to σ:

$$-\frac{1}{r^2}\frac{dr}{d\sigma} = \alpha\sin\theta\,\frac{d\theta}{d\sigma}. \qquad (8.5.28)$$

As $r \to \infty$, before the scattering,

$$\frac{dr}{d\sigma} = -\frac{v_0}{\sqrt{1 - v_0^2}} \quad\text{and}\quad r^2\frac{d\theta}{d\sigma} = L = \pm\frac{bv_0}{\sqrt{1 - v_0^2}},$$

the $\pm$ sign depending on the initial sign of $\dot\theta$, which without loss of generality we take to be positive. α is then given by $\alpha = 1/b\sin\theta_i$ and (8.5.27) becomes

$$\frac{1}{r} = \frac{2M_\odot G}{b^2} - \frac{\cos\theta}{b\sin\theta_i} \qquad (8.5.29)$$

where θ_i is the initial direction of θ from the scatterer, or with $\xi = r/b$

$$\frac{1}{\xi} = \epsilon - \frac{\cos\theta}{\sin\theta_i}, \qquad (8.5.30)$$

where $\epsilon = 2M_\odot G/b$ is a small dimensionless number.
For $\xi \to \infty$,

$$\epsilon = +\frac{\cos\theta}{\sin\theta_i}; \qquad (8.5.31)$$

the two solutions of (8.5.31), $\theta = \pm\theta_i$, give θ before and after the scattering. For $\epsilon = 0$, $\theta_i = \pm\pi/2$, so the particle comes in along the $\pm y$-axis, goes out along the $\pm y$-axis, and is undeflected. The scattering angle is therefore the difference between $2\theta_i$ and π. Since ϵ is small, the equation for θ_i can be solved:

$$0 = \epsilon - \cot\theta_i$$

and with

$$\theta_i = \left[\frac{\pi}{2} - \delta\right] \tag{8.5.32}$$

we have

$$\epsilon - \delta = 0. \tag{8.5.33}$$

The scattered angle is therefore, without paying attention to sign,

$$|\theta_{sc}| = 2\epsilon = \frac{4M_\odot G}{b},$$

or in ordinary units

$$\theta_{sc} = \frac{4M_\odot G}{bc^2}. \tag{8.5.34}$$

To find the nature of the trajectory (attraction or repulsion), we have to trace out the orbit from (8.5.29). We write again

$$\frac{1}{\xi} = \epsilon - \frac{\cos\theta}{\sin\theta_i}. \tag{8.5.35}$$

We start with $\theta_i \sim \pi/2$, $\dot\theta > 0$, $1/\xi \cong \epsilon - \cos\theta$ so that (see Figure 8.1)

$$\epsilon - \cos\theta_i \cong 0, \qquad \theta_i \cong \frac{\pi}{2} - \epsilon.$$

Clearly, θ must increase from θ_i to keep $1/\xi > 0$. Thus, the trajectory circles the origin, ending at $\theta_f = -\pi/2 + \epsilon$. The particle is attracted to the scatterer.

We may compare (8.5.31) with a naive (sometimes called Newtonian) application of the equivalence principle, which would have a light wave packet accelerate under gravity according to the local gravitational field. Thus, for a small deflection, we can calculate the transverse change in

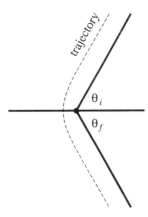

Figure 8.1.

velocity $\Delta \mathbf{v}$ from the unperturbed straight-line motion of the light. That is,

$$\Delta \mathbf{v} = \int_{-\infty}^{\infty} dt \, \mathbf{a}_{\perp}(t)$$

with

$$\mathbf{a}_{\perp} = - M_{\odot} \frac{G \mathbf{r}_{\perp}}{r^3}.$$

The result for impact parameter b is

$$\Delta \theta \approx \frac{\Delta v}{c} \approx \frac{2 M_{\odot} G}{c^2 b},$$

half the correct result.

We take up next the gravitational redshift. We note first that it follows directly and simply from the equivalence principle, as is shown in Problem 8.1.

A general argument for the effect is based on the tensor $g_{\mu\nu}$ itself. Consider, in a static gravitational field, a stationary clock at point A with an intrinsic period τ_A and a stationary clock at point B with an intrinsic period τ_B. Communication between them is by a fixed-frequency light wave. This is possible because $g_{\mu\nu}$ is time-independent. Call the period of the light wave T.

Now consider two events at A: two clicks of the stationary clock. The light wave will resonate with the clock if T is the time (*not* the proper time) between ticks of the clock, that is, if $T = \Delta y^0$, and

$$\sqrt{g_{\mu\nu}\Delta y^\mu \Delta y^\nu} = \tau_A.$$

Since the clock is stationary, our resonance condition is

$$T\sqrt{g_{00}(A)} = \tau_A.$$

Similarly, at B

$$T\sqrt{g_{00}(B)} = \tau_B.$$

Therefore,

$$\frac{\tau_B}{\tau_A} = \sqrt{\frac{g_{00}(B)}{g_{00}(A)}}$$

and the frequencies ω_B and ω_A have the ratio

$$\frac{\omega_B}{\omega_A} = \sqrt{\frac{g_{00}(A)}{g_{00}(B)}}.$$

Thus, if A is deep in a gravitational potential and B is not, $\omega_B/\omega_A < 1$. That is, the atom at A appears to be red-shifted.

We can also give an explicit mechanism for this to happen. We assume a $g_{\mu\nu}$ that is time-independent and imagine an atom in the sun and a different atom on Earth exchanging a light signal of definite frequency ω_0. We then calculate the resonant frequency ω_0 of the two atoms in terms of the gravitational potentials at the two positions and the field-free resonant frequencies ω_R. We will be solving an atomic equation at each site, so that the variation of the gravitational potential at each site will be negligible and may be ignored. Thus, we imagine $g_{\mu\nu}$ to be constant at each site, but different from one site to the other.

The atomic equation will be (for a scalar particle with charge e)

$$[(\partial_\mu - ieA_\mu)g^{\mu\nu}(\partial_\nu - ieA_\nu) + m^2]\psi = 0, \tag{8.5.36}$$

where we have replaced covariant derivatives with ordinary ones, as discussed above. The Klein–Gordon equation (8.5.36) with constant $g^{\mu\nu}$ and time-independent A_μ will permit a time dependence

$$\psi = \chi\, e^{-iEx^0} \tag{8.5.37}$$

where χ is time-independent. The result is an eigenvalue equation for E; the differences of E between two states of the same atom will be the exchanged frequency ω_0.

Before writing the equation, we transform to a coordinate system in which the three g_{i0}'s are zero. This can be done by shifting the time origin:

$$x^0 = x'^0 + \Lambda(\mathbf{x}'), \qquad \mathbf{x} = \mathbf{x}' \tag{8.5.38}$$

so that

$$0 = g'_{i0} = \frac{\partial x^\mu}{\partial x'^i} \frac{\partial x^\nu}{\partial x'^0} g_{\mu\nu}$$

$$= \frac{\partial x^\mu}{\partial x'^i} g_{\mu 0}$$

$$= \frac{\partial x^j}{\partial x'^i} g_{j0} + \frac{\partial x^0}{\partial x'^i} g_{00}$$

$$= g_{i0} + \frac{\partial \Lambda}{\partial x'^j} g_{00} \tag{8.5.39}$$

and

$$\frac{\partial \Lambda}{\partial x'^i} = - \frac{g_{i0}}{g_{00}}. \tag{8.5.40}$$

This equation can be solved—always neglecting the variation of g_{i0}/g_{00}. The solution is

$$\Lambda = - \frac{g_{i0}x'^i}{g_{00}}. \tag{8.5.41}$$

Note also that (8.5.38) leaves g_{00} unchanged.

We now return to (8.5.36), with the insertion (8.5.37) and the space components of the vector potential equal to zero. The equation is

$$[-g^{00}(E + eA_0)^2 + \partial_i \partial_j g^{ij} + m^2]\chi = 0. \tag{8.5.42}$$

The vector potential A^0 will be given (in non-rationalized units) by

$$-\partial_i g^{ij} \partial_j A^0 = 4\pi j^0 \tag{8.5.43}$$

where j^0 is the charge density of—as a model—a heavy proton:

$$j^0 = e_p \frac{\delta^3(\mathbf{x})}{\sqrt{g}}$$

$$= \frac{e_p \delta^3(\mathbf{x})}{\sqrt{g_{00}} \sqrt{g_3}} \tag{8.5.44}$$

where $-g_3$ is the determinant of the three-dimensional matrix g_{ij}:

$$g_3 = - \det g_{ij} \tag{8.5.45}$$

Since both (8.5.42) and (8.5.43) are spatially covariant equations, we can transform them to the form

$$\left[-g^{00}(E + eA_0)^2 - \nabla'^2 + m^2 \right] \chi = 0 \tag{8.5.46}$$

and

$$\nabla'^2 A^0 = 4\pi e_p \frac{\delta^3(\mathbf{x}')}{\sqrt{g_{00}}} \tag{8.5.47}$$

so that

$$A^0 = - \frac{e_p}{\sqrt{g_{00}} r'} \quad \text{and} \quad A_0 = - \sqrt{g_{00}} \frac{e_p}{r'}. \tag{8.5.48}$$

The eigenvalue equation is now

$$\left(-\left(\sqrt{g^{00}} E - \frac{ee_p}{r'} \right)^2 - \nabla'^2 + m^2 \right) \chi = 0 \tag{8.5.49}$$

and $\sqrt{g^{00}} E = E_R$, the solution of the eigenvalue problem in the absence of a gravitational field. Taking the difference of two states, we find, with $\Delta E = \hbar \omega_0$,

$$\sqrt{g^{00}} \omega_0 = \omega_R, \tag{8.5.50}$$

so $\sqrt{g_{00}} \omega_R = \omega_0$, for the two atoms that are exchanging the signal. Thus, in the sun, where

$$g_{00} = 1 + 2\phi, \tag{8.5.51}$$

with ϕ the gravitational potential, $g_{00} < 1$, whereas on Earth g_{00} will be much closer to 1. Therefore, the light emitted by the solar atom will have a lower frequency ω_0 than ω_R, the frequency that would be observed for the same atom in a field-free region. This calculation shows how general covariance produced a mechanism for the gravitational red shift.

8.6. CURVATURE

We are now ready to confront the inconsistency of the linear field theory with which we have been working. The clue is the equivalence principle. We have seen that the equivalence principle for particle motion or electromagnetic field equations could be guaranteed by making the equations generally covariant with respect to the second-rank tensor $g_{\mu\nu}$. However, for $g_{\mu\nu}$ to transform like a tensor, it must satisfy covariant equations. The problem is therefore to construct a covariant function $H^{\mu\lambda\cdots}$ of the g's and their derivatives and set it equal to the assumed source—presumably still the correctly calculated stress tensor for matter, $T_M^{\mu\nu}$. If so, the equation must be something like

$$H^{\mu\nu}\left(g, \frac{\partial g}{\partial x}, \frac{\partial^2 g}{\partial x^2}, \dots\right) = \lambda T_M^{\mu\nu}. \tag{8.6.1}$$

The immediate problem is that the obvious way—introducing the covariant derivatives $g_{\mu\nu;\lambda}$—does not work, since they are all zero.

There is, however, a tensor function of the g's that we can construct. We take advantage of the covariant derivatives of a vector field V^μ:

$$V^\mu_{;\nu} = \partial_\nu V^\mu + \Gamma^\mu_{\nu\lambda} V^\lambda \tag{8.6.2}$$

and

$$V^\mu_{;\nu;\lambda} = \partial_\lambda V^\mu_{;\nu} + \Gamma^\mu_{\lambda\sigma} V^\sigma_{;\nu} - \Gamma^\tau_{\lambda\nu} V^\mu_{;\tau} \tag{8.6.3}$$

so that the tensor $V^\mu_{;\nu;\lambda} - V^\mu_{;\lambda;\nu}$ is given by

$$V^\mu_{;\nu;\lambda} - V^\mu_{;\lambda;\nu} = \partial_\lambda\left(\partial_\nu V^\mu + \Gamma^\mu_{\nu\eta} V^\eta\right) + \Gamma^\mu_{\lambda\sigma}\left(\partial_\nu V^\sigma + \Gamma^\sigma_{\nu\tau} V^\tau\right)$$

$$- \partial_\nu\left(\partial_\lambda V^\mu + \Gamma^\mu_{\lambda\eta} V^\eta\right) - \Gamma^\mu_{\nu\sigma}\left(\partial_\lambda V^\sigma + \Gamma^\sigma_{\lambda\tau} V^\tau\right)$$

$$= \left(\partial_\lambda \Gamma^\mu_{\nu\eta} - \partial_\nu \Gamma^\mu_{\lambda\eta}\right) V^\eta + \left(\Gamma^\mu_{\lambda\sigma} \Gamma^\sigma_{\nu\eta} - \Gamma^\mu_{\nu\sigma} \Gamma^\sigma_{\lambda\eta}\right) V^\eta$$

$$\equiv R^\mu_{\;\eta\nu\lambda} V^\eta \tag{8.6.4}$$

where

$$R^\mu_{\;\eta\nu\lambda} = \partial_\lambda \Gamma^\mu_{\nu\eta} - \partial_\nu \Gamma^\mu_{\lambda\eta} + \Gamma^\mu_{\lambda\sigma} \Gamma^\sigma_{\nu\eta} - \Gamma^\mu_{\nu\sigma} \Gamma^\sigma_{\lambda\eta}. \tag{8.6.5}$$

Since $V^{\mu}{}_{;\nu;\lambda}$ is a tensor and V^{η} a vector, $R^{\mu}{}_{\eta\nu\lambda}$ is a tensor. It is called the curvature tensor. It has the following properties.

First, if the tensor $g_{\mu\nu}$ describes a space-time metric, and any component of $R^{\mu}_{\eta\nu\lambda}$ is different from zero in a finite region, then no coordinate change can transform $g_{\mu\nu}$ to pseudo-Euclidean form. This is true since if it could, and since all the components of $R^{\mu}_{\eta\nu\lambda}$ vanish in the pseudo-Euclidean system, and since $R^{\mu}_{\eta\nu\lambda}$ is a tensor, all the components of $R^{\mu}_{\eta\nu\lambda}$ would have had to vanish in the original coordinate system.

Second, the converse, which we shall not prove, also holds: If all components of $R^{\mu}_{\eta\nu\lambda}$ are zero in a finite region of space, then it is possible to introduce a coordinate system that transforms $g_{\mu\nu}$ to pseudo-Euclidean form in that entire region.

Third, the translation to the gravitational field $g_{\mu\nu}$ follows: If any component of $R^{\mu}_{\eta\nu\lambda}$ is different from zero in some region of space-time, there is no coordinate transformation that eliminates the gravitational field in that entire region. Conversely, if all the components of $R^{\mu}_{\eta\nu\lambda}$ are zero, one can find a pseudo-Euclidean coordinate system, in which $g_{\mu\nu} = \eta_{\mu\nu}$ in the entire region, the gravitational field $\phi_{\mu\nu}$ is zero, and the gravitation-free laws of special relativity hold. The curvature tensor $R^{\mu}_{\eta\nu\lambda}$ carries the invariant reality of the gravitational field.

In order to decide on the equations to be satisfied by $R^{\mu}_{\eta\nu\lambda}$, we need to know some of its general properties. These are most easily studied in the local coordinate system that makes $g_{\mu\nu} = \eta_{\mu\nu}$ and $\partial g_{\mu\nu}/\partial x^{\lambda} = 0$. In that coordinate system, at the chosen point,

$$R^{\lambda}{}_{\mu\nu\kappa} = R^{(0)\lambda}_{\mu\nu\kappa} = \partial_{\kappa}\Gamma^{\lambda}_{\mu\nu} - \partial_{\nu}\Gamma^{\lambda}_{\mu\kappa}$$

$$= \frac{1}{2}\eta^{\lambda\tau}\partial_{\kappa}(\partial_{\mu}g_{\tau\nu} + \partial_{\nu}g_{\tau\mu} - \partial_{\tau}g_{\mu\nu}) - \{\kappa \leftrightarrow \nu\} \qquad (8.6.6)$$

The covariant curvature

$$R_{\lambda\mu\nu\kappa} = g_{\lambda\tau}R^{\tau}{}_{\mu\nu\kappa},$$

is

$$R^{(0)}_{\lambda\mu\nu\kappa} = \frac{1}{2}(\partial_{\kappa}\partial_{\mu}g_{\lambda\nu} + \partial_{\nu}\partial_{\lambda}g_{\mu\kappa} - \partial_{\kappa}\partial_{\lambda}g_{\mu\nu} - \partial_{\nu}\partial_{\mu}g_{\lambda\kappa}). \qquad (8.6.7)$$

We can read off from (8.6.7) the symmetry properties of $R^{(0)}_{\lambda\mu\nu\kappa}$; since $R_{\lambda\mu\nu\kappa}$ is a tensor, they will hold in general. These are:

1. $R_{\lambda\mu\nu\kappa}$ is antisymmetric in exchange of ν and κ.
2. $R_{\lambda\mu\nu\kappa}$ is antisymmetric in exchange of λ and μ.
3. $R_{\lambda\mu\nu\kappa}$ is symmetric in exchange of the pair $(\lambda\mu)$ with $(\nu\kappa)$.
4. There is one more algebraic relation:

$$R_{\lambda\mu\nu\kappa} + R_{\lambda\kappa\mu\nu} + R_{\lambda\nu\kappa\mu} = 0. \tag{8.6.8}$$

Equation (8.6.8) only adds one constraint (in four dimensions) since if any two of the four indices are equal, the sum is identically zero by the symmetry properties 1–3.

We can now calculate the number of independent components of $R_{\lambda\mu\nu\kappa}$ by considering $R_{(\lambda\mu)(\nu\kappa)}$ as a symmetric 6×6 matrix—six being the number of values the antisymmetric pair $(\lambda\mu)$ and the pair $(\nu\kappa)$ can take. This number of components is

$$\frac{6 \times 5}{2} + 6 = 21.$$

Subtracting the single constraint (8.6.8), we get exactly the expected number, 20 (see the discussion in Section 7.7).

There is in addition one more identity, but this time it is differential rather than algebraic, the Bianchi identity,

$$R_{\lambda\mu\nu\kappa;\tau} + (\text{cyclic permutation of } \nu\kappa\tau) = 0 \tag{8.6.9}$$

or in the locally flat coordinate system

$$R^{(0)}_{\lambda\mu\nu\kappa,\tau} + (\text{cyclic permutations of } \nu\kappa\tau) = 0 \tag{8.6.10}$$

which is easily verified from (8.6.7).

Two more tensors can be constructed from $R_{\lambda\mu\nu\kappa}$ by tracing. Note that the symmetry properties of $R_{\lambda\mu\nu\kappa}$ permit only one trace (to within a sign):

$$R_{\mu\kappa} = g^{\lambda\nu} R_{\lambda\mu\nu\kappa} \tag{8.6.11}$$

so that

$$R^{(0)}_{\mu\kappa} = \eta^{\lambda\nu} R^0{}_{\lambda\mu\nu\kappa}$$

$$= \frac{1}{2}(\partial_\kappa\partial_\mu g + \partial_\nu\partial^\nu g_{\mu\kappa} - \partial_\kappa\partial^\nu g_{\mu\nu} - \partial_\mu\partial^\nu g_{\kappa\nu}) \tag{8.6.12}$$

which is symmetric in (μ, κ); a second trace gives the curvature scalar

$$R = g^{\mu\kappa} R_{\mu\kappa}$$

$$= \eta^{\mu\kappa} R^{(0)}_{\mu\kappa} = \partial_\kappa\partial^\kappa g - \partial^\kappa\partial^\mu g_{\kappa\mu}. \tag{8.6.13}$$

Here, $g = \eta^{\mu\lambda} g_{\mu\lambda}$.

8.7. THE EINSTEIN FIELD EQUATIONS AND THE PRECESSION OF ORBITS

We recognize in (8.6.12) and (8.6.13) the two components of the equations for $\phi_{\mu\nu}$ that we found in (8.2.21) (setting $\mu = 0$). That is, remembering $g_{\mu\nu} = \eta_{\mu\nu} + 2\lambda\phi_{\mu\nu}$, we found

$$R^{(0)}{}_{\mu\kappa} - \frac{\eta_{\mu\kappa}}{2} R^{(0)} = -\lambda^2 T^{(0)}_{\mu\kappa} \tag{8.7.1}$$

where $T^{(0)}{}_{\mu\kappa}$ is the matter stress tensor in the locally Minkowskian coordinate system. As in Section 8.4, the left-hand side of (8.7.1) is identically conserved:

$$\partial^\mu \left(R^{(0)}_{\mu\kappa} - \frac{\eta_{\mu\kappa}}{2} R^{(0)} \right) = 0. \tag{8.7.2}$$

Notice that now, however, (8.7.1) is consistent, because the right-hand side $T^{(0)}_{\mu\kappa}$ is also conserved (the Γ's are all zero!).

The covariant equation that follows from (8.7.1) in a general coordinate system is

$$R_{\mu\kappa} - \frac{g_{\mu\kappa}}{2} R = -\lambda^2 T_{\mu\kappa}. \tag{8.7.3}$$

The conservation law (8.7.2) becomes

$$\left(R^{\mu\kappa} - \frac{g^{\mu\kappa}}{2} R \right)_{;\mu} = 0, \tag{8.7.4}$$

which, of course, is the reflection of the Bianchi identity on the properties of the second-rank tensor $R^{\mu\kappa}$. However, the right-hand side of (8.7.3) also has vanishing covariant divergence, since it is constructed from the covariant matter (including electromagnetic) Lagrangian as discussed in Section 7.8. The equations are therefore formally[9] consistent.

Note that in arriving at the field equation (8.7.2), the weak field assumption has not been made, although it is clearly the case that the weak field (small $\phi_{\mu\nu}$) limit of (8.7.3) reproduces (8.2.21); however, in this limit, the right- and left-hand sides are no longer consistent with the

---

[9]They are only formally consistent since the same kind of point singularity that occurs in other relativistic field theories occurs here.

particle equations and must be supplemented by the higher-order terms coming from the nonlinearities in $R_{\mu\nu}$ and $T_{\mu\nu}$. We turn now to that problem, with the goal of finding the next approximation to the gravitational potential produced by the sun, and in particular of finding the correct value of the coefficient C in (8.5.22).

We wish to expand (8.7.3) in powers of the potential $V = GM_\odot/r$, which by the virial theorem is $\sim v^2$, and therefore small for $v/c \ll 1$. We see from (8.5.21) and (8.5.12), since $W \sim 1$, that we need g_{ij} only to first order in V, but g_{00} to the next order. Thus, we must find the next correction to ϕ_{00}.

We write

$$R_{\mu\kappa} = R_{\mu\kappa}^{(1)} + R_{\mu\kappa}^{(2)} + \cdots$$

where $R_{\mu\kappa}^{(1)}$ is linear in ϕ, $R^{(2)}$ quadratic, etc. In turn, $g_{\mu\nu}$ is calculated as

$$g_{\mu\nu} = g_{\mu\nu}^{(0)} + g_{\mu\nu}^{(1)} + \cdots,$$

where $g_{\mu\nu}^{(0)} = \eta_{\mu\nu}$, $g_{\mu\nu} - g_{\mu\nu}^{(0)} = 2\lambda\phi_{\mu\nu}$, and

$$\phi_{\mu\nu} = \phi_{\mu\nu}^{(1)} + \phi_{\mu\nu}^{(2)} + \cdots \qquad (8.7.5)$$

with $\phi_{\mu\nu}^{(1)}$ given by (8.4.27):

$$\phi_{\mu\nu}^{(1)} = -\frac{\lambda M_\odot}{8\pi} \frac{t_{\mu\nu}}{|\mathbf{x} - \mathbf{x}_p|} \qquad (8.7.6)$$

with $t_{00} = t_{11} = t_{22} = t_{33} = 1$, all other components zero. We expand by expressing $R^{(1)}$ as a function of $\phi^{(2)}$ and $R^{(2)}$ as a function of the known $\phi^{(1)}$; we then solve the resulting inhomogeneous equation for $\phi^{(2)}$.

Before carrying out that procedure, we observe that (8.7.3) provides us with an exactly conserved, symmetric tensor $\tau^{\mu\nu}$, which is a nonlinear function of the ϕ field and a suitable candidate for the stress tensor of the coupled system. Since

$$R_{\mu\nu}^{(1)} - \frac{g_{\mu\nu}^{(0)}}{2} R^{(1)} = G_{\mu\nu}^{(0)}$$

is identically conserved [i.e., $\partial^\mu G_{\mu\nu}^{(0)} = 0$, where ∂^μ is *not* the covariant derivative, but the ordinary derivative], we must have that

$$\tau^{\mu\nu} = -\lambda^2 T^{\mu\nu} - \left(R^{\mu\nu} - R^{(1)\mu\nu} - \left(\frac{g^{\mu\nu}}{2} R - \frac{\eta^{\mu\nu}}{2} R^{(1)} \right) \right) \qquad (8.7.7)$$

is also exactly conserved. Further, since $T^{\mu\nu}$ includes contributions from all nongravitational sources (point particles, electromagnetic field, etc.) that reduce to known forms in the ξ coordinate system, the conserved tensor $\tau^{\mu\nu}$ correctly describes the exchange of energy and momentum between gravitational and other degrees of freedom. The tensor $\tau^{\mu\nu}$ therefore makes it possible to calculate conventional energy, momentum, and angular momentum fluxes of gravitational radiation. In the present application, it makes it possible, by using (8.7.3), to calculate the next approximation to g_0 [as defined in (8.5.9)] and from that and (8.5.22) the precession of the planetary orbits.

We choose a gauge (coordinate system) in which $\phi^{(2)}$ satisfies the same linear gauge condition as $\phi^{(1)}$, that is,

$$\partial^\mu \phi^{(2)}_{\mu\nu} = \frac{1}{2}\partial_\nu \phi^{(2)} \tag{8.7.8}$$

where $\phi = \phi^\lambda_\lambda$ and all contractions are carried out with $\eta_{\mu\nu}$. Then

$$R^{(1)}_{\nu\kappa} = \frac{1}{2}\partial_\nu\partial^\nu g_{\mu\kappa} = \lambda\partial_\nu\partial^\nu \phi_{\mu\kappa} \tag{8.7.9}$$

and

$$R^{(1)} = \frac{1}{2}\partial_\nu\partial^\nu g = \lambda\partial_\nu\partial^\nu \phi. \tag{8.7.10}$$

Equation (8.7.3) is therfore

$$\lambda\left(\partial_\nu\partial^\nu \phi_{\mu\kappa} - \frac{1}{2}\eta_{\mu\kappa}\partial_\nu\partial^\nu \phi\right) = -\lambda^2 T_{\mu\kappa} - \left(R^{(2)}_{\mu\kappa} - \frac{(g_{\mu\kappa}R)^{(2)}}{2}\right) + 0(\phi_1^3)$$

$$+ 0(\phi_2\phi_1). \tag{8.7.11}$$

[Note that (8.7.11) would become an exact equation if instead of $R^{(2)}$, we wrote $R - R^{(1)}$.] Proceeding systematically, we rewrite (8.7.3) as

$$R^{(1)}_{\mu\nu} - \frac{\eta_{\mu\nu}}{2}R^{(1)} + R^{(2)}_{\mu\nu} - \frac{\eta_{\mu\nu}}{2}R^{(2)} - \frac{\overset{(1)}{g_{\mu\nu}}}{2}R^{(1)} = -\lambda^2 T_{\mu\nu} + \cdots, \tag{8.7.12}$$

where

$$\overset{(0)}{g_{\mu\nu}} = \eta_{\mu\nu}, \qquad \overset{(1)}{g_{\mu\nu}} = 2\lambda\phi^{(1)}_{\mu\nu}, \qquad \text{etc.}$$

We note that $R^{(1)}$ is zero outside the source to lowest order. Since contributions inside the source will give corrections to the $1/r$ potential, which causes no precession, they can be ignored. Then the equation we have to solve is

$$R_{\mu\nu}^{(1)}\big(\phi^{(2)}\big) - \frac{\eta_{\mu\nu}}{2} R^{(1)}\big(\phi^{(2)}\big) + R_{\mu\nu}^{(2)}\big(\phi^{(1)}\big) - \frac{\eta_{\mu\nu}}{2} R^{(2)}\big(\phi^{(1)}\big) = 0 \qquad (8.7.13)$$

or

$$R_{\mu\nu}^{(1)}\big(\phi^{(2)}\big) = - R_{\mu\nu}^{(2)}\big(\phi^{(1)}\big) \qquad (8.7.14)$$

and, in the chosen gauge,

$$\lambda \partial_\sigma \partial^\sigma \phi_{\mu\nu}^{(2)} = - R_{\mu\nu}^{(2)}\big(\phi^{(1)}\big). \qquad (8.7.15)$$

We first find a general formula for $R^{(2)}$ and then specialize to the case at hand:

$$R_{\mu\eta} = \partial_\eta \Gamma_{\mu\lambda}^{\lambda} - \partial_\lambda \Gamma_{\mu\eta}^{\lambda} + \Gamma_{\mu\lambda}^{\sigma}\Gamma_{\sigma\eta}^{\lambda} - \Gamma_{\mu\eta}^{\sigma}\Gamma_{\sigma\lambda}^{\lambda} \qquad (8.7.16)$$

and

$$\begin{aligned}
\frac{R_{\mu\eta}^{(2)}}{4\lambda^2} &= -\frac{1}{2}\phi^{\lambda\sigma}\{\partial_\eta \partial_\mu \phi_{\sigma\lambda} - \partial_\lambda \partial_\mu \phi_{\sigma\eta} - \partial_\lambda \partial_\eta \phi_{\mu\sigma} + \partial_\lambda \partial_\sigma \phi_{\mu\eta}\} \\
&\quad - \frac{1}{2}\partial_\eta \phi^{\lambda\sigma}\partial_\mu \phi_{\sigma\lambda} + \left(\partial_\lambda \frac{\phi^{\lambda r}}{2} - \frac{\partial^\sigma \phi_\lambda^{\lambda}}{4}\right)(\partial_\mu \phi_{\sigma\eta} + \partial_\eta \phi_{\sigma\mu} - \partial_\sigma \phi_{\mu\eta}) \\
&\quad + \frac{1}{4}\left(\partial_\mu \phi_\lambda^{\sigma} + \partial_\lambda \phi_\mu^{\sigma} - \partial^\sigma \phi_{\mu\lambda}\right)\left(\partial_\sigma \phi_\eta^{\lambda} + \partial_\eta \phi_\sigma^{\lambda} - \partial^\lambda \phi_{\sigma\eta}\right). \qquad (8.7.17)
\end{aligned}$$

In (8.7.17), ϕ stands for $\phi^{(1)}$, and all raising is done via $\eta^{\mu\nu}$. However, note that with this convention

$$g^{\mu\nu} = \eta^{\mu\nu} -- 2\lambda\phi^{\mu\nu} \qquad (8.7.18)$$

although

$$g_{\mu\nu} = \eta_{\mu\nu} + 2\lambda\phi_{\mu\nu},$$

since

$$g_{\mu\nu}g^{\nu\sigma} = \delta_\mu^\sigma.$$

We now specialize to the case at hand: We set $\mu = \eta = 0$, $\partial_\mu = \partial_\eta = 0$ since we are looking for time-independent solutions. This leaves (note the gauge condition on $\phi^{(1)}$),

$$\frac{R_{00}^{(2)}}{4\lambda^2} = -\frac{1}{2}\phi^{ij}\partial_i\partial_j\phi_{00} + \frac{1}{4}(\partial_\lambda \phi_0^{\sigma} - \partial^\sigma \phi_{0\lambda})(\partial_\sigma \phi_0^{\lambda} - \partial^\lambda \phi_{\sigma 0}). \qquad (8.7.19)$$

The first term in (8.7.19) will be proportional to $\nabla^2\phi_{00}$ and, hence, confined to the source; it therefore does not contribute to the precession. There remains

$$R_{00}^{(2)} = -2\lambda^2 \, \partial_\lambda \phi_0^0 \, \partial^\lambda \phi_{00} = 2\lambda^2 (\nabla \phi_{00})^2 \tag{8.7.20}$$

and (8.7.15) becomes

$$-\nabla^2 \lambda \phi_{00}^{(2)} = -2\lambda^2 (\nabla \phi_{00})^2$$

or

$$\nabla^2 \phi_{00}^{(2)} = \frac{\lambda^3}{32\pi^2} \frac{M_\odot^2}{r^4} \tag{8.7.21}$$

and

$$\phi_{00}^{(2)} = \frac{\lambda^3 M_\odot^2}{64\pi^2 r^2} + \frac{A}{r} + B \tag{8.7.22}$$

where A and B are integration constants that do not affect the precession. Our formula for g_0 is then

$$g_0 = 1 - \frac{2\lambda^2 M_\odot}{8\pi r} + \frac{\lambda^4 M_\odot^2}{32\pi^2 r^2} = 1 + 2V + 2V^2 \tag{8.7.23}$$

where

$$V = -\frac{\lambda^2 M_\odot}{8\pi r} = -\frac{GM_\odot}{r}$$

is the nonrelativistic potential of the sun.

We now return to (8.5.20) for the orbit equation. Setting $W \sim 1 - \epsilon$, we find

$$\left(\frac{du}{d\theta}\right)^2 + u^2 = \frac{1}{L^2}(1 - 2V)\left(\frac{(1-\epsilon)^2}{1 + 2V + 2V^2} - 1\right)$$

$$\cong -\frac{2\epsilon}{L^2} - \frac{2V}{L^2} + \frac{6V^2}{L^2}. \tag{8.7.24}$$

The nonrelativistic equation would be

$$\left(\frac{du}{d\theta}\right)^2 + u^2 = \frac{2(E - V)}{L^2}$$

where E is the total energy, $E = (mv^2/2) + V$. Therefore, the equivalent potential for our problem is $V_{eff} = V - 3V^2$, and the precession is given by the solution

$$u \cong B + u_0 \cos\left(1 - 3\left(\frac{GM_\odot}{L}\right)^2\right)\theta \qquad (8.7.25)$$

which returns to the same value of u at

$$\theta = 2\pi\left(1 + 3\left(\frac{GM_\odot}{L}\right)^2\right) \qquad (8.7.26)$$

for a forward precession of $6\pi(GM_\odot/L)^2$ per revolution.

Note that $L \sim rv$ and $GM_\odot/r \sim v^2$, so that the precession angle is of order $(v/c)^2$, equivalent to the scale of fine structure in atomic physics. (See Problem 8.8.)

Note also that the second-order contribution to g_0 was $-2V^2$ out of a final $6V^2$. However, that division is not gauge-invariant; transformations of the form $r' = f(r)$ can shift contributions between different orders, since both first- and second-order potentials give precession of order $(v/c)^2$.

We close this section by constructing an action integral for (8.7.3). The Lagrangian density for the field $g_{\mu\nu}$ must be an invariant. The simplest guess is, of course, R itself. We therefore try the action

$$\frac{S_g}{\alpha} = \int d^4x\sqrt{g}\, R, \qquad (8.7.27)$$

where α is to be determined from the known equation of motion, with

$$S_{\text{matter}} = -\sum_p \frac{m_p}{2} \int d\sigma_p \frac{dx_p^\mu}{d\sigma_p} g_{\mu\nu}(x_p) \frac{dx_p^\nu}{d\sigma_p}. \qquad (8.7.28)$$

We can verify our guess most simply by transforming, at a given point in space, to the coordinate system that is pseudo-Euclidean in the immediate neighborhood of that point. It is easily seen that $\sqrt{g}\, R$ differs by a derivative (or equivalently by an integration by parts) from a function of the $g_{\mu\nu}$'s which is bilinear in the first derivatives of $g_{\mu\nu}$'s. Therefore, at the chosen point, we may set $g_{\mu\nu} = \eta_{\mu\nu}$ in the Lagrangian, since the variation in $g_{\mu\nu}$ will be multiplied by derivatives of the $g_{\mu\nu}$'s that themselves vanish. We may *not* set the derivatives equal to zero in the Lagrangian, since *this* variation will lead to second derivatives, which do not vanish, in the equations of motion. However, since the local equations in this coordinate system are given by the weak field equations (8.2.21) (with $\mu^2 = 0$), the Lagrangian must be given by the weak field Lagrangian,

(8.3.9). One verifies easily that the equality holds with $1/\alpha = 16\pi G$. Note that the sign of the curvature scalar is an odd function of the sign of $g_{\mu\nu}$. Therefore, choosing $g_{\mu\nu}$ to have positive space components would result in a negative sign for α.

8.8. GRAVITATIONAL RADIATION

The equation satisfied by the gravitational potential $\phi^{\mu\nu}$ is given by (8.4.14). In the harmonic gauge, it is given by (8.4.22):

$$\partial_\lambda \partial^\lambda \phi^{\mu\nu} = -\lambda S^{\mu\nu} \tag{8.8.1}$$

where

$$S^{\mu\nu} = T^{\mu\nu} - \frac{1}{2}\eta^{\mu\nu}T \tag{8.8.2}$$

and $T^{\mu\nu}$ is supplemented, as in (8.7.7), to make it exactly conserved: $\partial_\mu T^{\mu\nu} = 0$; this leads to the harmonic gauge condition on $S^{\mu\nu}$:

$$\partial_\mu S^{\mu\nu} = \frac{1}{2}\partial^\nu S, \tag{8.8.3}$$

where

$$S = S^\lambda_\lambda = \eta_{\lambda\alpha}S^{\alpha\lambda}. \tag{8.8.4}$$

As usual, we choose the retarded solution, to which we return shortly. First, we discuss the radiation itself.

The propagation equation (for finite μ, which we consider first) is

$$\left(\partial_\lambda \partial^\lambda + \mu^2\right)\phi_{\rho\nu} - \partial^\lambda \partial_\rho \phi_{\lambda\nu} - \partial^\lambda \partial_\nu \phi_{\lambda\rho} + \partial_\rho \partial_\nu \phi$$
$$- \eta_{\rho\nu}\left(\left(\partial_\lambda \partial^\lambda + \mu^2\right)\phi - \partial^\alpha \partial^\beta \phi_{\alpha\beta}\right) = 0. \tag{8.8.5}$$

By applying the operator ∂^ρ to the left-hand side of (8.8.5), we find the constraint equation

$$\mu^2(\partial^\rho \phi_{\rho\nu} - \partial_\nu \phi) = 0$$

or, since $\mu^2 \neq 0$,

$$\partial^\rho \phi_{\rho\nu} = \partial_\nu \phi. \tag{8.8.6}$$

Equation (8.8.5) can therefore be rewritten as

$$\left(\partial_\lambda \partial^\lambda + \mu^2\right) \phi_{\rho\nu} - \partial_\rho \partial_\nu \phi - \eta_{\rho\nu} \mu^2 \phi = 0. \tag{8.8.7}$$

The trace of (8.8.7) gives

$$\mu^2 \phi = 0 \qquad \text{or} \qquad \phi = 0, \tag{8.8.8}$$

and, thus, from (8.8.6)

$$\partial^\rho \phi_{\rho\nu} = 0. \tag{8.8.9}$$

Equations (8.8.8) and (8.8.9) were the starting point of our investigation of tensor fields, so it should not be a surprise that we have recovered them. The propagation equations are thus

$$\left(\partial_\lambda \partial^\lambda + \mu^2\right) \phi_{\mu\nu} = 0 \tag{8.8.10}$$

which with the five constraint equations leaves five propagating modes; one sees from the Fourier transform constraint equation (with $k_0 = \omega$ and $k_1 = k_2 = 0$)

$$\omega \phi_{00} + k \phi_{30} = 0, \tag{8.8.11}$$

$$\omega \phi_{0i} + k \phi_{3i} = 0, \tag{8.8.12}$$

and

$$\phi_{00} - \phi_{33} - \phi_{11} - \phi_{22} = 0, \tag{8.8.13}$$

that one can conveniently take as dynamical variables ϕ_{12}, ϕ_{13}, ϕ_{23}, ϕ_{33}, and $\phi_{11} - \phi_{22}$. The other components can then be found from (8.8.11), (8.8.12), and (8.8.13).

We turn next to the gravitational field, with $\mu = 0$. We note first that the four constraint equations (8.8.6) cannot be derived and, in fact, will not usually hold. There are, however, four constrained variables. We see directly from (8.8.5) that no second time derivatives of ϕ_{00} and ϕ_{0i} occur, so that ϕ_{00} and ϕ_{0i} are constrained variables. The second derivatives that do occur are of the six linearly independent components ϕ_{12}, ϕ_{13}, ϕ_{23}, $\phi_{11} + \phi_{22}$, $\phi_{11} + \phi_{33}$, and $\phi_{22} + \phi_{33}$. We still have the four freedoms of gauge choice, so that we can eliminate four more components, leaving us with two propagating modes. We now proceed to carry out the reduction. The wave equation (8.8.5) in wave number space is

$$k^2 \phi_{\mu\nu} - k^\lambda k_\mu \phi_{\lambda\nu} - k^\lambda k_\nu \phi_{\lambda\mu} + k_\mu k_\nu \phi - \eta_{\mu\nu}(k^2 \phi - k^\alpha k^\beta \phi_{\alpha\beta}) = 0. \tag{8.8.14}$$

Taking the trace of (8.8.14) shows that the $\eta_{\mu\nu}$ term can be dropped, leaving

$$k^2 \phi_{\mu\nu} - k^\lambda k_\mu \phi_{\lambda\nu} - k^\lambda k_\nu \phi_{\lambda\mu} + k_\mu k_\nu \phi = 0. \tag{8.8.15}$$

Consider first $k^2 \neq 0$. Then

$$\phi_{\mu\nu} = k^\lambda \frac{k_\mu}{k^2} \phi_{\lambda\nu} + k^\lambda k_\nu \frac{\phi_{\lambda\mu}}{k^2} - \frac{k_\mu k_\nu}{k^2} \phi \tag{8.8.16}$$

and the gauge transformation

$$\phi_{\mu\nu} \rightarrow \phi_{\mu\nu} + k_\mu \xi_\nu + k_\nu \xi_\mu \tag{8.8.17}$$

with

$$\xi_\nu = -\frac{k^\lambda \phi_{\lambda\nu}}{k^2} - \frac{1}{2} \frac{k_\nu \phi}{k^2} \tag{8.8.18}$$

reduces every component to zero.

Therefore, only $k^2 = 0$ modes propagate. They, however, must be in the harmonic gauge, as we now show! Since $k^2 = 0$, (8.8.15) becomes

$$k_\mu \psi_\nu + k_\nu \psi_\mu = k_\mu k_\nu \phi \tag{8.8.19}$$

where $\psi_\mu = k^\lambda \phi_{\mu\lambda}$.

We solve (8.8.19) by letting

$$\psi_\nu = \frac{k_\nu \phi}{2} + \epsilon_\nu \tag{8.8.20}$$

so that

$$k_\mu \epsilon_\nu + k_\nu \epsilon_\mu = 0. \tag{8.8.21}$$

Clearly, the only solution of (8.8.21) is $\epsilon_\mu = 0$. For example, go to a Lorentz system where no component of k_μ is zero. Then $k_\mu \epsilon_\mu$ (no sum) must be zero, as must ϵ_μ; since ϵ_μ is a four-vector, it is zero in general.

So,

$$\psi_\nu = \frac{k_\nu \phi}{2}, \tag{8.8.22}$$

which is the harmonic gauge.

Within the harmonic gauge, we can eliminate the four $\phi_{0\mu}$'s by the gauge transformation

$$\phi_{0\mu} \to \phi_{0\mu} + \xi_\mu k_0 + \xi_0 k_\mu \qquad (8.8.23)$$

with

$$\xi_0 = -\frac{\phi_{00}}{2k_0}$$

and

$$\xi_i = -\frac{1}{k_0}\left(\phi_{0i} - \frac{\phi_{00}k_i}{2k_0}\right). \qquad (8.8.24)$$

This leaves (8.8.22) as a further constraint:

$$k^\mu \phi_{\mu\nu} = \frac{k_\nu}{2}\phi. \qquad (8.8.25)$$

First, take $\nu = 0$. This gives $k^\mu \phi_{\mu 0} = k_0 \phi/2$, so that $\phi = 0$ and, thus,

$$k^\mu \phi_{\mu i} = k^j \phi_{ji} = \frac{k_i}{2}\phi = 0.$$

From $k^j \phi_{ji} = 0$, we see that for $i = 3$, $\phi_{33} = 0$. Since ϕ and ϕ_{00} are also zero, so is $\phi_{11} + \phi_{22}$. We are thus left with the two propagating components $\phi_{11} - \phi_{22}$ and ϕ_{12}.

We can understand these degrees of freedom by studying their transformation properties under a rotation about the z-axis:

$$x = x' \cos\theta - y' \sin\theta, \qquad y = x' \sin\theta + y' \cos\theta \qquad (8.8.26)$$

so

$$\frac{\partial x}{\partial x'} = \cos\theta, \qquad \frac{\partial x}{\partial y'} = -\sin\theta, \qquad \frac{\partial y}{\partial x'} = \sin\theta, \qquad \text{and} \qquad \frac{\partial y}{\partial y'} = \cos\theta.$$

Thus, since $\phi_{\mu\nu}$ is a tensor,

$$\phi'_{\mu\nu} = \frac{\partial x^\sigma}{\partial x'^\mu} \frac{\partial x^\lambda}{\partial x'^\nu} \phi_{\sigma\lambda} \qquad (8.8.27)$$

and

$$\phi'_{11} = \cos^2\theta \phi_{11} + \sin^2\theta \phi_{22} + 2\sin\theta \cos\theta \phi_{12}, \qquad (8.8.28)$$

$$\phi'_{22} = \cos^2\theta \phi_{22} + \sin^2\theta \phi_{11} - 2\sin\theta \cos\theta \phi_{12}, \qquad (8.8.29)$$

and

$$\phi'_{12} = -\cos\theta\sin\theta\phi_{11} + \cos\theta\sin\theta\phi_{22} + \left(\cos^2\theta - \sin^2\theta\right)\phi_{12} \quad (8.8.30)$$

so that

$$\frac{\phi'_{11} - \phi'_{22}}{2} = \cos 2\theta\left(\frac{\phi_{11} - \phi_{22}}{2}\right) + \sin 2\theta\,\phi_{12}$$

and

$$\phi'_{12} = \cos 2\theta\,\phi_{12} - \sin 2\theta\,\frac{\phi_{11} - \phi_{22}}{2}. \quad (8.8.31)$$

Evidently, the linear combinations

$$\frac{\phi_{11} - \phi_{22}}{2} \pm i\phi_{12}$$

transform by a phase 2θ:

$$\left(\frac{\phi_{11} - \phi_{22}}{2} \pm i\phi_{12}\right)' = e^{\mp 2i\theta}\left(\frac{\phi_{11} - \phi_{22}}{2} \pm i\phi_{12}\right). \quad (8.8.32)$$

These components are said to have helicity ± 2 ($h = \pm 2$), respectively. We note that the helicity should be proportional to the projection of the wave angular momentum in the direction of propagation. The normalization for a classical plane wave would be

$$h = \frac{J_z\omega}{W}. \quad (8.8.33)$$

This is analogous to the case of the electromagnetic field that we studied in Section 3.8. There we found $h = \pm 1$ for the propagating modes, and we explicitly demonstrated the relationship (8.8.33) for a suitable wave packet.

We return to (8.8.1) and solve for the retarded gravitational radiation emitted by a known source $S^{\mu\nu}$:

$$\phi^{\mu\nu}(x) = -\frac{\lambda}{4\pi}\int d^4x'\, G_R(x - x')\, S^{\mu\nu}(x'). \quad (8.8.34)$$

G_R is given in (4.2.14). For a monochromatic source,

$$S^{\mu\nu} = S^{\mu\nu}(\mathbf{r}, \omega) e^{-i\omega x_0} + \text{c.c.},$$

we have, for large r,

$$\phi^{\mu\nu} = \epsilon^{\mu\nu} \frac{e^{i\omega(r-t)}}{4\pi r} + \text{c.c.} \tag{8.8.35}$$

with

$$\epsilon^{\mu\nu} = -\lambda \int d\mathbf{r}' \, e^{-i\mathbf{k}\cdot\mathbf{r}'} S^{\mu\nu}(\mathbf{r}', \omega) \equiv S^{\mu\nu}(k), \tag{8.8.36}$$

and $k_\mu S^{\mu\nu}(k) = \frac{1}{2} k^\nu S^\lambda_\lambda(k)$.

The gravitational stress tensor $\tau^g_{\mu\kappa}$ is calculated to lowest order from the last term in (8.7.11):

$$\tau^g_{\mu\kappa} = \frac{1}{\lambda^2} \left[R^{(2)}_{\mu\kappa} - \frac{(g_{\mu\kappa}R)^{(2)}}{2} \right]. \tag{8.8.37}$$

Now $(g_{\mu\kappa}R)^{(2)} = \eta_{\mu\kappa}R^{(2)} + 2\lambda\phi_{\mu\kappa}R^{(1)}$; since $R^{(1)}(\phi^{(1)}) = 0$, we drop it. Our $\tau^{\mu\nu}_g$ is then given by

$$\tau^g_{\mu\kappa} = \frac{1}{\lambda^2} \left(R^{(2)}_{\mu\kappa} - \frac{\eta_{\mu\kappa}}{2} R^{(2)} \right), \tag{8.8.38}$$

with $R^{(2)}_{\mu\kappa}$ given by (8.7.17).

As usual, we average $\tau_{\mu\kappa}$ over a cycle of the radiation. In evaluating the expression (8.7.17) for $\tau_{\mu\kappa}$, we may then substitute $\epsilon^*_{\mu\nu}$ for $\phi_{\mu\nu}$ on the left and $\epsilon_{\mu\nu}$ for $\phi_{\mu\nu}$ on the right, finally adding the complex conjugate. With $\partial_\mu \to ik\mu$, we have, in harmonic gauge, with $k_\alpha k^\alpha = 0$

$$\frac{(4\pi)^2 r^2 R^{(2)}_{\mu\eta}}{4\lambda^2} = \frac{1}{2} \epsilon^{*\lambda\sigma} \left(\epsilon_{\lambda\sigma} k_\mu k_\eta - k_\lambda k_\mu \epsilon_{\sigma\eta} - k_\lambda k_\eta \epsilon_{\mu\sigma} + k_\lambda k_\sigma \epsilon_{\mu\eta} \right)$$

$$- \frac{1}{2} k_\mu k_\eta \epsilon^{*\lambda\sigma} \epsilon_{\lambda\sigma} + \frac{1}{4} \left(k_\mu \epsilon^{\sigma*}_\lambda + k_\lambda \epsilon^{\sigma*}_\mu - k^\sigma \epsilon^*_{\mu\lambda} \right)$$

$$\times \left(k_\sigma \epsilon^\lambda_\eta + k_\eta \epsilon^\lambda_\sigma - k^\lambda \epsilon_{\sigma\eta} \right) + \text{c.c.} \tag{8.8.39}$$

$$= \frac{1}{4} k_\mu k_\eta \left\{ \epsilon_{\lambda\sigma} \epsilon^{*\lambda\sigma} - \frac{\epsilon^* \epsilon}{2} \right\} + \text{c.c.} \tag{8.8.40}$$

Evidently, the trace $R^{(2)}$ of (8.8.40) is zero. Thus, we have for $\tau^{\mu\eta}$

$$\tau^{\mu\eta} = \frac{2k^\mu k^\eta}{(4\pi)^2 r^2} \left\{ \epsilon_{\lambda\sigma} \epsilon^{*\lambda\sigma} - \frac{|\epsilon|^2}{2} \right\}. \tag{8.8.41}$$

The expression for $\tau^{\mu\eta}$ is easily seen to be gauge-invariant (for $\epsilon^{\mu\nu}$ in the harmonic gauge). Since we can choose $\epsilon = 0$ and $\epsilon_{0i} = 0$, (8.8.41) shows that the energy of the free field is positive.

The gravitational Poynting vector $\mathcal{P}_g$ follows from (8.8.41). Since the only amplitudes that differ from zero are ϵ_{11}, $\epsilon_{22} = -\epsilon_{11}$ and ϵ_{21}, we have

$$\mathcal{P}_g = \frac{2k^0\mathbf{k}}{(4\pi)^2 r^2} \left\{ |\epsilon_{11}|^2 + |\epsilon_{22}|^2 + 2|\epsilon_{12}|^2 \right\} \tag{8.8.42}$$

$$= \frac{2k^0\mathbf{k}}{(4\pi)^2 r^2} \left\{ \frac{|\epsilon_{11} - \epsilon_{22}|^2}{2} + 2|\epsilon_{12}|^2 \right\} \tag{8.8.43}$$

and the energy radiated per unit time and solid angle in the three-direction is

$$\frac{dW}{dt\, d\Omega} = \frac{2k^0\mathbf{k}}{(4\pi)^2 \mathbf{r}^2} \left\{ \frac{|\epsilon_{11} - \epsilon_{22}|^2}{2} + 2|\epsilon_{12}|^2 \right\}. \tag{8.8.44}$$

To find the angular distribution of the emitted radiation, we must express $dW/dt\, d\Omega$ in invariant terms: We find

$$\frac{dW}{dt\, d\Omega} = \frac{2(k^0)^2}{(4\pi)^2} \left\{ -\frac{1}{2}|\epsilon_{ii}|^2 + |\epsilon_{ij}|^2 + \frac{1}{2}|\epsilon_{ij}\hat{k}_i\hat{k}_j|^2 - 2\epsilon_{ij}\hat{k}_j\epsilon^*_{i\ell}\hat{k}_\ell \right.$$

$$\left. + \epsilon_{ii}\hat{k}_\ell\hat{k}_j\epsilon^*_{\ell j} + \epsilon^*_{ii}\hat{k}_\ell\hat{k}_j\epsilon_{\ell j} \right\} \tag{8.8.45}$$

which, with $\mathbf{k}$ in the three-direction, yields (8.8.44). The total rate is given by integrating (8.8.45) over $d\Omega$. With

$$\int d\Omega\, \hat{k}_i\hat{k}_j = \frac{4\pi}{3}\delta_{ij} \tag{8.8.46}$$

and

$$\int d\Omega\, \hat{k}_i\hat{k}_j\hat{k}_\ell\hat{k}_m = \frac{4\pi}{15}\left(\delta_{ij}\delta_{\ell m} + \delta_{i\ell}\delta_{jm} + \delta_{im}\delta_{j\ell}\right) \tag{8.8.47}$$

the result is

$$\frac{dW}{dt} = \frac{2(k^0)^2}{4\pi} \times \frac{6}{15}\left(\epsilon_{ij}\epsilon_{ij}^* - \frac{1}{3}|\epsilon_{ii}|^2\right). \tag{8.8.48}$$

We note here [as is already clear from (8.8.44)] that (8.8.48) responds only to the quadrupole (i.e., traceless) components of ϵ_{ij}, since, with

$$\epsilon_{ij} = q_{ij} + \delta_{ij}\epsilon_{\ell\ell}/3 \tag{8.8.49}$$

and $q_{ii} = 0$,

$$\epsilon_{ij}\epsilon_{ij}^* - \frac{1}{3}|\epsilon_{ii}|^2 = q_{ij}q_{ij}^*. \tag{8.8.50}$$

Therefore, turning to (8.8.36), we have

$$\epsilon_{ij} = -\lambda \int d\mathbf{r}'\, e^{-i\mathbf{k}\cdot\mathbf{r}'} T_{ij}(\mathbf{r}', \omega). \tag{8.8.51}$$

For $kr' \ll 1$, we can express the integrals over T_{ij} in terms of moments of the energy density T_{00}. From the conservation of $T_{\mu\nu}$,

$$\partial^\mu T_{\mu\nu} = 0, \tag{8.8.52}$$

follows

$$-(k^0)^2 T_{00} = \partial_i \partial_j T_{ij} \tag{8.8.53}$$

and

$$-\int d\mathbf{r}\, T_{ij} = \frac{(k^0)^2 Q_{ij}}{2} \tag{8.8.54}$$

where

$$Q_{ij} = \int d\mathbf{r}\, T_{00} x_i x_j, \tag{8.8.55}$$

with or without a subtracted trace. Also, although Q_{ij} evidently depends on the choice of origin of the coordinate system, it only depends on it through the constant total energy, $\int d\mathbf{r}\, T_{00} = W$, and the center of energy coordinate, $\int d\mathbf{r}\, T_{00}\,\mathbf{x}$, which depends linearly on t; since we are considering a finite frequency, neither of these terms will survive.

Putting it all together, we find for the total radiation rate:

$$\frac{dW}{dt} = \frac{2}{5} G(k^0)^6 \left(Q_{ij} Q_{ij}^* - \frac{1}{3} |Q_{ii}|^2 \right). \tag{8.8.56}$$

At this point, we reinstate the velocity of light c:

$$\frac{dW}{dt} = \frac{2}{5} \frac{G\omega^6}{c^5} \left\{ Q_{ij} Q_{ij}^* - \frac{1}{3} |Q_{ii}|^2 \right\}, \tag{8.8.57}$$

with ω the circular frequency of the radiation.

CHAPTER 8 PROBLEMS

8.1 The weak field gravitational red shift follows directly (both classically and in quantum theory) from the equivalence principle. Show this by considering the emission and absorption of light, as follows:

(a) Consider an emitter A and absorber B, both at rest, separated by a distance L along a uniform gravitational field. Describe the emission by A and absorption by B of a light signal from an appropriately accelerated coordinate system. In that coordinate system, the velocity of B when the light is absorbed will differ from the velocity of A when the light is emitted; the light will therefore be Doppler-shifted. Show that the Doppler shift agrees with the expected gravitational red shift.

(b) Consider an atom on the ground in a state of definite mass M_1. The atom emits a photon of frequency ω and goes to a state of mass M_2. The photon, now with frequency ω', is absorbed by an atom with mass M_3 at a height L. The atom now goes to a state of mass M_4. Using energy conservation, show that ω' has the correct red shift.

8.2 Examine the effect of a rotating source on the gravitational field.

Consider a set of particles of equal mass m making up a rigid sphere that is rotating with angular velocity $\boldsymbol{\omega}$. The linear velocity of a particle in the sphere is then

$$\mathbf{v}_a = \boldsymbol{\omega} \times \mathbf{r}_a,$$

and the first-order (in $\boldsymbol{\omega}$) addition to the stress tensor is an off-diagonal component T^{0i}.

For a spherically symmetric distribution of matter, the angular momentum of matter **J** of the sphere is given by

$$\mathbf{J} = I\boldsymbol{\omega}$$

where I is the moment of inertia of the sphere

$$I = \frac{2}{3} \int d\mathbf{r}\, \rho(\mathbf{r})\, \mathbf{r}^2$$

with ρ the spherically symmetric density of matter in the sphere.

Now find the new gravitational potential outside the sphere. It is given by

$$\phi^{0j} = -\frac{\lambda}{8\pi} \left(\frac{\mathbf{J} \times \mathbf{r}}{r^3} \right)^j.$$

*8.3 (a) One suggested way of detecting the new potential would be to measure the precession of the spin of a gyroscope in an orbit around Earth.

Calculate the effect, starting from the last equation in Problem 8.2, keeping only the leading terms in v/c and ϕ_{0i} in the equation for $d^2\mathbf{x}/dt^2$. Proceed by considering each point mass b in the gyroscope to be at a position $\mathbf{x}_b = \mathbf{x} + \mathbf{y}_b$, where $\Sigma_b\, \mathbf{y}_b = 0$.

Expand in $\mathbf{y}_b$ to include only linear terms in $\mathbf{y}_b$ (this is legitimate, since

$$\frac{y}{r} \sim \frac{\text{size of gyroscope}}{\text{radius of orbit}} \ll 1 \quad \text{and} \quad \frac{\dot{y}}{\dot{r}} = \frac{\text{spin velocity of gyroscope}}{\text{velocity in orbit}} \ll 1.$$

Now sum over b. The $\mathbf{y}_b$ and $\dot{\mathbf{y}}_b$ terms disappear and the internal forces cancel by momentum conservation, leaving an equation of motion for the center of the gyroscope. Subtracting this equation from the original ones gives an equation for each $d^2\mathbf{y}_b/dt^2$ in terms of the external gravitational field (Earth's field) and the internal forces.

Now calculate

$$\frac{d\mathbf{L}}{dt} = \sum_b \mathbf{y}_b \times m_b\, \frac{d^2\mathbf{y}_b}{dt^2} = \frac{d}{dt} \sum m_b \mathbf{y}_b \times \frac{d\mathbf{y}_b}{dt}.$$

The internal forces holding the gyroscope together cancel again, this time by angular momentum conservation, leaving a right-hand side having terms bilinear in **y** and $\dot{\mathbf{y}}$, and terms quadratic in **y**. Assuming a spherically symmetric gyroscope, express the first in terms of the angular momentum **L** itself; then show that the second is a time derivative. You should find

$$\frac{d\mathbf{L}}{dt} = \mathbf{\Omega} \times \mathbf{L} - \frac{I}{2}\frac{d\mathbf{q}}{dt}$$

where I is the moment of inertia of the sphere,

$$\mathbf{q} = 2G\nabla \times \frac{(\mathbf{J} \times \mathbf{r})}{r^3}$$

and $\mathbf{\Omega} = \mathbf{q}/2$.

The time derivative makes no contribution in a periodic orbit (or, for that matter, in any confined orbit, averaged over a long time.) The final result is thus

$$\frac{d\mathbf{L}}{dt} = \mathbf{\Omega} \times \mathbf{L},$$

showing that $\mathbf{L}$ precesses about the vector $\mathbf{\Omega}$ with angular velocity $\mathbf{\Omega}$. This precession is called the Lense–Thirring effect.

Remember: The spherical symmetry of the gyroscope yields the identities

$$\sum_a x_i^a x_j^a \propto \delta_{ij}$$

and

$$\frac{d}{dt}\sum_a x_i^a x_j^a = 0.$$

(b) Calculate the precession rate Ω in relevant units for a satellite experiment: arc-sec/year. Take the satellite at one Earth radius in a polar orbit.

The effect you have just calculated is of special interest, since it is distantly related to Mach's principle, which to some extent is embodied in Einstein's gravitational theory. Mach proposed that inertial frames are those frames that are at rest, or in uniform motion, with respect to the total matter in the universe. Thus, proximity to a large mass must distort the choice of inertial frame; indeed, we know that to be the case, since the curvature tensor is not zero near a gravitational source.

Near a large rotating source, the transformation to an inertial frame should involve a rotation, so that a gyroscope would be expected to precess about the rotation vector of the source. Your result shows that this is not exactly true, since the precession vector Ω is not in the direction of $\mathbf{J}$; however, the smallness of the effect shows that the mass of the earth is much too small to compete with the large-scale matter in the universe. Presumably, a large enough mass, properly configured and rotating, would drag the inertial system around with it.

*8.4 (a) Starting from the equation of motion for a particle in a gravitational field,

$$\frac{d^2x^\mu}{d\sigma^2} + \Gamma^\mu_{\nu\lambda} \frac{dx^\lambda}{d\sigma} \frac{dx^\nu}{d\sigma} = 0, \qquad d\sigma = \sqrt{g_{\mu\nu}\, dx^\mu dx^\nu} \quad (1)$$

transform to a new independent variable $d\tau$. With $v^\mu = dx^\mu/d\tau$, show that the new equation of motion is

$$\frac{dv^\mu}{d\tau} - \frac{v^\mu v^\beta g_{\alpha\beta}}{v^\sigma g_{\sigma\eta} v^\eta} \frac{dv^\alpha}{d\tau} = \frac{v^\mu}{2} \frac{v^\alpha v^\beta v^\gamma \partial_\gamma g_{\alpha\beta}}{v^\sigma g_{\sigma\eta} v^\eta} - \Gamma^\mu_{\nu\lambda} v^\nu v^\lambda. \qquad (2)$$

(b) Expand $dv^\mu/d\tau$ in characteristic vectors of the matrix $M^\mu_\alpha = v^\mu v^\beta g_{\alpha\beta}/v^\sigma g_{\sigma\eta} v^\eta$. One of these is v^α, with a characteristic value of 1; call the three others, with a characteristic value of 0, η^α_i.

(c) Show that the η^α_i can be chosen as

$$\eta^\alpha_i = g^{\alpha\beta} u_{i\beta}$$

where the three u_i's are orthogonal to v^μ and to each other:

$$v^\mu u_{i\mu} = 0$$

$$u^\mu_j u_{i\mu} = -\delta_{ij} \qquad \text{(note minus sign!)}$$

where $u^\mu_j = \eta^{\mu\nu} u_{j\nu}$.

Equation (2) now becomes, with $dv^\mu/d\tau = gv^\mu + \Sigma_i b_i \eta_i^\mu$,

$$\sum_i b_i \eta_i^\mu = \frac{v^\mu}{2} \frac{v^\alpha v^\beta v^\gamma \partial_\gamma g_{\alpha\beta}}{v^\sigma g_{\sigma\eta} v^\eta} - \Gamma^\mu_{\nu\lambda} v^\nu v^\lambda, \qquad (3)$$

with g remaining undetermined until a definite choice for $d\tau$ has been made.

The free choice of g arises from the circumstance that the four components of (1) are not independent:

$$\frac{dx^\mu}{d\sigma} g_{\mu\nu} \frac{dx^\nu}{d\sigma} = \text{constant} = 1.$$

Since g may be freely chosen, only three of the resulting equations are independent.

(d) Two obvious choices for τ are time and proper time. Show how g must be determined for each case. We will then continue with τ chosen as the proper time, since that will give us a covariant set of equations—the gravitational equivalent of the Lorentz force law.

(e) Show how to project the coefficients b_i from (3) and find (after some work) the new equation of motion:

$$\frac{dv^\mu}{d\tau} = -(g^{\mu\alpha} - v^\mu g^{\alpha\sigma} v_\sigma)\Gamma_\alpha, \tag{4}$$

where

$$\Gamma_\sigma = \frac{1}{2}(\partial_\nu g_{\sigma\lambda} + \partial_\lambda g_{\sigma\nu} - \partial_\sigma g_{\lambda\nu})v^\nu v^\nu. \tag{5}$$

Note that the tensor $g^{\mu\alpha}$ in (4) is *not* the tensor $g_{\mu\alpha}$ raised with the Minkowski metric $\eta^{\mu\nu}$. It is the matrix inverse of $g_{\mu\nu}$:

$$g^{\mu\nu}g_{\alpha\nu} = \delta^\mu_\alpha.$$

Remember for future applications. For weak fields,

$$g_{\mu\nu} = \eta_{\mu\nu} + 2\lambda\phi_{\mu\nu} \qquad \text{but} \qquad g^{\mu\nu} = \eta^{\mu\nu} - 2\lambda\phi^{\mu\nu}.$$

8.5 (a) The covariant curvature tensor $R_{\lambda\mu\nu\eta}$ is antisymmetric in $\lambda \leftrightarrow \mu$ and $\nu \leftrightarrow \eta$; it is symmetric in $\lambda\mu \leftrightarrow \nu\eta$. Show that these conditions make $Q_{\lambda\mu\nu\eta} = R_{\lambda\mu\nu\eta} + R_{\lambda\eta\mu\nu} + R_{\lambda\nu\eta\mu}$ antisymmetric in any pair of indices. Therefore, the equation $Q_{\lambda\mu\nu\eta} = 0$ constitutes an extra constraint on R but only in dimensions four or higher.

(b) From the symmetry properties of $R_{\mu\nu\lambda\eta}$, find the number of independent components of $R_{\mu\nu\lambda\eta}$ in three, two, and one dimension(s).

(c) Show that in two dimensions $R_{\lambda\mu\nu\eta}$ can be expressed as

$$R_{\lambda\mu\nu\eta} = \frac{(g_{\lambda\nu}g_{\mu\eta} - g_{\lambda\eta}g_{\mu\nu})}{-g} R_{1212},$$

and, therefore, the curvature $R = 2R_{12}/-g$ and $R_{\lambda\mu\nu\eta} = (g_{\lambda\nu}g_{\mu\eta} - g_{\lambda\eta}g_{\mu\nu})R/2$.

8.6 (a) Extend the work of Section 7.7 to an arbitrary dimensionality d. Calculate the number of conditions N_c that must be imposed in transforming the tensor $g_{\mu\nu}$ at $x = x_0$ to the form $g_{\mu\nu} = \eta_{\mu\nu} + O((x - x_0)^2)$. Now count the number of parameters N_p available to carry out the transformation. Show that $N_c = N_p - N_L$, where N_L is the number of parameters of a Lorentz transformation in d dimensions.

(b) Now, in one, two, and three dimensions, compare the number N_c' of conditions imposed in transforming $g_{\mu\nu}$ to the form $g_{\mu\nu} = \eta_{\mu\nu} + O((x - x_0)^3)$ with the number of parameters N_p' available to carry out the transformation. Show that $N_c' = N_p' - N_L + N_R$, where N_R is the number of independent components of the curvature tensor that you found in Problem 8.1 above.

*8.7 Verify the statement made following (8.7.28) that justifies the formula (8.7.27) for the gravitational action.

*8.8 Consider an electron bound to a proton in a relativistic Coulomb orbit. The proton generates a potential $A_0 = - e/r$, $\mathbf{A} = 0$. Following the technique of Section 8.5, find the orbit equation $r = f(\theta L, e^2, m)$, where L is the electron's angular momentum.

8.9 **(a)** Show that in one dimension the radius of curvature of a curve $y = f(x)$ at a point x is $R = (1 + (y')^2)^{3/2}/y''$, where a negative R signifies that the curve is concave downward. The radius of curvature is here defined as the radius of the tangent circle that has the same second derivative at the matching point.

(b) A two-dimensional surface is embedded in three-dimensional space. It is described by an equation $x_3 = f(x_1, x_2)$. Choose $\partial f/\partial x_1$ and $\partial f/\partial x_2$ to be zero at $x_1 = x_2 = 0$. Then $f(x_1, x_2)$, near $x, x_2 = 0$, can be expanded (with proper choice of axes) as

$$f(x_1, x_2) = \frac{x_1^2}{2R_1} + \frac{x_2^2}{2R_2} + O(x^3).$$

Calculate the metric tensor g_{ij}. (Note that $g_{ij} = g^{ij} = \delta_{ij}$ at $x_1 = x_2 = 0$.)

(c) Calculate the curvature tensor R_{1212} at $x_1 = x_2 = 0$ and from it the curvature scalar R.

8.10 Define

$$g_{\mu\nu} = \frac{\partial \xi^\alpha}{\partial x^\mu} \eta_{\alpha\beta} \frac{\partial \xi^\beta}{\partial x^\nu} \quad \text{and} \quad g^{\mu\nu} = \frac{\partial x^\mu}{\partial \xi^\alpha} \eta^{\alpha\beta} \frac{\partial x^\nu}{\partial \xi^\beta},$$

where η is the usual $1, -1, -1, -1$. Show that $g^{\mu\nu}$ is the inverse matrix of $g_{\mu\nu}$.

8.11 Consider a two-dimensional space, with metric

$$(d\tau)^2 = (dx_1)^2 + f(x_1)(dx_2)^2,$$

so

$$g_{11} = 1, \qquad g_{22} = f(x_1), \qquad g_{12} = g_{21} = 0.$$

Find the value of $\Gamma^\mu_{\sigma\lambda}$ for all μ, σ, λ.

8.12 **(a)** Consider a metric $g_{11} = 1$, $g_{22} = f(x_1)$. Calculate the curvature tensor R_{1212} and from it the scalar curvature R.

(b) Verify that for $f(x) = \sin x$, R is constant. Verify that for $f(x) = x$, R is zero.

8.13 In the harmonic gauge, $\partial_\mu \phi^{\mu\nu} = \frac{1}{2}\partial^\nu \phi$, the equation coupling $\phi^{\mu\nu}$ to $T^{\mu\nu}$ is

$$\partial_\lambda \partial^\lambda \phi^{\mu\nu} - \frac{\eta^{\mu\nu}}{2}\partial_\lambda \partial^\lambda \phi = -\lambda T^{\mu\nu} \qquad \text{or} \qquad \partial_\lambda \partial^\lambda \phi^{\mu\nu} = -\lambda S^{\mu\nu}$$

$$(1)$$

where

$$S^{\mu\nu} = T^{\mu\nu} - \frac{\eta^{\mu\nu}}{2}T.$$

Show that (1) is consistent with the choice of gauge, in that

$$\partial_\mu S^{\mu\nu} = \frac{1}{2}\partial^\nu S.$$

APPENDIX A

Vectors and Tensors

A.1. UNIT VECTORS AND ORTHOGONAL TRANSFORMATIONS

A given orthogonal coordinate system determines an orthonormal set of unit vectors in terms of which an arbitrary vector can be expanded. In three dimensions, with $\hat{\mathbf{e}}_1$, $\hat{\mathbf{e}}_2$, and $\hat{\mathbf{e}}_3$ as the unit vectors,

$$\mathbf{A} = \hat{\mathbf{e}}_1 A_1 + \hat{\mathbf{e}}_2 A_2 + \hat{\mathbf{e}}_3 A_3. \tag{A.1.1}$$

The numbers A_1, A_2, and A_3 are called the 1, 2, 3 components of the vector $\mathbf{A}$.

A useful simplification is achieved by introducing tensor notation, in which a subscript takes on successively the values 1, 2, 3. Thus, (A.1.1) would be written as

$$\mathbf{A} = \sum_i \hat{\mathbf{e}}_i A_i \tag{A.1.2}$$

or

$$\mathbf{A} = \hat{\mathbf{e}}_i A_i \tag{A.1.3}$$

where in (A.1.3) the summation convention has been introduced. This convention requires that a repeated index be summed, unless otherwise stated. Notice that the dimensionality is now implicit. The tensor formalism allows us to deal simply with any (finite) dimension.

The sum of two vectors $\mathbf{A}$ and $\mathbf{B}$ is defined as

$$\mathbf{A} + \mathbf{B} = \hat{\mathbf{e}}_i (A_i + B_i).$$

The product of a scalar c and vector $\mathbf{A}$ is defined as

$$c\mathbf{A} = \hat{\mathbf{e}}_i(cA_i).$$

The unit vectors are orthonormal. That is, we define the dot, or inner, product by

$$\hat{\mathbf{e}}_i \cdot \hat{\mathbf{e}}_j = \delta_{ij} \tag{A.1.4}$$

where the Kronecker delta δ_{ij} is defined by

$$\delta_{ij} = \begin{cases} 1 & \text{if } i = j \\ 0 & \text{otherwise} \end{cases}. \tag{A.1.5}$$

The dot, or inner, product of two vectors is defined to be distributive under addition. That is,

$$\mathbf{A} \cdot \mathbf{B} + \mathbf{A} \cdot \mathbf{C} = \mathbf{A} \cdot (\mathbf{B} + \mathbf{C}) \tag{A.1.6}$$

for any three vectors $\mathbf{A}$, $\mathbf{B}$, and $\mathbf{C}$. Therefore (remember the summation convention),

$$\begin{aligned} \mathbf{A} \cdot \mathbf{B} &= \hat{\mathbf{e}}_i A_i \cdot \hat{\mathbf{e}}_j B_j \\ &= \delta_{ij} A_i B_j \\ &= A_i B_i. \end{aligned} \tag{A.1.7}$$

Although we have introduced vectors as abstract objects, in two or three dimensions we can visualize them as directed line segments in our very own Euclidian space. Then, in two dimensions, we would write

$$\mathbf{A} = A_x \hat{\mathbf{e}}_x + A_y \hat{\mathbf{e}}_y \tag{A.1.8}$$

and with θ the angle between $\mathbf{A}$ and the x-axis, $\hat{\mathbf{e}}_x$,

$$\begin{aligned} A_x &= A \cos \theta \\ A_y &= A \sin\theta \end{aligned} \tag{A.1.9}$$

and

$$A^2 = A_x^2 + A_y^2 = \mathbf{A} \cdot \mathbf{A}. \tag{A.1.10}$$

A similar expansion for $\mathbf{B}$,

$$\mathbf{B} = B_x \hat{\mathbf{e}}_x + B_y \hat{\mathbf{e}}_y, \tag{A.1.11}$$

with

$$B_x = B \cos \psi \quad \text{and} \quad B_y = B \sin \psi \tag{A.1.12}$$

gives

$$\mathbf{A} \cdot \mathbf{B} = AB \left(\cos \theta \cos \psi + \sin \theta \sin \psi \right)$$

$$= AB \cos(\theta - \psi), \tag{A.1.13}$$

the well-known relation of the inner product to the magnitudes of two vectors and the angle between them.

We may consider a second orthogonal coordinate system with unit vectors $\hat{\mathbf{e}}_{i'}$. The new unit vectors can be expanded in terms of the old ones:

$$\hat{\mathbf{e}}_{i'} = \mathbf{O}_{i'i}\hat{\mathbf{e}}_i. \tag{A.1.14}$$

Since the $\hat{\mathbf{e}}_{i'}$ must be orthonormal,

$$\delta_{i'j'} = \hat{\mathbf{e}}_{i'} \cdot \hat{\mathbf{e}}_{j'} = O_{i'i}O_{j'j}\delta_{ij} = O_{i'i}O_{j'i}. \tag{A.1.15}$$

In matrix notation, (A.1.15) can be written more compactly as

$$OO^T = I, \tag{A.1.16}$$

where the ij matrix element of I is δ_{ij} and the matrix elements of the matrix O are $O_{i'i}$, and of its transpose O^T are

$$(O^T)_{ij'} = O_{j'i}. \tag{A.1.17}$$

A matrix satisfying (A.1.16) is called orthogonal, since it connects two orthogonal coordinate systems.

The determinant of an orthogonal matrix must be ± 1. To see this, we take the determinant of (A.1.16). This yields

$$\det I = \det(OO^T)$$

$$= (\det O)(\det O^T)$$

$$= (\det O)^2 \tag{A.1.18}$$

so that $\det O = \pm 1$.

It follows immediately that O has a unique inverse (as it must, since we could have expressed the $\hat{\mathbf{e}}_i$'s in terms of the $\hat{\mathbf{e}}_{i'}$'s)

$$O^{-1} = O^T \tag{A.1.19}$$

so that also

$$O^T O = I. \tag{A.1.20}$$

It is geometrically evident that any reorientation of the three unit coordinate vectors, keeping their relative positions unchanged, can be generated by a rotation. Since any rotation can be generated by a succession of small rotations, the determinant of the transformation matrix O describing the rotation can never jump from 1 to -1. Therefore, when $\det O = 1$, O describes a rotation, whereas when $\det O = -1$, O describes a rotation followed by a single reflection, for example,

$$\hat{\mathbf{e}}_{x'} = \hat{\mathbf{e}}_x, \qquad \hat{\mathbf{e}}_{y'} = \hat{\mathbf{e}}_y, \qquad \text{and} \qquad \hat{\mathbf{e}}'_z = -\hat{\mathbf{e}}_z. \tag{A.1.21}$$

A.2. TRANSFORMATION OF VECTOR COMPONENTS

When we transform the base vector $\hat{\mathbf{e}}_i$, we must transform the vector components A_i in such a way that the vector $\mathbf{A}$ itself remains unchanged:

$$\mathbf{A} = \hat{\mathbf{e}}_i A_i = \hat{\mathbf{e}}_{i'} A_{i'}. \tag{A.2.1}$$

From (A.1.14) we see that

$$\hat{\mathbf{e}}_i A_i = O_{i'i} \hat{\mathbf{e}}_i A_{i'} \tag{A.2.2}$$

so that

$$A_i = O_{i'i} A_{i'} \tag{A.2.3}$$

or, in matrix notation, with A as a column vector,

$$A = O^T A' \tag{A.2.4}$$

or

$$A' = OA \tag{A.2.5}$$

so that the components A_i transform exactly like the unit vectors $\hat{\mathbf{e}}_i$. This is a special property of orthogonal transformations. In general, the base vectors transform differently from the vector components. Thus, if the transformations on the $\hat{\mathbf{e}}_i$'s is P, that is,

$$\hat{\mathbf{e}}_{i'} = P_{i'i} \hat{\mathbf{e}}_i \tag{A.2.6}$$

and

$$\hat{\mathbf{e}}_{i'} A_{i'} = \hat{\mathbf{e}}_i A_i, \tag{A.2.7}$$

we again must have, as in (A.2.4),

$$A = P^T A' \qquad (A.2.8)$$

from which, in general,

$$A' = (P^T)^{-1} A \neq PA \qquad (A.2.9)$$

except for orthogonal matrices. The transformation law of the base vectors is called covariant, that of the vector components contravariant.

We define a vector under orthogonal transformations to be a set of objects A_i that transform according to (A.2.5). The prototype vector is formed by the x, y, and z components of a point in space referred to some origin.

The dot product of two vectors is independent of the coordinate system:

$$\mathbf{A'} \cdot \mathbf{B'} = A_{i'} B_{i'} = O_{i'j} O_{i'k} A_j B_k = A_i B_i = \mathbf{A} \cdot \mathbf{B}. \qquad (A.2.10)$$

We call $\mathbf{A} \cdot \mathbf{B}$ an invariant, or scalar. There are also, of course, trivial scalars, such as fixed numbers, mass ratios, etc. Note that the product SA_i of a scalar S and a vector A_i is a vector.

The importance of the transformation properties of vector components for physics is that a linear relation between vectors is preserved under coordinate transformations. Thus, if a physical law is

$$A_i = B_i, \qquad (A.2.11)$$

then obviously $O_{i'i} A_i = O_{i'i} B_i$, or

$$A_{i'} = B_{i'} \qquad (A.2.12)$$

and the law is the same expressed in the new coordinate system. Therefore, no experiment can tell us which coordinate system we are using. We say this law of Nature is invariant (strictly, covariant) under rotations and reflections. A simple way of describing what this means is the following: Suppose we go to sleep in our laboratory and a playful genie turns our laboratory around. There is no way that we can on awakening tell whether and by how much we have turned. (Of course, there must be no nearby unturned objects that might serve as references.) Physicists believe that this property is exactly true of rotations: Space is isotropic.

A reflection of our laboratory puts more of a strain on our genie: She would have to tear apart the laboratory and then reconstruct a mirror image of it and everything in it (including us). Then, if the laws of Nature are covariant with respect to reflections as well as to rotations, we would not be able to tell whether the deconstruction had taken place or not. Put more simply and avoiding the painful deconstruction, we could not tell by watching an experiment whether we were looking at the real world, or at a reflection of the real world in a mirror. Unlike the case of rotations, reflection invariance holds to a very good approximation for most phenomena, but not exactly for any, and not even approximately for the weak interactions.

A simple example of a vector law is Newton's second law describing the acceleration of a particle of mass m at position x_i in the gravitation field of a second particle of mass M at position y_i:

$$m\frac{d^2x_i}{dt^2} = -mM\frac{G(x_i - y_i)}{[(x_k - y_k)^2]^{3/2}} \tag{A.2.13}$$

where m, M, G, t, and $[(x_k - y_k)^2]^{3/2}$ are scalars, and x_i and y_i are vectors. Note that the summation convention defines $(x_k - y_k)^2$ to be

$$\begin{aligned}(x_k - y_k)^2 &= (x_k - y_k)(x_k - y_k)\\ &= \sum_k (x_k - y_k)(x_k - y_k) = (\mathbf{x} - \mathbf{y})^2,\end{aligned} \tag{A.2.14}$$

the square of the distance between x and y. Note also that $x_i(t)$ and $x_i(t + \Delta t)$ are vectors and, thus,

$$\frac{dx_i}{dt} = \lim_{\Delta t \to 0} \frac{x_i(t + \Delta t) - x_i(t)}{\Delta t}$$

is also a vector, as is d^2x_i/dt^2, etc.

A.3. TENSORS

We define an nth-rank tensor to be a set of objects that transform like products of n vectors. Thus if $T_{i_1 \dots i_n}$ is an nth-rank tensor,

$$T_{i'_1 \dots i'_n} = O_{i'_1 i_1} \dots O_{i'_n i_n} T_{i_1, \dots, i_n}. \tag{A.3.1}$$

A simple example of such a tensor is, in fact, a product of n vectors:

$$T_{i_1, \dots, i_n} = A_{i_1} B_{i_2} \dots Z_{i_n} \tag{A.3.2}$$

which will obviously have the property (A.3.1). Note that a scalar is a tensor of rank zero and a vector a tensor of rank one.

Symmetry properties of tensors are invariant. Thus, if $T_{i_1 i_2 \ldots i_n}$ is symmetric (antisymmetric) under an exchange of i_1 and i_2, then $T_{i'_1 i'_2 \ldots i'_n}$ will have the same property under exchange of i'_1 and i'_2.

We recall from (A.2.10) that the inner product $A_i B_i$ of two vectors A_i and B_i is invariant. The generalization of this rule to tensors is that the trace with respect to any two indices lowers the rank of the tensor by 2. The trace with respect to two indices is defined as

$$\text{Tr}_{i,j}(T_{i_1 \ldots i \ldots j \ldots k_n}) = T_{i_1 \ldots i \ldots i \ldots i_n} \tag{A.3.3}$$

where, as always, the repeated index i is summed.

Laws of nature that are linear relations between tensors have the same property as those between vectors, in that they do not permit experiments to differentiate between coordinate systems.

We give a simple example of a tensor law. Suppose for simplicity that in (A.2.13) we let M be a heavy immobile object at $y = 0$. Then (A.2.13) can be written as

$$\frac{m d^2 x_i}{dt^2} = -\frac{m M G x_i}{(x_k^2)^{3/2}}. \tag{A.3.4}$$

Now multiply (A.3.4) by x_j:

$$m x_j \frac{d^2 x_i}{dt^2} = -m M \frac{G x_i x_j}{(x_k^2)^{3/2}}, \tag{A.3.5}$$

interchange i and j, and subtract. The right-hand side of (A.3.5) vanishes, and we are left with

$$O = m \left(x_j \frac{d^2 x_i}{dt^2} - x_i \frac{d^2 x_j}{dt^2} \right)$$

$$= \frac{d}{dt} m \left(x_j \frac{dx_i}{dt} - x_i \frac{dx_j}{dt} \right) \tag{A.3.6}$$

and we have a tensor conservation law:

$$\frac{d}{dt} L_{ji} = 0 \tag{A.3.7}$$

where

$$L_{ji} = m\left(x_j \frac{dx_i}{dt} - x_i \frac{dx_j}{dt}\right). \tag{A.3.8}$$

Equation (A.3.7) is a new way of writing the conservation of angular momentum as the constancy of a second-rank antisymmetric tensor.

A.4. PSEUDOTENSORS

The connection between the second-rank antisymmetric tensor L_{ij} (A.3.8) and the usual angular momentum vector

$$\mathbf{L} = m\mathbf{x} \times \mathbf{v} \tag{A.4.1}$$

is given by the equation

$$L_i = \frac{1}{2} \epsilon_{ijk} L_{jk} \tag{A.4.2}$$

where ϵ_{ijk} is totally antisymmetric in exchange of any two indices, and $\epsilon_{123} = 1$. This uniquely determines all components of ϵ:

$$1 = \epsilon_{123} = \epsilon_{312} = \epsilon_{231} = -\epsilon_{213} = -\epsilon_{321} = -\epsilon_{132}; \tag{A.4.3}$$

all other components are zero.

One might suppose that ϵ is a third-rank totally antisymmetric tensor. To check, we calculate

$$\bar{\epsilon}_{i'j'k'} = O_{i'i}O_{j'j}O_{k'k}\epsilon_{ijk}. \tag{A.4.4}$$

Since $\bar{\epsilon}_{i'j'k'}$ is totally antisymmetric in exchange of any two indices, $\bar{\epsilon}$ must be a multiple of ϵ. We can evaluate the multiple by calculating $\bar{\epsilon}_{123}$:

$$\bar{\epsilon}_{123} = O_{1i}O_{2j}O_{3k}\epsilon_{ijk} = \det O \tag{A.4.5}$$

so that ϵ is a tensor under rotations but has an extra change of sign under reflections. It is called a pseudotensor. In contrast, as shown in Problem (A.1), δ_{ij} is a tensor.

Returning to (A.4.2), we see that L_i transforms like a vector under rotations, but has an extra sign change under inversions. It is called a pseudovector, or axial vector. This property corresponds exactly to that of an ordinary vector cross-product, as in (A.4.1): Note that an inversion $(\mathbf{x} \to -\mathbf{x})$ takes $\mathbf{v} \to -\mathbf{v}$, but $\mathbf{L} \to \mathbf{L}$, which is not what a true vector should do.

The pseudotensor ϵ_{ijk} evidently makes possible the construction of pseudotensors of any rank. For example, if $\mathbf{A}$, $\mathbf{B}$, and $\mathbf{C}$ are vectors, then

$$\mathbf{A} \cdot \mathbf{B} \times \mathbf{C} = \epsilon_{ijk} A_i B_j C_k \qquad (A.4.6)$$

is a pseudoscalar.

A.5. VECTOR AND TENSOR FIELDS

In the previous sections of this appendix, we have considered single tensors, vectors, and scalars. Examples are the angular momentum, the velocity and the energy of a body at some time, or the instantaneous electric field at a given point in space.

We often wish to consider collectively objects associated with all of space, for example, the density $\rho(\mathbf{x})$ of matter at a given time. Since mass and volume are scalars, $\rho(\mathbf{x})$, the mass per unit volume, is also a scalar. We call $\rho(\mathbf{x})$ a scalar field. Similarly, we can consider vector fields—such as the electric field—and, in general, tensor and pseudotensor fields of higher rank.

Let us start with a scalar field $\rho(\mathbf{x})$. The transformation rule for $\rho(\mathbf{x})$ is

$$\rho'(\mathbf{x}') = \rho(\mathbf{x}) = \rho\big(\mathbf{x}(x')\big) \qquad (A.5.1)$$

That is, the densities measured *at the same point* in both the primed and unprimed coordinate systems are equal. In the primed system, we express the density as a function of the primed coordinate $\mathbf{x}'$. Therefore, the new function $\rho'(\mathbf{x}')$ is not the same function of $\mathbf{x}'$ as ρ is of $\mathbf{x}$. Hence, the notation $\rho'(\mathbf{x}')$ in (A.5.1).

Differentiation of a scalar field produces a vector field. To see this, let ρ be the scalar field. Then consider the gradient

$$A_i(\mathbf{x}) = \frac{\partial}{\partial x_i} \rho(\mathbf{x}). \qquad (A.5.2)$$

The instructions for calculating the gradient in the primed coordinate system would be, by the chain rule for partial differentiation,

$$A_{i'}(\mathbf{x}') = \frac{\partial}{\partial x_{i'}} \rho'(\mathbf{x}')$$

$$= \frac{\partial \rho}{\partial x_j} \frac{\partial x_j}{\partial x_{i'}}. \tag{A.5.3}$$

Recall that $x_i' = O_{i'j} x_j$, so

$$x_j = (O^T)_{ji'} x_{i'},$$

$$\frac{\partial x_j}{\partial x_{i'}} = (O^T)_{ji'} = O_{i'j},$$

and therefore,

$$A_{i'} = O_{i'j} A_j \tag{A.5.4}$$

which is the correct transformation law for a vector.

In general, differentiation of a tensor of rank n produces a tensor of rank $n + 1$.[1] For example,

$$B_{ji} = \frac{\partial}{\partial x_j} A_i$$

where A_i is a vector produces a second-rank tensor B_{ji}. Note that the antisymmetric tensor

$$F_{ji} = B_{ji} - B_{ij} = \frac{\partial}{\partial x_j} A_i - \frac{\partial}{\partial x_i} A_j$$

is related to the curl of A:

$$\frac{1}{2} \epsilon_{ijk} F_{jk} = (\nabla \times \mathbf{A})_i. \tag{A.5.5}$$

Since $(\partial/\partial x_j) A_i$ is a tensor, its trace, $(\partial/\partial x_j) A_j$, is a scalar. Similarly, the divergence of a tensor of rank n produces one of rank $n - 1$:

$$\frac{\partial}{\partial x_{i_1}} F_{i_1 \ldots i_n} = B_{i_2 \ldots i_n}. \tag{A.5.6}$$

---

[1]*Warning*: We have shown this to be true for orthogonal transformations. It is specifically *not* true for general coordinate transformations. (See Section 7.6.)

A.6. SUMMARY OF RULES OF THREE-DIMENSIONAL VECTOR ALGEBRA AND ANALYSIS

We list here, without proof, the most important of these.[2]

1. For any vectors **A** and **B**

$$\mathbf{A} \cdot \mathbf{B} = AB \cos \theta \qquad (A.6.1)$$

where A is the magnitude of **A**,

$$A = (\mathbf{A}^2)^{1/2},$$

B the magnitude of **B** and θ the angle between them.

2. $$\mathbf{A} \times \mathbf{B} = \hat{\mathbf{n}} AB \sin \theta \qquad (A.6.2)$$

where $\hat{\mathbf{n}}$ is normal to the plane containing **A** and **B**, and in a direction given by a right-handed screw going from **A** to **B** (through the smallest angle between them). Right-handed is defined here by the coordinate system in which (A.4.2) is written: $\hat{\mathbf{e}}_1 \times \hat{\mathbf{e}}_2 = \hat{\mathbf{e}}_3$.

3. $$\mathbf{A} \times (\mathbf{B} \times \mathbf{C}) = \mathbf{A} \cdot \mathbf{C} \mathbf{B} - \mathbf{A} \cdot \mathbf{B} \mathbf{C} \qquad (A.6.3)$$

provided **B** and **C** commute.

4. $$\mathbf{A} \cdot \mathbf{B} \times \mathbf{C} = \mathbf{A} \times \mathbf{B} \cdot \mathbf{C}$$
$$= \det \begin{pmatrix} A_x & A_y & A_z \\ B_x & B_y & B_z \\ C_x & C_y & C_z \end{pmatrix}. \qquad (A.6.4)$$

5. Stokes' theorem:

$$\oint_C d\boldsymbol{\ell} \cdot \mathbf{A} = \int_S d\mathbf{S} \cdot \boldsymbol{\nabla} \times \mathbf{A}. \qquad (A.6.5)$$

Here, C is a closed path along which we take the line integral of

---

[2]For a simple discussion, see H. M. Schey, *Div, Grad, Curl and All That*, New York: W. W. Norton, 1973.

A, S is a surface bounded by C, and $d\mathbf{S} = dS\hat{\mathbf{n}}$, with dS the surface element and $\hat{\mathbf{n}}$ a unit vector normal to the surface, in a direction given by a right-handed screw going around the circuit in the direction specified by $d\boldsymbol{\ell}$.

6. Gauss' theorem:

$$\int_V d\mathbf{r}\nabla \cdot \mathbf{A} = \int_S d\mathbf{S} \cdot \mathbf{A} \tag{A.6.6}$$

where V is a volume bounded by S, and $d\mathbf{S} = dS\hat{\mathbf{n}}$, as before, with $\hat{\mathbf{n}}$ the outward normal.

7. A vector $\mathbf{E}$ with $\nabla \times \mathbf{E} = 0$ is the gradient of a scalar:

$$\mathbf{E} = \nabla\phi. \tag{A.6.7}$$

8. A vector $\mathbf{B}$ with $\nabla \cdot \mathbf{B} = 0$ is the curl of a vector:

$$\mathbf{B} = \nabla \times \mathbf{A}. \tag{A.6.8}$$

9. The gradient operator in a general orthogonal coordinate system q_1, q_2, and q_3 with unit vectors $\hat{\mathbf{e}}_1$, $\hat{\mathbf{e}}_2$, and $\hat{\mathbf{e}}_3$ is given by

$$\nabla = \frac{\hat{\mathbf{e}}_1}{h_1}\frac{\partial}{\partial q_1} + \frac{\hat{\mathbf{e}}_2}{h_2}\frac{\partial}{\partial q_2} + \frac{\hat{\mathbf{e}}_3}{h_3}\frac{\partial}{\partial q_3}$$

where $h_i\, dq_i$ is the increment of length in the q_i direction. For explicit calculation, one must remember that the h's and $\hat{\mathbf{e}}$'s are, in general, functions of the q's. In particular, the Laplacian operator ∇^2 is given by

$$\nabla^2 = \frac{1}{h_1 h_2 h_3}\left(\frac{\partial}{\partial q_1}\left(\frac{h_2 h_3}{h_1}\frac{\partial}{\partial q_1}\right) + \frac{\partial}{\partial q_2}\left(\frac{h_1 h_3}{h_2}\frac{\partial}{\partial q_2}\right) + \frac{\partial}{\partial q_3}\left(\frac{h_1 h_2}{h_3}\frac{\partial}{\partial q_3}\right)\right).$$
$$\tag{A.6.9}$$

In spherical coordinates r, θ, φ, $h_r = 1$, $h_\theta = r$, and $h_\varphi = r\sin\theta$.

APPENDIX A: PROBLEMS

A.1 Show that δ_{ij} is a second-rank tensor.

A.2 Construct a second-rank tensor T_{ij} that is symmetric, has zero trace, and is a function of only one vector and the Kronecker delta.

A.3 Repeat Problem A.2, but for a third-rank tensor.

A.4 Repeat Problem A.2, but for a fourth-rank tensor.

A.5 Construct the most general third-rank tensor T_{ijk} that is a linear function of a vector y, a bilinear function of a vector x, and antisymmetric under the exchange of i and j.

A.6 Construct a third-rank tensor T_{ijk} that is a function of two vectors, x_i and y_i, antisymmetric under exchange of i and j, and symmetric under exchange of x and y.

A.7 Show that the triple scalar product, $\mathbf{A} \cdot \mathbf{B} \times \mathbf{C}$, is given by

$$\mathbf{A} \cdot \mathbf{B} \times \mathbf{C} = \det \begin{Bmatrix} A_x & A_y & A_z \\ B_x & B_y & B_z \\ C_x & C_y & C_z \end{Bmatrix},$$

from which you must show that $\mathbf{A} \cdot \mathbf{B} \times \mathbf{C}$ is invariant under an interchange of the dot and cross-product, that is,

$$\mathbf{A} \cdot \mathbf{B} \times \mathbf{C} = \mathbf{A} \times \mathbf{B} \cdot \mathbf{C}.$$

A.8 Using (A.6.3), show that the triple vector product is not, in general, associative, that is,

$$\mathbf{A} \times (\mathbf{B} \times \mathbf{C}) \neq (\mathbf{A} \times \mathbf{B}) \times \mathbf{C},$$

and find under what circumstances the inequality becomes an equality.

A.9 Given a tensor T_{ijkl}, symmetric (or antisymmetric) in i and j, prove that the transformed tensor $T_{i'j'k'l'}$ has the same property in i' and j'.

A.10 Let T_{ijk} be a third-rank tensor. Show that T_{ijj} is a first-rank tensor.

A.11 Show that $\epsilon_{ijk}\epsilon_{ilm} = \delta_{jl}\delta_{km} - \delta_{jm}\delta_{kl}$.

A.12 Use the result of Problem A.11 to prove the formula (A.6.3) for the triple vector product.

A.13 Use the definition of $\mathbf{A} \times \mathbf{B}$:

$$(\mathbf{A} \times \mathbf{B})_i = \epsilon_{ijk} A_j B_k,$$

to show that $\mathbf{A} \times \mathbf{B}$ is orthogonal to $\mathbf{A}$ and $\mathbf{B}$.

A.14 Use the result of Problem A.11 to show that the magnitude of $\mathbf{A} \times \mathbf{B}$ is given correctly by (A.6.2).

A.15 Show that the direction of $\mathbf{A} \times \mathbf{B}$, as defined in Problem A.13, satisfies the right-hand rule.

A.16 Suppose a body is spinning about a fixed point with instantaneous angular velocity $\boldsymbol{\omega}$. That is, every point $\mathbf{r}_p$ in the body has instantaneous velocity $\mathbf{v}_p = \boldsymbol{\omega} \times \mathbf{r}_p$. The angular momentum $\mathbf{L}$ of the body is

$$\mathbf{L} = \sum m_p \mathbf{r}_p \times (\boldsymbol{\omega} \times \mathbf{r}_p).$$

Show that the components of $\mathbf{L}$ are linear functions of the components of $\boldsymbol{\omega}$:

$$L_i = I_{ij}\omega_j$$

where I_{ij} is called the moment of inertia tensor. Show that I_{ij} is a second-rank symmetric tensor.

A.17 Using the fact [derived in (A.1.13)] that $\mathbf{A} \cdot \mathbf{B}$ for any two vectors is given by $AB \cos \theta$, where θ is the angle between them, show that

$$\cos \theta = \cos \theta_A \cos \theta_B + \sin \theta_A \sin \theta_B \cos(\varphi_A - \varphi_B)$$

where θ_A and θ_B and φ_A and φ_B are the spherical coordinates of the vectors A and B, with θ their polar angle and φ their azimuth.

A.18 Derive a formula equivalent to the one in Problem A.17 for the angle between two vectors in a four-dimensional Euclidean space, with polar angles θ, ψ, and φ for each vector, that is, $A_4 = A \cos \theta$, $A_3 = A \sin \theta \cos \psi$, $A_2 = A \sin \theta \sin \psi \cos \phi$, and $A_1 = A \sin \theta \sin \psi \times \sin \phi$.

A.19 From Gauss' theorem, (A.6.6), prove

(a)
$$\int_V d^3x\, \boldsymbol{\nabla}\phi = \int_S dS\phi$$

(b)
$$\int_V d^3x\, \boldsymbol{\nabla} \times \mathbf{A} = \int_S d\mathbf{S} \times \mathbf{A}$$

for any function ϕ and vector $\mathbf{A}$.

A.20 From Stokes' theorem, show that if a vector $\mathbf{A}(\mathbf{x})$ has zero curl, then $\mathbf{A}(\mathbf{x})$ is the gradient of a scalar $\psi(\mathbf{x})$, with $\psi(\mathbf{x})$ given by

$$\psi(\mathbf{x}) = \int_{\mathbf{x}_0}^{\mathbf{x}} d\boldsymbol{\ell}' \cdot \mathbf{A}(\mathbf{x}')$$

independent of the path from $\mathbf{x}_0$ to $\mathbf{x}$.

A.21 Let V be a volume bounded by a surface S. From Gauss' theorem, show that

$$\int_V (\boldsymbol{\nabla}\psi)^2 \, d^3x = \int_S \psi \boldsymbol{\nabla}\psi \cdot d\mathbf{S} - \int_V \psi \boldsymbol{\nabla}^2\psi \, d^3x$$

from which prove that a vector whose divergence and curl are both zero, and which vanishes sufficiently rapidly at ∞, must be zero everywhere. Give the criteria for "sufficiently rapidly."

A.22 From Gauss' theorem applied to a small cube, derive the expression (A.6.9).

Spherical Harmonics and Orthogonal Polynomials

The simple potential function

$$\phi(\mathbf{x} - \mathbf{x}') = \frac{1}{[(\mathbf{x} - \mathbf{x}')^2]^{1/2}} \tag{B.1.1}$$

can be expanded for small r'/r in a power series in r'/r, and for small r/r', in a power series in that variable. In order to avoid confusion with the x component of $\mathbf{x}$, we here denote the magnitude of $\mathbf{x}$ by r:

$$r = (\mathbf{x}^2)^{1/2}. \tag{B.1.2}$$

We can test for the radius of convergence of the series by finding the zeros of the function

$$D = (\mathbf{x} - \mathbf{x}')^2 = r^2 + r'^2 - 2rr' \cos\theta, \tag{B.1.3}$$

where θ is the angle between the vectors $\mathbf{x}$ and $\mathbf{x}'$.

If we let $r'/r = u$, the equation $D = 0$ becomes

$$1 + u^2 - 2u \cos\theta = 0 \tag{B.1.4}$$

whose solutions are

$$u = \cos\theta \pm \sqrt{\cos^2\theta - 1} = e^{\pm i\theta}. \tag{B.1.5}$$

The function ϕ, considered a function of the complex variable u, is therefore analytic in u for $|u| < 1$ so that the power series in u is convergent for $u < 1$. Thus, we may write

$$\phi = \frac{1}{(r^2 + r'^2 - 2rr' \cos \theta)^{1/2}} = \frac{1}{r'} \sum_{\ell=0}^{\infty} \left(\frac{r}{r'}\right)^{\ell} P_{\ell}(\cos \theta) \quad (B.1.6)$$

for $r < r'$, and

$$\phi = \frac{1}{r} \sum_{\ell=0}^{\infty} \left(\frac{r'}{r}\right)^{\ell} P_{\ell}(\cos \theta) \quad (B.1.7)$$

for $r > r'$.

The function $P_{\ell}(\cos \theta)$ defined by (B.1.6) and (B.1.7) is clearly a polynomial in $\cos \theta$ of degree ℓ, even or odd in $\cos \theta$ according to whether ℓ is even or odd. It is called a Legendre polynomial. It is also clear that the function $r^{\ell} P_{\ell}(\cos \theta)$ is a polynomial of degree ℓ in the components x, y, z of the vector $\mathbf{x}$, with $\mathbf{x}'$ held fixed. In particular, if we choose $\mathbf{x}'$ as the polar axis of the $\mathbf{x}$ coordinate system, then $r \cos \theta = z$ and $r^2 = x^2 + y^2 + z^2$, so that $r^{\ell} P_{\ell}(\cos \theta)$ is a polynomial function only of z and r^2. The function ϕ with the property (B.1.6) is called a generating function for the P_{ℓ}'s.

We study some properties of the P_{ℓ}'s:

1. Since

$$\nabla^2 \frac{1}{|\mathbf{x} - \mathbf{x}'|} = 0 \quad (\mathbf{x} \neq \mathbf{x}') \qquad \text{and} \qquad \nabla'^2 \frac{1}{|\mathbf{x} - \mathbf{x}'|} = 0 \quad (\mathbf{x} \neq \mathbf{x}'),$$
$$(B.1.8)$$

(B.1.6) shows that the polynomial function of $\mathbf{x}$, $r^{\ell} P_{\ell}(\cos \theta)$, is harmonic; that is, it satisfies the equation

$$\nabla^2 r^{\ell} P_{\ell}(\cos \theta) = 0; \quad (B.1.9)$$

Similarly,

$$\nabla^2 \frac{1}{r^{\ell+1}} P_{\ell}(\cos \theta) = 0. \quad (B.1.10)$$

2. In the forward and backward directions, the expansion (B.1.6) becomes

$$\phi = \frac{1}{|r \mp r'|} = \frac{1}{r'} \sum_{\ell=0}^{\infty} (\pm 1)^{\ell} \left(\frac{r}{r'}\right)^{l} \quad (B.1.11)$$

so that

$$P_{\ell}(\pm 1) = (\pm 1)^{\ell}. \quad (B.1.12)$$

3. Harmonic polynomials of different order are orthogonal. That is, if

$$\nabla^2 p_\ell = \nabla^2 q_{\ell'} = 0 \qquad\qquad (B.1.13)$$

where p_ℓ and $q_{\ell'}$ are polynomials of order ℓ and ℓ' in x, y, z, then the integral over solid angle, $d\Omega$,

$$\int d\Omega\, p_\ell q_{\ell'} = 0, \qquad \ell \neq \ell'. \qquad\qquad (B.1.14)$$

Proof: Integrate over a spherical volume:

$$\int (p_\ell \nabla^2 q_{\ell'} - q_{\ell'} \nabla^2 p_\ell)\, d\mathbf{r} = 0 \qquad\qquad (B.1.15)$$

so that

$$\int_S (p_\ell \nabla q_{\ell'} - q_{\ell'} \nabla p_\ell) \cdot d\mathbf{S} = 0 \qquad\qquad (B.1.16)$$

integrated over the spherical surface bounding the volume. Since $d\mathbf{S} \cdot \nabla = r^2\, d\Omega(\partial/\partial r)$, (B.1.16) becomes, with

$$p_\ell = r^\ell Y_\ell(\Omega) \qquad \text{and} \qquad q_{\ell'} = r^{\ell'} Z_{\ell'}(\Omega) \qquad (B.1.17)$$

$$0 = r^{\ell + \ell' + 1}(\ell' - \ell) \int Y_\ell(\Omega)\, Z_{\ell'}(\Omega)\, d\Omega \qquad (B.1.18)$$

and the result is proved. In particular, Legendre polynomials $P_\ell(w)$ and $P_{\ell'}(w)$ with $\ell \neq \ell'$ are orthogonal when integrated over w. Here, $w = \cos\theta$, $dw = \sin\theta\, d\theta$, and $d\Omega = d\varphi\, dw$, where φ goes from θ to 2π and w now goes from -1 to 1.

4. The normalization integral for P_ℓ can be found by integrating $1/|\mathbf{r} - \mathbf{r}'|^2$:

$$\int \frac{d\Omega}{|\mathbf{r} - \mathbf{r}'|^2} = 2\pi \int_{-1}^{1} dw\, \frac{1}{(r^2 + r'^2 - 2rr'w)}$$

$$= \frac{2\pi}{2rr'} \log\frac{(r + r')^2}{(r - r')^2}$$

$$= \frac{2\pi}{rr'} [\log(r' + r) - \log(r' - r)] \qquad (B.1.19)$$

for $r' > r$. Here, the symbol $\log x$ stands for the natural logarithm of x. Expanding in powers of r/r', we have

$$\int \frac{d\Omega}{|\mathbf{r} - \mathbf{r}'|^2} = \frac{2\pi}{rr'}\left[\log\left(1 + \frac{r}{r'}\right) - \log\left(1 - \frac{r}{r'}\right)\right]$$

$$= \frac{2\pi}{rr'}\left[\frac{r}{r'} - \frac{1}{2}\left(\frac{r}{r'}\right)^2 + \frac{1}{3}\left(\frac{r}{r'}\right)^3 + \cdots\right.$$

$$\left. - \left(-\frac{r}{r'} - \frac{1}{2}\left(\frac{r}{r'}\right)^2 - \frac{1}{3}\left(\frac{r}{r'}\right)^3 \cdots\right)\right]$$

$$= \frac{4\pi}{r'^2}\left[1 + \frac{1}{3}\left(\frac{r}{r'}\right)^2 + \frac{1}{5}\left(\frac{r}{r'}\right)^5 + \cdots\right]. \quad \text{(B.1.20)}$$

On the other hand,

$$\int \frac{d\Omega}{|\mathbf{r} - \mathbf{r}'|^2} = \int d\Omega\left(\frac{1}{r'}\sum_{\ell=0}^{\infty}\left(\frac{r}{r'}\right)^\ell P_\ell(w)\right)\left(\frac{1}{r'}\sum_{\ell'=0}^{\infty}\left(\frac{r}{r'}\right)^{\ell'} P_{\ell'}(w)\right)$$

$$= \frac{2\pi}{r'^2}\sum_{\ell=0}^{\infty}\left(\frac{r}{r'}\right)^{2\ell}\int dw P_\ell(w)\, P_\ell(w) \qquad \text{(B.1.21)}$$

so that, in all,

$$\int P_\ell(w)\, P_{\ell'}(w)\, dw = \frac{2}{2\ell + 1}\,\delta_{\ell\ell'}. \qquad \text{(B.1.22)}$$

5. The P_ℓ's are orthogonal polynomials of parity $(-1)^\ell$ and of order ℓ in w, with a weight function 1 on the interval w between ± 1 and a normalization $P_\ell(1) = 1$. Evidently, the P_ℓ's can be sequentially constructed from these rules. Thus,

$$P_0(w) = 1 \qquad \text{(B.1.23)}$$

$$P_1(w) = w \qquad \text{(B.1.24)}$$

$$P_2(w) = \alpha + \beta w^2, \qquad \text{(B.1.25)}$$

etc.

P_2 is automatically orthogonal to P_1; orthogonality to P_0

requires

$$\int_{-1}^{1} dw(\alpha + \beta w^2) = 0$$

or

$$\alpha = -\frac{\beta}{3}. \qquad (B.1.26)$$

Together with $P_2(1) = 1$, or $\alpha + \beta = 1$, (B.1.26) determines P_2 as

$$P_2(w) = \frac{3}{2} w^2 - \frac{1}{2}. \qquad (B.1.27)$$

This procedure can be continued to construct any finite-order Legendre polynomial.

B.2. SPHERICAL HARMONICS

We return to (B.1.1) and expand directly in powers of x, y, and z using the three dimensional Taylor's theorem. That is, for $r < r'$,

$$\frac{1}{|\mathbf{x} - \mathbf{x}'|} = \sum_{\ell=0}^{\infty} \frac{(-1)^{\ell}}{\ell!} (x \cdot \nabla')^{\ell} \frac{1}{r'}, \qquad (B.2.1)$$

or, in tensor notation,

$$\phi(\mathbf{x} - \mathbf{x}') = \sum_{\ell=0}^{\infty} \frac{(-1)}{\ell!} x_{i_1} \dots x_{i_\ell} \frac{\partial}{\partial x'_{i_1}} \dots \frac{\partial}{\partial x'_{i_\ell}} \frac{1}{r'}. \qquad (B.2.2)$$

We define multipole fields of order ℓ as

$$\phi_{i_1 \dots i_\ell}^{(\ell)}(\mathbf{x}') \equiv \frac{\partial}{\partial x'_{i_1}} \dots \frac{\partial}{\partial x'_{i_\ell}} \frac{1}{r'}. \qquad (B.2.3)$$

$\phi^{(\ell)}(\mathbf{x}')$ is a symmetric tensor of rank ℓ. It is also traceless, since contracting on any pair of indices in (B.2.3) produces a ∇'^2, which in turn gives zero acting on $1/r'$.

At a given point in space, we can count the number of independent components of $\phi^{(\ell)}$ [i.e., the number of independent numbers we must specify to determine all the components of $\phi^{(\ell)}$]. That number is $2\ell + 1$, as we now show.

We count first the number $S(\ell, N)$ of components of an ℓth-rank symmetric tensor in N dimensions. We call the number of x_1 components n_1, of x_2 components n_2, etc. The number of independent components is then

$$S(\ell, N) = \sum_{n_1} \cdots \sum_{n_N} \cdot 1 .$$

$$(\Sigma^{n_i = \ell})$$

(B.2.4)

The sum can be deduced from

$$S(z, N) = \sum_{\ell=0}^{\infty} S(\ell, N) \, z^\ell$$

$$= \sum_{n_1=0}^{\infty} z^{n_1} \cdots \sum_{n_N=0}^{\infty} z^{n_N}$$

$$= \left(\frac{1}{1-z} \right)^N$$

(B.2.5)

which is convergent for $|z| < 1$. We can recover the ℓth term in the series by integrating the complex function $1/(2\pi i z^{\ell+1}) \, S(z, n)$ around the origin:

$$S(\ell, N) = \oint \frac{dz}{2\pi i z^{\ell+1}} S(z, N)$$

(B.2.6)

$$= \frac{1}{\ell!} \frac{\partial^\ell}{\partial z^\ell} \left(\frac{1}{1-z} \right)^N \bigg|_{z=0}$$

$$= \frac{N \cdot (N+1) + \cdots + (N+\ell-1)}{\ell!}$$

$$= \frac{(N+\ell-1)!}{(N-1)!\ell!}$$

(B.2.7)

which is the desired formula. For $N = 3$, we have

$$S(\ell, 3) = \frac{(\ell+2)!}{2!\ell!} = \frac{(\ell+1)(\ell+2)}{2} .$$

(B.2.8)

The number $S_0(\ell, 3)$ of independent components of a traceless symmetric tensor in three dimensions is

$$S_0(\ell, 3) = S(\ell, 3) - S(\ell - 2, 3) = 2\ell + 1$$

(B.2.9)

since the symmetry of $S(\ell, 3)$ assures that all traces are identical and since the trace of an ℓth-rank symmetric tensor is an $\ell - 2$th-rank symmetric tensor.

We return to (B.2.2), which we may write as

$$\phi(\mathbf{x} - \mathbf{x}') = \sum_{\ell=0}^{\infty} \frac{(-1)^\ell}{\ell!} \overline{P}_{i_1 \ldots i_\ell}^{(\ell)} \phi_{i_1 \ldots i_\ell}^{(\ell)}(\mathbf{x}') = \sum_{\ell=0}^{\infty} \frac{(-1)^\ell}{\ell!} x_{i_1} \ldots x_{i_\ell} \phi_{i_1 \ldots i_\ell}^{(\ell)}(\mathbf{x}') \tag{B.2.10}$$

where

$$\overline{P}_{i_1 \ldots i_\ell}^{(\ell)} = x_{i_1} \ldots x_{i_\ell} \tag{B.2.11}$$

is a symmetric ℓth-rank tensor.

Since there are only $2\ell + 1$ independent $\phi^{(\ell)}$'s, there must be only $2\ell + 1$ independent components of the $\overline{P}^{(\ell)}$'s that matter. We exhibit this, making $\overline{P}_\ell$ traceless by subtracting Krönecker deltas. These have no effect on ϕ, since

$$\delta_{ij} \phi_{i_1 \ldots i \ldots j \ldots i_l}^{(\ell)} = 0. \tag{B.2.12}$$

We illustrate the subtraction procedure for $\ell = 2$. Define

$$P_{ij}^{(2)} = \overline{P}_{ij}^{(2)} - \alpha \delta_{ij} \tag{B.2.13}$$

such that $P_{ii}^{(2)} = 0$; that is,

$$3\alpha = \overline{P}_{ii}^{(2)}$$

and

$$P_{ij}^{(2)} = \overline{P}_{ij}^{(2)} - \frac{1}{3} \delta_{ij} \overline{P}_{kk}^{(2)} \tag{B.2.14}$$

$$= x_i x_j - \frac{1}{3} \delta_{ij} \mathbf{x}^2.$$

The above procedure, applied to $\overline{P}^\ell$ of arbitrary rank, produces an ℓth-rank, symmetric, traceless tensor, $P_{i_1 \ldots i_\ell}^{(\ell)}(\mathbf{x})$, with $2\ell + 1$ independent components.

Since $\nabla^2 1/|\mathbf{x} - \mathbf{x}'| = 0$ for $(\mathbf{x} \neq \mathbf{x}')$, the polynomial product $P_{i_1 \ldots i_\ell}^{(\ell)}(\mathbf{x}) \phi_{i_1 \ldots i_\ell}^{(\ell)}(\mathbf{x}')$ must satisfy Laplace's equation; since there are $2\ell + 1$ independent $\phi^{(\ell)}$'s and $2\ell + 1$ independent $P^{(\ell)}$'s, each $P^{(\ell)}$ must satisfy Laplace's equation, that is,

$$\nabla^2 P_{i_1 \ldots i_\ell}^{(\ell)} = 0. \tag{B.2.15}$$

We have thus shown that there exist at least $2\ell + 1$ harmonic polynomials of degree ℓ: the $P_{i_1 \ldots i_\ell}^{(\ell)}$'s. And we may write

$$\phi = \sum_{\ell=0}^{\infty} \frac{(-1)^{\ell}}{\ell!} P_{i_1 \ldots i_\ell}^{(\ell)} \phi_{i_1 \ldots i_\ell}^{(\ell)}. \tag{B.2.16}$$

We show next that there are exactly $2\ell + 1$ independent harmonic polynomials of degree ℓ. We proceed by writing an arbitrary polynomial of degree ℓ as

$$\psi^{(\ell)} = \sum_{m_1+m_2+m_3=\ell} (x + iy)^{m_1}(x - iy)^{m_2}z^{m_3}a(m_1, m_2, m_3) \tag{B.2.17}$$

where the coefficient $a(m_1, m_2, m_3)$ is to be determined.

Suppose $m_1 > m_2$; we then write

$$(x + iy)^{m_1}(x - iy)^{m_2} = (x + iy)^{m_1 - m_2}(x^2 + y^2)^{m_2}; \tag{B.2.18}$$

if $m_1 < m_2$, we write

$$(x + iy)^{m_1}(x - iy)^{m_2} = (x - iy)^{m_2 - m_1}(x^2 + y^2)^{m_1}. \tag{B.2.19}$$

We can expand

$$\psi^{(\ell)} = \sum_m \psi^{(\ell,m)} \tag{B.2.20}$$

where, for fixed $m = m_1 - m_2 > 0$,

$$\psi^{(\ell,m)} = (x + iy)^m \sum_{2m_2+m_3=\ell-m} (x^2 + y^2)^{m_2}z^{m_3}a(m_1, m_2, m_3). \tag{B.2.21}$$

Note that m may take on values from 0 to ℓ. The m_2 and m_1 values with $m_2 > m_1$ add ℓ functions proportional to $(x - iy)^{|m|}$, so that in all we have functions $\psi^{(\ell,m)}$ whose azimuthal φ dependence (in spherical coordinates) is $e^{im\varphi}$, with m taking on $2\ell + 1$ values between $\pm\ell$. Note that functions with different m values are orthogonal:

$$\int \psi^{(\ell,m)}\psi^{*(\ell',m')} \, d\Omega = 0, \qquad m \neq m'. \tag{B.2.22}$$

We already know from (B.1.14) that $\psi^{\ell,m}$'s of different ℓ are orthogonal.

Finally, we show that orthogonality between different ℓ values determines $\psi^{(\ell,m)}$ to within a constant multiple; normalization determines it to within a phase.

We first note from (B.2.17) that

$$\psi^{(\ell)}(-\mathbf{x}) = (-1)^{m_1 + m_2 + m_3}\psi^{\ell}(\mathbf{x})$$
$$= (-1)^{\ell}\psi^{\ell}(\mathbf{x}). \qquad (B.2.23)$$

The transformation $\mathbf{x} \to -\mathbf{x}$ in spherical coordinates is given by

$$\varphi \to \varphi + \pi, \qquad \theta \to \pi - \theta \qquad (B.2.24)$$

so that, since $e^{im\varphi} \to (-1)^m e^{im\varphi}$ under the transformation (B.2.24), the residual function in (B.2.21)

$$\sum (x^2 + y^2)^{m_2} z^{m_3} a(m_1, m_2, m_3), \qquad (B.2.25)$$

must be multiplied by $(-1)^{\ell-m}$ under the transformation $z \to -z$.

Factoring the r^{ℓ} dependence of $\psi^{(\ell,m)}$, we define

$$\psi^{(\ell,m)} = r^{\ell} \sin^{|m|}\theta\, e^{im\varphi} \overline{P}_{\ell,m}(\cos\theta) \qquad (B.2.26)$$

where $\overline{P}_{\ell,m}(\cos\theta)$ is a polynomial of degree $\ell - m$ in $\cos\theta$.[1] Under the transformation $\mathbf{r} \to -\mathbf{r}$, then,

$$\overline{P}_{\ell,m}(-\cos\theta) = (-1)^{\ell-m}\overline{P}_{\ell,m}(\cos\theta).$$

The functions (B.2.26) are already orthogonal for different m values. If they are to be harmonic, they must also be orthogonal for different ℓ values. That is,

$$\int \psi^{(\ell,m)*}\psi^{(\ell',m')}\, d\Omega = 0, \qquad \ell \neq \ell' \quad \text{or} \quad m \neq m'. \qquad (B.2.27)$$

Following (B.2.27), we construct, for each m, a sequence of orthogonal polynomials of degrees $\ell - |m|$ in z, starting with $\ell = |m|$ and containing only even or odd powers of z according to the parity $(-1)^{\ell-m}$. The weight function for the sequence is $(\sin^{|m|}\theta)^2 = (1 - \cos^2\theta)^{|m|}$.

---

[1]We introduce the bar in the symbol $\overline{P}_{\ell,m}$ because there exists a conventionally defined symbol $P_{\ell,m}$ with a different normalization.

TABLE B.1.

$m\backslash\ell$	0	1	2	3	4
4					$\sin^4\theta\, e^{4i\varphi}$
3				$\sin^3\theta\, e^{3i\varphi}$	$\sin^3\theta\, e^{3i\varphi}\cos\theta$
2			$\sin^2\theta\, e^{2i\varphi}$	$\sin^2\theta\, e^{2i\varphi}\cos\theta$	$\sin^2\theta\, e^{2i\varphi}(g\cos^2\theta+h)$
1		$\sin\theta\, e^{i\varphi}$	$\sin\theta\, e^{i\varphi}\cos\theta$	$\sin\theta\, e^{i\varphi}(c\cos^2\theta+d)$	$\sin\theta\, e^{i\varphi}\cos\theta(j\cos^2\theta+k)$
0	1	$\cos\theta$	$a\cos^2\theta+b$	$\cos\theta(e\cos^2\theta+f)$	.
−1		$\sin\theta\, e^{-i\varphi}$	$\sin\theta\, e^{-i\varphi}\cos\theta$	.	
.			$\sin^2\theta\, e^{-2i\varphi}$	.	

This is illustrated in Table B.1.

Evidently, there are just enough orthogonality relations to determine all the coefficients to within a constant multiple. If we normalize these functions to one, they are determined up to a constant phase. For some purposes, it is convenient to choose

$$\overline{P}_{\ell,m} = (-1)^m \overline{P}_{\ell,-m}, \tag{B.2.28}$$

and we shall generally do so.

We have now constructed for each m a unique orthonormal set of polynomials of degree ℓ; since m ranges from $-\ell$ to ℓ, there are, for each ℓ, $2\ell + 1$ polynomials. Harmonic polynomials may be chosen to have a fixed m, since the Laplacian operator ∇^2 does not mix values of m, that is,

$$\nabla^2 e^{im\varphi} F(\theta) = e^{im\varphi} G(\theta). \tag{B.2.29}$$

In addition, harmonic polynomials must be orthogonal for different ℓ, a property by which they are uniquely determined, as we have just shown. Therefore, our orthogonality procedure has determined the $2\ell + 1$ harmonic polynomials of each order.

The notation we use for the normalized functions is

$$Y_{\ell,m}(\theta, \varphi) = \sin^{|m|}\theta\, e^{im\varphi} \overline{P}_{\ell,m}(\omega). \tag{B.2.30}$$

B.3. COMPLETENESS OF THE $Y_{\ell,m}$

We show first that the Legendre polynomials are complete, that is, that

$$\sum_{\ell} (2\ell + 1) \int P_{\ell}(\hat{r}\cdot\hat{r}') f(\mathbf{r}')\, d\Omega' = 4\pi f(\mathbf{r}) \tag{B.3.1}$$

for $f(\mathbf{r})$ a sufficiently well-behaved function of angle, and where the vectors $\mathbf{r}'$ and $\mathbf{r}$ have the same length: $|\mathbf{r}'| = |\mathbf{r}| = r$.

We proceed by considering the integral

$$F = - \int_{S'} dS' \cdot \left(\nabla' \frac{1}{|\mathbf{r}' - \mathbf{r}_i|} - \nabla' \frac{1}{|\mathbf{r}' - \mathbf{r}_0|} \right) f(\mathbf{r}') \qquad \text{(B.3.2)}$$

where the surface S' is a sphere of radius r. The vector $\mathbf{r}_i$ is inside the sphere, $\mathbf{r}_0$ outside the sphere; for simplicity we take them to be colinear. We will take the limit $\mathbf{r}_i \to \mathbf{r}$ and $\mathbf{r}_0 \to \mathbf{r}$ in two different ways.

Except for the singularities in the integrand (B.3.2) at $\mathbf{r}' = \mathbf{r}_i$ and $\mathbf{r}' = \mathbf{r}_0$, the limit would be zero. We take the singularity properly into account by expanding $f(\mathbf{r}')$ about $\mathbf{r}_i$ in the first term of (B.3.2) [as $f(\mathbf{r}') = f(\mathbf{r}_i) + (\mathbf{r}' - \mathbf{r}_i) \cdot \nabla f + \cdots$] and about $\mathbf{r}_0$ in the second. The terms with $\mathbf{r}' - \mathbf{r}_i$ and $\mathbf{r}' - \mathbf{r}_0$ to the first and higher powers remove the singularity and, hence, in the limit can be dropped. There remains

$$F = - \int dS \cdot \nabla \left(\frac{1}{|\mathbf{r} - \mathbf{r}_i|} f(\mathbf{r}_i) - \frac{1}{|\mathbf{r} - \mathbf{r}_0|} f(\mathbf{r}_0) \right)$$

+ terms that go to zero, so that

$$\lim_{\substack{\mathbf{r}_i \to \mathbf{r}^- \\ \mathbf{r}_0 \to \mathbf{r}^+}} F = 4\pi f(\mathbf{r}). \qquad \text{(B.3.3)}$$

On the other hand, expanding $1/|\mathbf{r} - \mathbf{r}_i|$ in powers of r_i/r and $1/|\mathbf{r} - \mathbf{r}_0|$ in powers of r/r_0, we have, from (B.3.2),

$$F = -r^2 \sum_\ell \int d\Omega' \frac{d}{dr'} \left(\frac{r_i^\ell}{r'^{\ell+1}} - \frac{r'^\ell}{r_0^{\ell+1}} \right) P_\ell(\hat{\mathbf{r}}_i \cdot \hat{\mathbf{r}}') f(\mathbf{r}')$$

or

$$F = \sum_\ell \int d\Omega' \left[(\ell + 1) \left(\frac{r_i}{r} \right)^\ell + \ell \left(\frac{r}{r_0} \right)^{\ell+1} \right] P_\ell(\hat{\mathbf{r}}_i \cdot \hat{\mathbf{r}}') f(\mathbf{r}'). \qquad \text{(B.3.4)}$$

It is now safe to take the limit, which yields, as claimed,

$$4\pi f(\mathbf{r}_i) = \sum_\ell (2\ell + 1) \int P_\ell(\hat{\mathbf{r}}_i \cdot \hat{\mathbf{r}}') f(\mathbf{r}') \, d\Omega'. \qquad \text{(B.3.1)}$$

Next, we note that since $r^\ell P_\ell(\hat{\mathbf{r}} \cdot \hat{\mathbf{r}}')$ is a harmonic polynomial of order

ℓ, as is $r'^{\ell}P_{\ell}(\hat{\mathbf{r}} \cdot \hat{\mathbf{r}}')$, we must be able to expand $P_{\ell}(\hat{\mathbf{r}} \cdot \hat{\mathbf{r}}')$ in terms of the $Y_{\ell,m}(\Omega)$'s and $Y_{\ell,m}(\Omega')$'s, that is,

$$P_{\ell}(\hat{\mathbf{r}} \cdot \hat{\mathbf{r}}') = \sum_{m,m'} a_{m,m'} Y_{\ell,m}(\theta, \varphi) \, Y^*_{\ell,m'}(\theta', \varphi'). \qquad (B.3.5)$$

Since $\hat{\mathbf{r}} \cdot \hat{\mathbf{r}}' = \cos\theta\cos\theta' + \sin\theta\sin\theta'\cos(\varphi - \varphi')$, (B.3.5) must depend only on $\varphi - \varphi'$ (not on $\varphi + \varphi'$). Therefore, $a_{m'm}$ must equal zero unless $m' = m$, and (B.3.5) becomes

$$P_{\ell}(\hat{r} \cdot \hat{r}') = \sum_{m} a_m Y_{\ell,m}(\theta, \varphi) \, Y^*_{\ell,m}(\theta', \varphi'). \qquad (B.3.6)$$

Note that since P_{ℓ} is real and invariant under interchange of $\hat{\mathbf{r}}$ and $\hat{\mathbf{r}}'$, a_m must be real. We perform two operations on (B.3.6):

1. Set $\hat{\mathbf{r}} = \hat{\mathbf{r}}'$ and integrate over $d\Omega$. We find

$$4\pi = \sum_{m} a_m. \qquad (B.3.7)$$

2. Now square (B.3.6):

$$P_{\ell}^2(\hat{r} \cdot \hat{r}') = \sum_{m_1} a_{m_1} Y_{\ell,m_1}(\Omega) \, Y^*_{\ell,m_1}(\Omega')$$
$$\times \sum_{m_2} a^*_{m_2} Y^*_{\ell,m_2}(\Omega) \, Y_{\ell,m_2}(\Omega'). \qquad (B.3.8)$$

3. Integrate over $d\Omega$. There results

$$\frac{4\pi}{2\ell + 1} = \sum_{m} |a_m|^2 \, |Y_{\ell,m}(\Omega')|^2. \qquad (B.3.9)$$

4. Finally, integrate over $d\Omega'$ to find

$$\frac{(4\pi)^2}{2\ell + 1} = \sum_{m} |a_m|^2. \qquad (B.3.10)$$

The unique solution of (B.3.7) and (B.3.10) is $a_m = 4\pi/(2\ell + 1)$. (See Problem B.11.) Our final result is therefore

$$P_{\ell}(\hat{\mathbf{r}} \cdot \hat{\mathbf{r}}') = \frac{4\pi}{2\ell + 1} \sum_{m=-\ell}^{\ell} Y_{\ell,m}(\Omega) \, Y^*_{\ell,m}(\Omega'). \qquad (B.3.11)$$

The completeness theorem (B.3.4) for the P_ℓ's then tells us that any function of angle can be expanded in the $Y_{\ell,m}$'s:

$$f(\Omega) = \sum_\ell \frac{(2\ell + 1)}{4\pi} \int P_\ell(\hat{r} \cdot \hat{r}') f(\Omega') \, d\Omega'$$

or

$$f(\Omega) = \sum_{\ell,m} Y_{\ell,m}(\Omega) \int Y_{\ell,m}^*(\Omega') f(\Omega') \, d\Omega'. \tag{B.3.12}$$

The $Y_{\ell,m}$'s are sometimes referred to as spherical tensors. Recall the Cartesian form of harmonic polynomials generated by the coefficients of $\phi^{(\ell)}$'s in (B.2.16):

$$P_{i_1 \ldots i_\ell} = (x_{i_1} \ldots x_{i_\ell} - \text{traces}). \tag{B.3.13}$$

These transform under rotations as ℓth-rank, symmetric traceless tensors.

The $Y_{\ell,m}$'s also have a simple transformation property, since a solution of the Laplace equation must remain a solution under rotations. Therefore,

$$r^\ell Y_{\ell,m}(\Omega_R) = \sum_{m'} D_{m,m'}^\ell(R) \, r^\ell \, Y_{\ell,m'}(\Omega) \tag{B.3.14}$$

where Ω_R takes the θ, φ of the original coordinate system (Ω) and changes them to the θ, φ of the same point with respect to the new coordinate system. The expansion coefficients $D_{m,m'}^\ell(R)$ define the transformation of a spherical tensor.

APPENDIX B PROBLEMS

B.1   From the definition (B.1.6) of the P_ℓ's, show that $P_\ell'(w = 1) = \ell(\ell + 1)/2$.

B.2   Again using (B.1.6), show that with $w = \cos\theta$,

$$P_\ell(w = 0) = 0 \qquad\qquad \text{for } \ell \text{ odd}$$

$$= (-1)^n \frac{\frac{1}{2} \cdots (n - \frac{1}{2})}{n!} \qquad \text{for } \ell \text{ even}$$

where $n = \ell/2$, and the product $\frac{1}{2} \cdots (n - \frac{1}{2}) = 1$ for $n = 0$.

B.3 Again using (B.1.6), show that the P_ℓ's satisfy the differential equation

$$-\frac{d}{dw}(1 - w^2)\frac{d}{dw}P_\ell(w) = \ell(\ell + 1)P_\ell(w).$$

B.4 Again using (B.1.6), show that

$$(\ell + 1)\, P_{\ell+1}(w) + \ell P_{\ell-1}(w) = w(2\ell + 1)\, P_\ell(w)$$

and from that, show that

$$\int_{-1}^{1} P_{\ell'} w P_\ell\, dw = 0$$

unless $\ell' = \ell \pm 1$ and that

$$\int_{-1}^{1} P_{\ell+1}(w)w\, P_\ell(w)\, dw = \frac{2(\ell + 1)}{(2\ell + 1)(2\ell + 3)}.$$

B.5 Use the orthonormality of the $Y_{\ell,m}$'s to construct all the $Y_{\ell,m}$'s for $\ell \leq 4$ as suggested by Table B.1.

B.6 Show that the function

$$P_\ell(\tau, \mathbf{r}) = \left(\tau(x + iy) - \frac{1}{\tau}(x - iy) - 2z\right)^\ell$$

for any τ is a harmonic polynomial of order ℓ and is a generating function for the $Y_{\ell,m}$'s:

$$P_\ell(\tau, \mathbf{r}) = r^\ell \sum_{m=-\ell}^{\ell} \tau^m c_{\ell,m}\, Y_{\ell,m}(\theta, \varphi)$$

where the $c_{\ell,m}$'s are constant coefficients.

B.7 Show from the results of Problem B.6 that

$$\left(\frac{\partial}{\partial x} + i\frac{\partial}{\partial y}\right)r^\ell Y_{\ell,m} = d_{\ell,m}\, r^{\ell-1}\, Y_{\ell-1,m+1},$$

that

$$\left(\frac{\partial}{\partial x} - i\frac{\partial}{\partial y}\right)r^\ell Y_{\ell,m} = d'_{\ell,m}\, r^{\ell-1}\, Y_{\ell-1,m-1},$$

and that

$$\frac{\partial}{\partial z} r^{\ell} Y_{\ell,m} = d''_{\ell,m} r^{\ell-1} Y_{\ell-1,m}$$

where d, d', and d'' are ℓ, m-dependent constants.

B.8 The charge distribution on a spherical surface is given by

$$\sigma = \mathbf{A} \cdot \mathbf{r},$$

where $\mathbf{A}$ is a constant vector.
    Find the potential

$$\phi(\mathbf{r}) = \int \frac{\sigma(\mathbf{r}')}{|\mathbf{r} - \mathbf{r}'|} d\Omega'$$

and the field

$$\mathbf{E} = -\nabla\phi$$

inside and outside of the sphere.

B.9 Verify by direct integration Newton's theorem that the potential outside a spherically symmetric charge distribution $\rho_s(r)$ is the same as it would be were all the charge concentrated at the center. That is, for r outside of the region in which ρ is nonzero,

$$\int d\mathbf{r}' \frac{\rho_s(r')}{|\mathbf{r} - \mathbf{r}'|} = \frac{Q}{r}$$

where $Q = \int d\mathbf{r}'\rho(r')$.

B.10 We know from (B.2.26) that $r^{\ell}Y_{\ell,m}$ is harmonic and of the form $r^{\ell} e^{im\varphi} f_{\ell,m}(\theta)$. Show from this that $f_{\ell,m}$ satisfies the equation

$$\left\{ \frac{1}{\sin\theta} \frac{d}{d\theta} \sin\theta \frac{d}{d\theta} - \frac{m^2}{\sin^2\theta} + \ell(\ell + 1) \right\} f_{\ell,m} = 0.$$

B.11 Prove the statement made following (B.3.10), that (B.3.7) and (B.3.10) imply $a_m = 4\pi/(2\ell + 1)$.

Index

421